Horizons in Clinical Nanomedicine

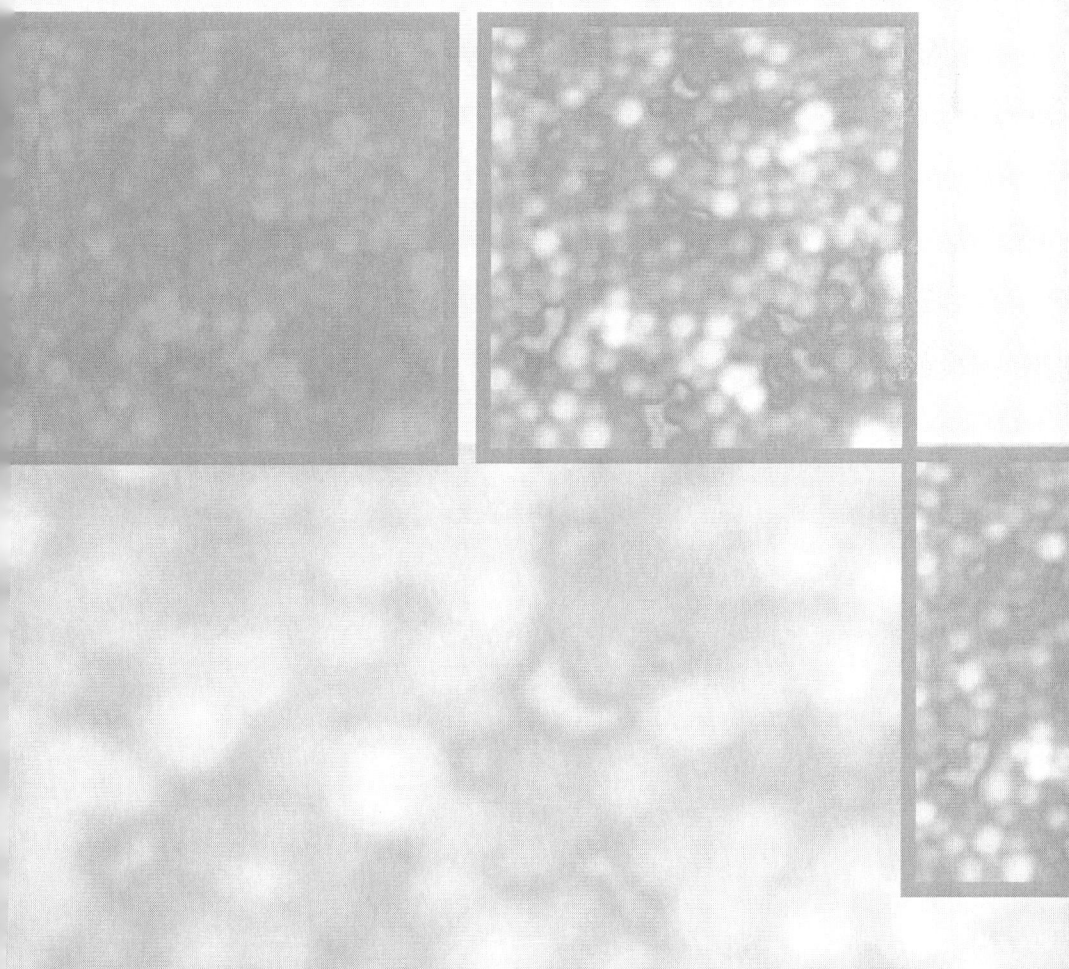

Horizons in Clinical Nanomedicine

edited by

Varvara Karagkiozaki
Stergios Logothetidis

Pan Stanford Publishing

Published by

Pan Stanford Publishing Pte. Ltd.
Penthouse Level, Suntec Tower 3
8 Temasek Boulevard
Singapore 038988

Email: editorial@panstanford.com
Web: www.panstanford.com

British Library Cataloguing-in-Publication Data
A catalogue record for this book is available from the British Library.

Horizons in Clinical Nanomedicine

ISBN 978-981-4411-56-1 (Hardcover)
ISBN 978-981-4411-57-8 (eBook)

Printed in the USA

Contents

Preface

Nanomedicine, a flourishing field of medical research, is expected to provide modern medicine with innovative solutions to the unmet and unsolved clinical needs. Nanostructured materials have the potential to revolutionize healthcare, due to their novel intrinsic physicochemical properties which can be exploited towards cutting-edge developments in the fields of diagnosing, treating, and preventing diseases, injuries, or genetic disorders. Thus, clinical nanomedicine holds great promise as a future powerful tool of medicine that improves human health.

This book presents a broad overview on nanomedicine tools, materials, and processes to be applied to different medical disciplines. It presents the broad spectrum of nanomedicines for the early, accurate diagnosis and effective treatment of human diseases. Taking into account the pillars of in vitro and in vivo nanodiagnostics, regenerative medicine, and nanopharmaceuticals, it deals with the unsolved medical problems in cardiovascular disease, AIDS, cancer, blood diseases, congenital defects, dermatology, dentistry, and orthopaedics, with a focus on personalized medicine. It addresses nanosafety and nanotoxicity issues to highlight the significance of nanomedicine applied into clinical practice for the benefit of the patient. The book will appeal to researchers, medical doctors, and graduate students who want to get in-depth knowledge of nanomedicine utilities for clinical applications.

Chapter 1

Introduction in Clinical Nanomedicine

Stergios Logothetidis
Lab for Thin Films—Nanosystems and Nanometrology,
Department of Physics, Aristotle University of Thessaloniki,
54124 Thessaloniki, Greece
logot@auth.gr

Nanotechnology represents the possibility of revolutionising many aspects of our lives. Nanomedicine, the application of nanotechnology to health, is one of the most promising fields of biomedical research, building up a novel research culture by encompassing the principles of traditional disciplinary boundaries (i.e., physics, chemistry, biology and engineering). Nanomedicine has the potential to give intelligent solutions to many of the current medical problems, by opening the door to a new generation of advanced drug delivery systems, improved diagnostic systems (in vitro and in vivo) and novel methods and materials for regenerative medicine. There are currently two families of therapeutic nanocarriers (i.e., liposomes and albumin nanoparticles) that have been approved and used in clinical settings, providing clinical benefit. Moreover, several nanocarriers are in clinical trials and even more are in pre-clinical

Horizons in Clinical Nanomedicine
Edited by Varvara Karagkiozaki and Stergios Logothetidis

ISBN 978-981-4411-56-1 (Hardcover), 978-981-4411-57-8 (eBook)
www.panstanford.com

phases. Despite the cutting-edge developments in nanomedicine, the process of converting basic research to viable products is expected to be long and ambitious. A crucial factor that should also be taken into consideration is the toxic effects of the novel therapeutical products in human health. Thus, a massive effort is required to translate laboratory innovation to the clinic and begin to change the landscape of medicine.

1.1 Introduction

Richard Feynman in 1959[1] was the first one to claim that 'there is plenty of room at the bottom', and since then, a booming interest in studying the nanoscale has emerged. The nano prefix comes from the Greek word 'Nano' meaning dwarf, with 1 nanometre (nm) being equal to one billionth of a metre (10^{-9} m), or 10 water molecules, or about the width of six carbon atoms. Atoms are smaller than 1 nm, whereas many molecules, including proteins, can range in size from 1 nm to several hundreds, as shown in Fig. 1.1.[2] Studying at such scales is of great importance, as the properties of matter differ significantly, especially due to quantum effects and the large surface-to-volume atom ratio. As a result, new findings arise, contributing to a better understanding of science.

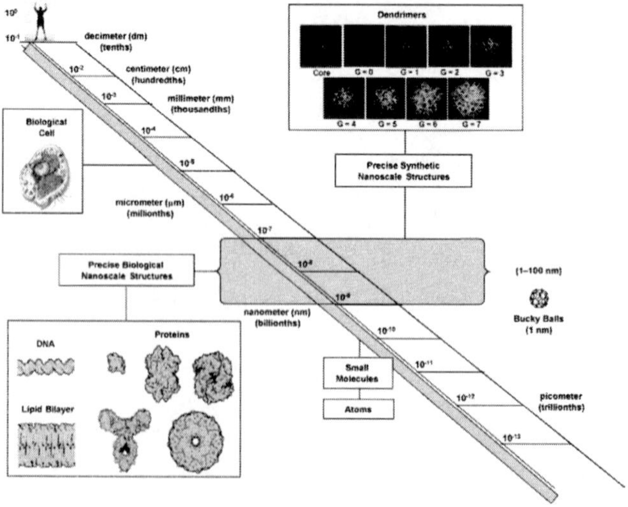

Figure 1.1 Nanoworld and Macroworld: the scale of natural objects.[3]

Nanoscience and Nanotechnology is a multidisciplinary field, derived from materials science, physics, chemistry, biology, medicine and engineering, covering a vast and diverse array of devices, with at least one dimension sized about 1 to 100 nm.

The primary aim of nanotechnology is the manipulation, synthesis, characterisation and production of novel materials, with nanoscale features, with better material properties, targeting to more challenging applications. Based on the processes involved in the fabrication of nanoscaled structures, two basic approaches exist. The *top down* process is a technique that uses nanofabrication tools, starting from larger dimensions, to create nanoscaled structures or devices of the desired shape and order. On the other hand, the bottom up approach involves the use of molecular self–assembly/self-organisation to achieve functional systems from the controlled deposition of atoms or molecules.[4] Biological systems especially utilise the bottom up approach, as it is crucial to the function of cells that exist in the organism. This is profound in the self-assembly of lipids to form the cellular membrane, the formation of the double helical DNA through hydrogen bonds of the individual strands, and the assembly of proteins.[5]

The use of the molecular knowledge of the human body as well as the molecular tools, such as the engineered nanodevices and nanostructures, is of great significance. The main fields of present and future applications of nanomedicine are nanodiagnostics, regenerative medicine and targeted drug delivery (Fig. 1.2).[6]

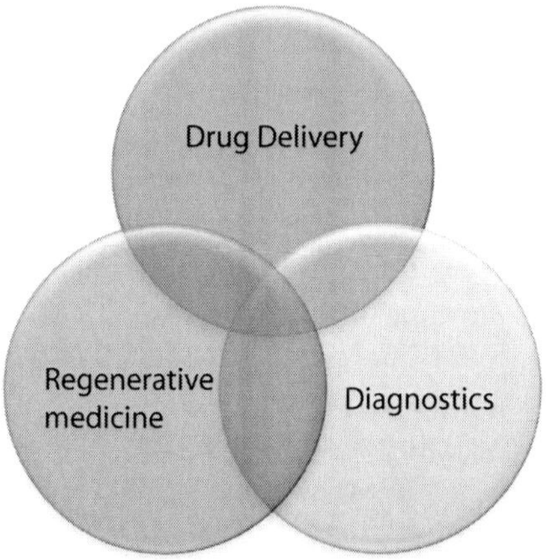

Figure 1.2 The three main pillars of nanomedicine.

In this direction, the application of nanotechnology to medicine and biology holds great promise towards the effective treatment of diseases that currently lack therapy. Nanomedicine can be defined as the science and technology of monitoring, repairing, constructing and controlling human biological systems at the molecular level, in order to preserve and improve human health.[7,8] Cancer, for instance, is a fatal disease that lacks therapy in many cases. A further analysis of this issue is presented in Chapter 3. Indeed, today extensive studies to understand the mechanism behind the tumour and cellular biology indicate that targeted therapy can be very promising compared to conventional chemotherapy.[9]

In addition to this, dramatic changes are taking place in diagnostics and imaging, which promise great benefits to nanomedicine. Nanoparticles and quantum dots (QDs) can be applied for early in vitro and in vivo diagnosis, as they are used as tracers or contrast agents. Unique nanodevices with nanomaterial features contribute to this aim. Also, several benefits of nano-imaging emerge, such as the early detection, and the monitoring of the disease stages. Bioelectronics, including biosensors, aim in that direction.[10] The optimistic goal of nanotechnology is to repair replace or regenerate cells, tissues and organs, upening up new possibilities to medicine that can definitely result in immense health benefits, providing a better and sufficient life to patients.

Although nanomedicine has become a very active and vital area of research, enabling evolutionary changes in several medical topics, great concern is put on the possible risks that rise from the use of nanomaterials, nanoparticles or nanotubes, because of their nanoscale dimensions.[11] Specifically, their extremely small size, makes them easy to penetrate cells and accumulate into vital organs. Haemocompatibility and activation of the human defence system are parameters that should also be taken into consideration,[12] in order to evaluate the cytotoxicity and the biocompatibility of such materials before they appear in the market.[13]

The importance of the nanoscale and how nanotechnology influences medicine, raising expectations of further applications in crucial medical issues, is presented in Section 1.1, in this chapter. Section 1.2 describes nanocarrier-based systems—a very promising and challenging field, especially in the domain of targeted drug delivery, referred to in Section 1.5. Regenerative medicine, is discussed in Section 1.3 as an idea of implementing nanosciences and nanotechnology in order to construct tissue substitutes. The third very promising tool of nanomedicine refers to in vitro and in vivo diagnostics for the early detection and prevention of diseases and is analysed in Section 1.4. Finally, the potential risk of nanomaterial toxicity is discussed in Section 1.6.

1.2 Nanomedicine and Diversity of Nanocarriers

This chapter focuses on the development of nanomaterials for enabling and improving the targeted delivery of therapeutic and diagnostic agents.

Nanotechnology - based carrier systems can specifically target the site of disease, either by targeted drug delivery, imaging, or by simultaneous strategic drug and gene delivery, reducing the drug dose and its harmful results to healthy tissues and organs.[14,15,16] Imaging contributes to a better outcome by the usage of multifunctionalized pharmaceutical nanocarrier systems that enhance the image contrast, by real-time bio-distribution. Several molecules or polymeric structures can be used (Fig. 1.3). Protein based nanocarriers represent an interesting approach, as their contribution to gene and drug delivery is very promising because of their low cytotoxicity, the abundant renewable sources, as well as their high drug binding capacity.[17]

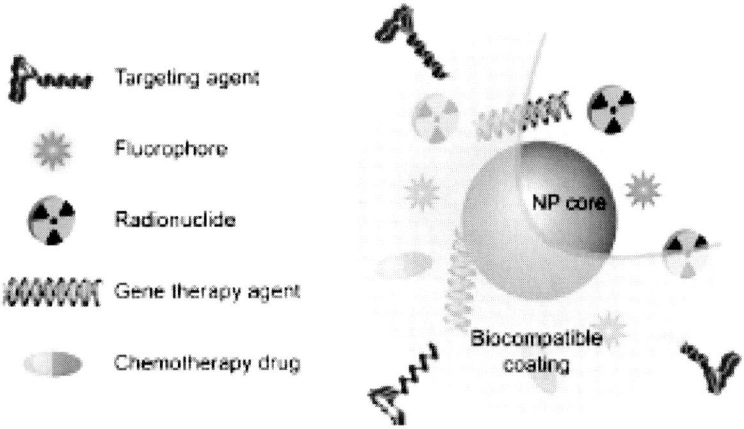

Figure 1.3 The schematic structure of the assembly of the multifunctional pharmaceutical nanocarrier.[18]

Nanocarriers can be divided into three basic categories: (a) *organic-based*, such as polymeric nanocarriers (i.e., dendrimers, micelles), liposomes and carbon-based nanocarriers (i.e., fullerenes and carbon nanotubes); (b) *inorganic-based* (i.e., metallic nanoparticles, silica nanoparticles, quantum dots); and (c) a *hybrid combination* of the above to develop a multi-functional carrier system.[19] Because of their size, nanocarrier systems represent a good perspective, as they can migrate through cell membranes and penetrate across physiological drug barriers.[20,21] Figure 1.4 depicts the relevant size of several types of nanocarriers with a virus, the platelet, the blood cells and the capillaries in the human body.

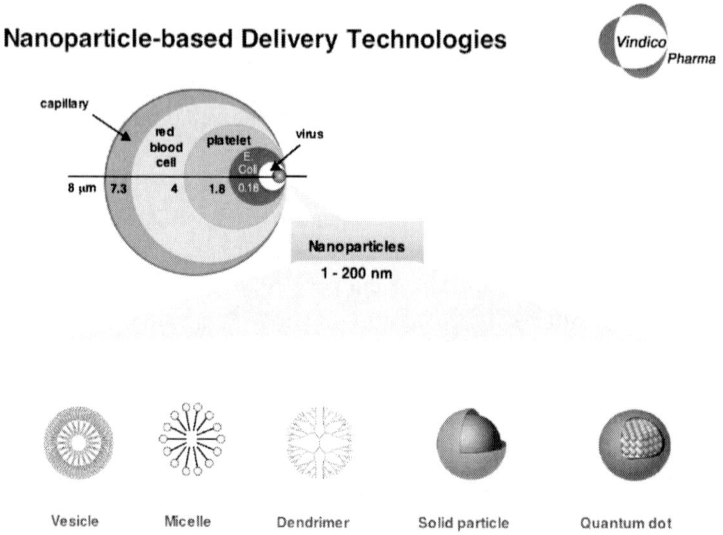

Figure 1.4 Size and shape diversity of nanocarrier-based systems.[22]

It is essential to highlight the fact that nanoparticles are about 100 to 10.000 times smaller than human cells and similar in size to large biological molecules (e.g., enzymes and cell receptors). Due to their small size, they can readily interact with biomolecules on the surface and inside the cells. More specifically, there are two possible mechanisms, which depend to a large extend on the nanocarrier's size. Nanocarriers smaller than 50 nm can easily enter most cells, either via passive diffusion or through active processes such as endocytocis and exocytosis, whereas those smaller than 20 nm follow the paracellular mechanism, as they can move out of blood vessels or pass through epithelial cells to circulate through the body.

The nanoparticle's *surface charge* is also considered to determine its behavior. For example, nanoparticles interact with oppositely charged cells, as well as cluster in the blood flow, or adhering to it. For targeted drug delivery, a prolonged circulation of nanocarriers in the body is needed so as to achieve drug efficacy and specificity compared to conventional drug approaches.

Nanoparticles are spherical structures, with sizes ranging from 1 to 100 nm, with the ability to encapsulate the drug. Nanoparticles also include nanospheres and nanocapsules. One obvious difference between these two is that in nanospheres the drug is distributed throughout the particles, whereas in nanocapsules, there is a polymeric membrane cavity where the drug is included.[23,24] Nanoparticles and other nanocarriers used in nanomedicine are sustained by proper surface treatment in order to be protected of reticuloendothelial system. This treatment involves the proper surface coating, usually with polyethylene glycol (PEG), a procedure called PEGylation. The outcome of this procedure is the increase of time they stay in the blood circulation, as carriers are not easily taken up by the macrophages. The significance of this is found in the fact that the longer the nanoparticles stay in the organism, the better the accumulation of the drug in a specific site, providing satisfactory results to drug delivery medicine. Also, nanocarriers through PEGylation exhibit a good solubility in aqueous solutions, flexibility of its polymer chain, low toxicity, immunogenicity, and antigenicity as well.[25]

Liposomes consist of one or more phospholipid bilayers, which are chemically active in order to enhance its efficient accumulation in the target site. A liposome encapsulates a region of aqueous solution, making it appear like a spherical vesicle.[21] In addition to this, electrostatic interaction plays a great role as the cell surface glycoproteines are negatively charged. The method of PEGylation, referred to above, is especially used for liposome - carriers.[26]

Viruses display a diversity of shape and sizes, varying between 20 and 300 nm. Virus particles form a hollow scaffold via self-assembly, in which their viral nucleic acid is encapsulated. They can be easily tailored at the genetic level, and appear to be biocompatible and easy to functionalise for several applications. So, many viruses (e.g., cowpea chlorotic mottle virus, cowpea mosaic virus, red clover necrotic mosaic virus, MS2 RNA-containing bacteriophage, the bacteriophage Qβ, M13 bacteriophage, etc.) have been studied for the potential to use them as nanocarrier systems for drug delivery applications.[27] Packing drug molecules into virus-like particles is based on supramolecular chemistry. The idea relies on the principle that the viral RNA or DNA packages into the virus, and this way, functional cargoes can also be packed through self-assembly and disassembly processes.

Dendrimers have tree-like structures with very small dimensions (1–10 nm). Their shape is globular and present unique features, compared to traditional polymers.[20] The core of these molecules consists of amino acids or sugars, whereas the branches are polymeric chains, developed by a series of polymerisation reactions, around the dendrimer's core. Thanks to their size, dendrimers can slip through openings in cell membrane in order to act as delivery particles. Drugs are attached to surface groups, which exist in internal cavities, via chemical modifications.

Micelles are polymeric structures, below 50 nm, dispersed in a liquid colloid. Micelles have a hydrophilic head, surrounded by solvent and a hydrophobic tail in their core. The drug is encapsulated into the core cavity formed by the hydrophobic tail. Polymeric micelles might have several advantages; some of them would be their size, the loading capacity, as well the reduced toxicity. The hydrophilic head provides a long circulation time in the bloodstream and their penetration ability through tissues and cells makes them candidates for targeted drug and gene delivery therapies. Until now, micelles have been studied for several antitumour drugs, both in pre-clinical and clinical trials. Results that derive from these endeavours are very promising.[14]

Nanotubes and fullerenes exhibit good electric, electronic, thermal and optical properties. Carbon nanotubes, cylindrical in shape, are useful carrier 'vehicles', which can be functionalised to act as a drug delivery system. As the drug is encapsulated into the nanotube, a drug-CNT complex develops, whose structural properties as well as the dynamic of the system drug-CNT.[28] Some recent surveys report that nanotubes can be loaded to the tumour site and then excited with radio waves, resulting in the heating up of the abnormal cells that would kill them.[29] Fullerenes form a sphere or an ellipsoid, inside which drug molecules can be incorporated. Compared to nanotubes, fullerenes also have some good properties. However, the real advantage of therapies based on fullerenes as compared to other targeted therapeutic agents is that fullerenes are expected to carry multiple drug loads.[30]

Quantum dots (QDs) are semiconductor nanocrystals, about 2–10 nm. QDs implications derive from the diagnostics segment, as they are vital for in vivo imaging, that provides real-time biodistribution and target accumulation of drug. They can also be easily attached to a variety of surface ligands and finally inserted into the body, enhancing the in vivo drug efficiency. By the illumination effect of ultraviolet light, QDs are used to localise cells or their activities. The physical principle behind the phenomenon relies on the fact that different sizes of the QDs give corresponding wavelengths. Specifically, smaller QDs result in larger energy jumps between the highest valence band and the lowest conduction band, and consequently more energy is needed in order to excite the dot. There are several reports demonstrating the advantages of QDs compared to fluorescent molecules, indicating that the intensity of the signal is brighter and present better photostability as well. Due to the better photostability, the acquisition of many consecutive focal-plane images is feasible, providing a three-dimensional image with high resolution.[31]

1.3 Nanomedicine in Regenerative Medicine

A very challenging and promising domain of implementing nanotechnology is regenerative medicine. Regenerative medicine, or tissue engineering, can be defined as the application of physical theories and principles in the design, construction, modification, development and maintenance of living tissues. It is considered to be a multi-disciplinary field, as many different scientists contribute to it with knowledge that derives from material science, molecular biology, engineering, medicine, etc. (Fig. 1.5).

The ultimate goal of regenerative medicine is to solve the problem of donor shortage of tissues and organs.[32] Patients with osteoarthritis (OA), for example, resulting from trauma or age-related disease, present a significant clinical challenge because of the limited repair of articular cartilage. Another implementation of tissue engineering can be on cardiovascular issues, finding solutions to clinical problems of heart valves, arteries and myocardium by developing tissue replacements in vitro or inside the human body.[33] Also, tissue engineering is supposed to work with the patient's repair mechanisms in order to prevent and treat chronic diseases such as diabetes and disorders of the central nervous system.[5] The concept of tissue engineering is based on three segments—cells, scaffold and bioactive factors—as shown in Fig. 1.5.

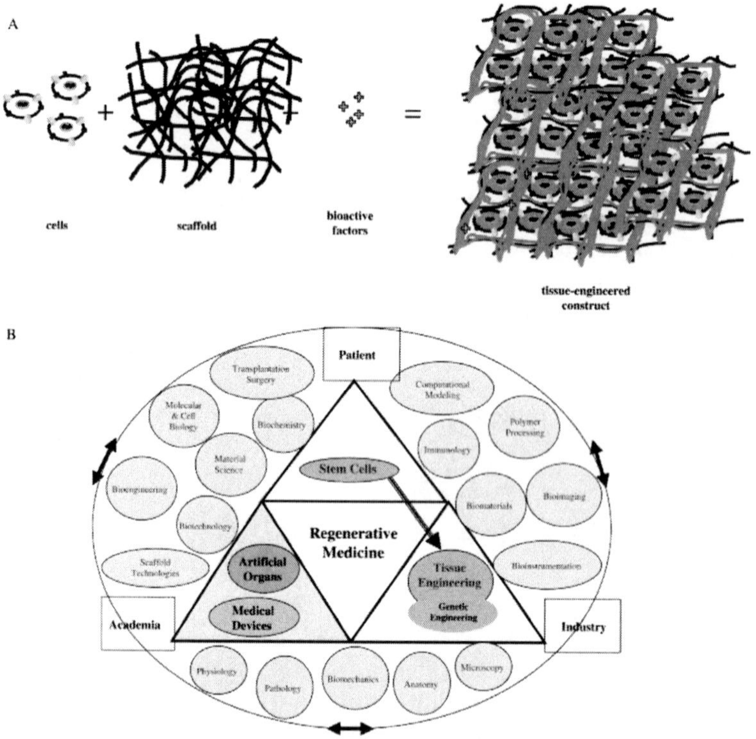

Figure 1.5 (A) Basic principles of tissue engineering: A scaffold or matrix, living cells and/or biologically active molecules are used in variable strategies to form a "tissue-engineered construct" promoting the construction of tissue. **(B)** Multidisciplinarity and complexity of combining tissue engineering and genetic engineering within the context of regenerative medicine.[34]

The first part of tissue engineering is the proper polymeric scaffold, either biodegradable or not, that will support the formation of tissue. Most mammalian cells have to adhere onto a proper substrate in order to proliferate and function properly. The ideal scaffold should be biocompatible and not cytotoxic, with good mechanical properties and a large surface-to-volume ratio that will promote cellular attachment, proliferation and differentiation.[35] There is a need of three-dimensional matrices as many surveys suggest that they allow cell ingrowth and organisation, resembling the native conditions of tissues found in vivo.[36] Today, a wide range of materials is studied with different scaffold architectures, such as hydrogel, sponge, mesh, etc. Also, among the vast majority of materials used for scaffolds, synthetic biopolymers appear to have an advantage over natural materials, as they can be tailored in order to have more predictable properties.[35]

The second part dealing with regenerative medicine involves the candidate cells that should be used to form tissue. Although it is not clear yet which cell type is the most suitable for in vitro implementation and in vivo transplantation, a number of studies have already proven the proper effect of live cells on bone or cartilage regeneration.[37,38,39,40] Chondrocytes,[36] (e.g., articular,[41,42,43] auricular,[36,44,45] costal,[36,46,47]) fibroblasts, and stem cells (e.g., bone-marrow derived, adipose-derived, muscle-derived, synovium-derived periosteum-derived, embryonic) as well as genetically modified cells have all been explored in order to demonstrate the mechanism of action—both in vitro and in vivo—for their potential as a viable cell source for cartilage implants. The ultimate aim is to find the ideal cell type, which is easily isolated and can be expanded via culturing processes, making large quantities available for use. Interestingly, mesenchymal stem cells (MSCs) represent an alternative approach, with great potential in osteogenic applications giving rise to the progress of tissue engineering.[37]

The third part, which also contributes to regenerative medicine, is the use of stimulating factors. Nowadays, there is a multitude of such agents exist, that are either capable to accelerate and/or enhance the tissue formation, or provide mechanical stability, preserving the scaffold's structure for proper cell proliferation. Growth factors usually include proteins, hormones that are vital, as they intend to control the activity of tissue specific cells, via regulation of the metabolism as well as differentiation, proliferation and expression of extracellular matrix proteins. However, it is important to highlight that growth factors may give different outcomes dependent on cell type and the culture's conditions.[36,37] As referred above, mechanical stability plays its role, also when dealing with bone or cartilage tissue. That is to say, hydrostatic pressure and dynamic compression[48] applied within physiological levels can be positive,[49] although it is of great significance in the loading regimen of the applied hydrostatic pressure and dynamic compression. Another objective of the stimulating factors is gene therapy, which seeks to advance the success of tissue engineering. Gene therapy is an alternative process, providing the ability to encapsulate bioactive molecules in scaffolds and through the expression of specific genes, release of proteins to provide local delivery into cells or tissues.

Mathematical models provide useful tools which help analyse, predict, and optimise the cellular response to scaffolds and growth factors.[50] Even though much has already been done regarding tissue substitutes-scaffolds, further studies are needed to investigate the mechanisms behind molecular biology, with material science providing a boost to tissue engineering science and techniques. There is also a need for developing new strategies to promote cell adhesion, proliferation and differentiation, too. Studying the contribution of different stimulating factors is vital, as well as the proper three-dimensional scaffold with the adequate support of cellular adhesion, proliferation and differentiation. For this, long term toxic impact of the implants to the organism is also an issue that should be taken into consideration, as well as the haemocompatibility and possible activation of the human defense system, especially due to the scaffolding materials used.

1.4 Nanomedicine in the Early Diagnosis of Diseases

Molecular diagnostic techniques are used for the early detection of diseases in order to enable their prevention and provide an immediate response. They are essential for several diseases such as cancer, cardiovascular diseases and neurological diseases. The use of nanotechnology in life sciences offers new options for clinical diagnostic procedures. This new field of science, which is called nanomolecular diagnostics[51] has an advantage over the molecular and biotechnological diagnostic methods. More specifically, nanomolecular procedures are faster and more sensitive because of the nanoparticles, which are used as labels for in vivo or in vitro imaging. Nanomolecular diagnostic technologies are classified into various categories, which are listed in Table 1.1.

Table 1.1 Classification of nanomolecular diagnostic techniques[52]

• Nanoscale visualisation (AFM)
• Nanoparticles biolabels
• Nanotechnology-based biochips/microarrays
• Nanobiosensors
• Nanoparticle-based immunoassays
• Nanoproteomic-based diagnostics
• DNA nanomachines for molecular diagnostics
• Nanopore technology
• Nanoparticle -based nucleic acid diagnostics

1.4.1 In vitro Diagnostics

In vitro diagnostic devices include nanobiosensors, microarrays, biochips of different elements (DNA, proteins, and cells) and lab-on-a-chip devices. The main advantage of these devices is that only small amounts of the sample are needed. The new devices, based on nanotechnology, are smaller, easier to use and cheaper than the conventional ones.[52]

Nanobiosensors are devices, based on nanotechnology, that are conjugated with biological molecules such as DNA, proteins, tissue, cells, or biomimetic molecules like aptamers and macrophage-inflammatory proteins.[53,54] These molecules are used as recognition elements. Figure 1.6 shows the label-free protein biosensors based on CNT-FETs that use antibodies and aptamers.[55] Interaction between the recognition elements and molecules of the sample causes changes in one or more physical-chemical properties (ion transfer, pH, heat, and optical properties). These physical-chemical changes produce an electronic signal, which evaluates the presence of the analyte of interest and its concentration in the sample. These sensors can be electronically gated to respond to the binding of a single molecule. Detection of nucleic acids, proteins and ions has been achieved using prototype sensors. These sensors can operate in the gas or liquid phase, and consequently can be used in a wide range of applications. Prototype sensors use inexpensive low-voltage measurement methods and detect binding events directly. So there is no need for costly, complicated and time-consuming chemical labelling; e.g., with fluorescent dyes, or for bulky and expensive optical detection systems. Consequently, as they are inexpensive to manufacture and portable,[51] they represent a very useful tool for in vitro diagnosis of various diseases, providing high-throughput screening and detection of a single disease in different samples, or of different diseases in a single sample. These devices can also offer multiple clinical parameters using an effective, accurate and simple test.[56]

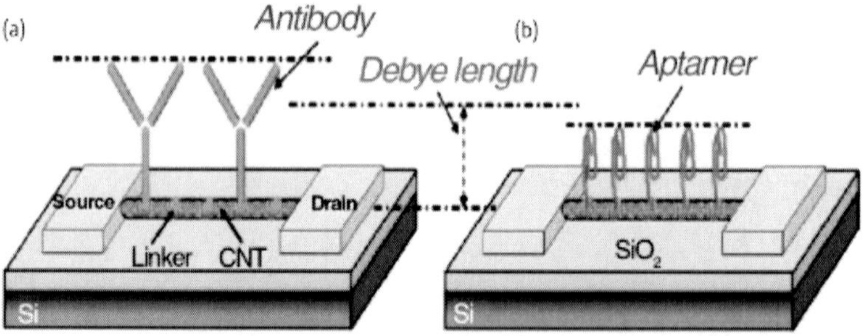

Figure 1.6 Schematic of label-free protein biosensors based on CNT-FETs: (a) antibody-modified CNT-FET (b) aptamer modified CNT-FET.[55]

Cantilever biosensors' technology is based on transformation of a reaction into a mechanical motion on the nanometer scale, approximately 10 nm, which can be measured directly by deflecting a light beam from the cantilever surface (Fig. 1.7). These cantilever biosensors are going to replace or complement the existing molecular diagnostic methods such as PCR and DNA -protein microarrays. There is no need to label or copy the target molecules. These biosensors provide fast, label-free recognition of specific DNA sequences for single-nucleotide polymorphisms, oncogenes, and genotyping. Also, cantilever biosensors provide real-time measurements and continuous monitoring of clinical parameters in personalised medicine.[52]

Viral Nanobiosensors: As it is already known, viruses can be considered as biological nanoparticles. Herpes simplex virus, hendra virus and adenovirus are used to promote the assembly of magnetic nanobeads as nanosensors for clinically relevant viruses.[58] These nanosensors can detect as few as five viral particles in a 10 ml serum sample. Also viruses can be used as sensors that utilise piezo-electric methods that include mass based biosensors, which are essentially based on atomic force microscopy (AFM). It is feasible to apply such electromechanical devices for virus detection, owing to the relatively high macromolecular mass of these entities (Fig. 1.7).[57] Contrary to molecular diagnostic techniques like ELISA and PCR, this diagnostic method is more sensitive, more efficient, cheaper, faster and with fewer artifacts.[52]

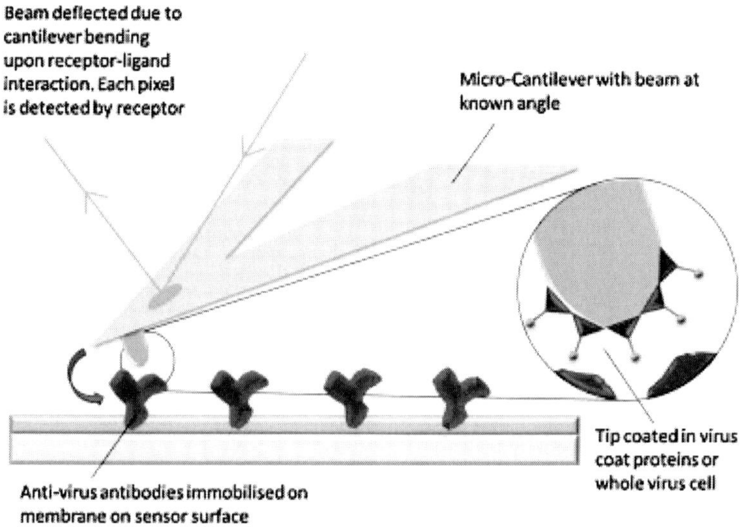

Figure 1.7 Action of cantilever using AFM topping mode is shown on the schematic. A laser is angled to reflect off the cantilever as it passes over the sensor surface. Upon virus–antibody interaction, the cantilever bends and causes the angle of the reflected beam to change. This change is then recorded and analysed.[57]

Optical Nanobiosensors: The most widely used optical biosensors are those that use the Surface Plasmon Resonance (SPR) technology. Optical-detectable tags can be formed by Surface Enhanced Raman Scattering (SERS) of active molecules at the glass-metal interface. Various small molecules are used for different types of tags. SERS bands are 1/50 the width of fluorescent bands. Also, the spectral intensity of SERS-based tags is linearly proportional to the number of particles. Consequently, it achieves a greater degree of multiplexing, allowing these tags to be used for multiplexed analyte quantification than current fluorescence-based quantification tags. SERS-based tags are stable, resistant to photodegradation and are coated with glass so that biomolecules such as proteins can be easily attached on the tags. The particles can be interrogated in the near-infrared range, enabling detection in blood and other tissues. A single test without interference from biological matrices, such as whole blood, can be measured up to 20 biomarkers and it is available by Nanoplex Biotags (Oxonica).[59]

Microarrays are important tools for early detection of diseases and for high-throughput analysis of biomolecules. For instance, microarrays have been widely applied in the study of various conditions, including atherosclerosis,[60] breast cancer,[61,62] colon cancer[63] and pulmonary fibrosis.[64] Several types of microarrays have been developed for different target materials (Table 1.2).[51]

Table 1.2 Types of microarrays

• DNA-microarrays
• cDNA-microarrays
• Single nucleotide polymorphism (SNP)-microarrays
• mRNA-microarrays
• Protein-microarrays
• Small molecules-microarrays
• Tissue -microarrays

The main drawbacks of this technique are the requirements for relatively large sample volumes and elongated incubation time, as well as the limit of detection. The best solution to these problems is to develop nanoarrays, particularly electronics-based nanoarrays. These novel nanoarrays are divided into three categories: (i) label-free nucleic acids analysis using nanoarrays, (ii) nanoarrays for protein detection by conventional optical fluorescence microscopy, as well as by novel label-free methods such as atomic force microscopy, and (iii) nanoarray for enzymatic-based assay. With further miniaturisation, higher sensitivity, and simplified sample preparation, nanoarrays could potentially be employed for biomolecular analysis in personal healthcare and monitoring of trace pathogens.[51] Figure 1.8 shows a schematic of the immunoassay format used to detect HIV-I p24 antigen with an anti-p24 antibody nanoarray.[65]

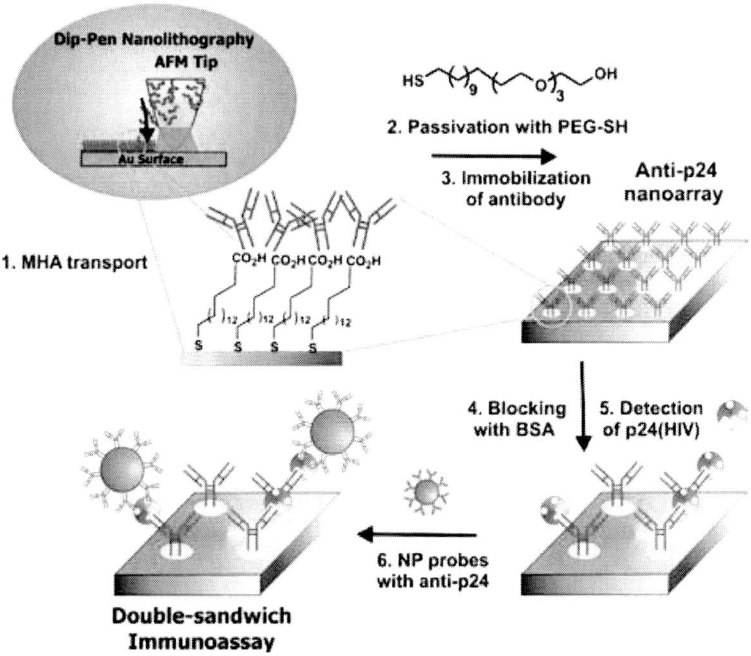

Figure 1.8 Schematic immunoassay format used to detect HIV-I p24 antigen with an anti-p24 antibody nanoarray.[65]

Lab-on-a-chip: The term lab-on-a-chip refers to a single device which integrates one or several laboratory functions on a single chip[66] (Fig. 1.9). These functions include sample preparation, purification, storage, mixing, and detection. These chips use pressure, electro-osmosis, electrophoresis and other mechanisms to move samples and reagents through microscopic channels and capillaries, some as small as a few dozen nanometers.[51] Lab-on-a-chip has many applications in in vitro diagnostics. Their main advantages are the extremely rapid analysis of small samples, the high degree of automation and the low cost due to the low consumption of reagents and samples.[51] Some of the applications of lab-on-a-chip are in real-time polymerase chain reaction[67] and immunoassays[68] to detect bacteria, viruses and cancers. It can also be used in blood sample preparation to crack cells and extract their DNA.[66]

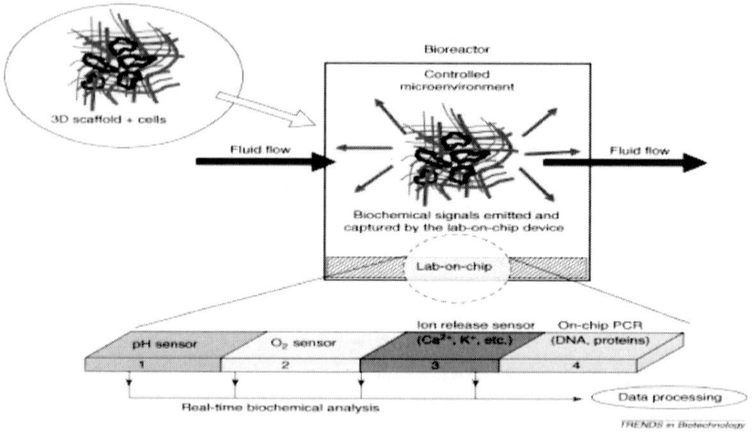

Figure 1.9 A 3D scaffold with cells is cultured within a bioreactor in which the microenvironment is controlled. The physical and chemical stimuli, which are produced with the application of a controlled fluid flow within the scaffold, induce biochemical signals that are captured by a lab-on-a-chip, in which several nanosensors can be integrated to detect pH, O_2, ions, DNA or proteins. These data can be analysed in real-time to regulate the microenvironments automatically through a feedback-loop mechanism.[69]

1.4.2 In vivo Diagnostics

The in vivo use of nanomaterials as diagnostic agents is of intense interest owing to their unique properties.[70] The most popular nanoparticles that are used for in vivo diagnostics are gold nanoparticles, quantumdots (QDs), and magnetic nanoparticles.[52] These nanoparticles have demonstrated their potential for in vivo imaging due to their size and unique optical properties.[71,72,73] For instance, their size ranges from 20 to 200 nm, so they can avoid renal filtration, leading to prolonged residence time in the blood stream that enables more effective targeting of diseased tissues.[69]

In vivo imaging

The use of nanotechnology for in vivo imaging includes techniques designed to obtain molecular data to identify the causes of the disease in vivo. Contrary to classic imaging diagnosis with computed tomography (CT), magnetic resonance imaging (MRI) or ultrasound, nano-imaging makes it possible to study human biochemical processes in different organs in vivo, opening up new horizons in instrumental diagnostic medicine. These systems, however, have emerged recently and only some of them are in clinical and pre-clinical use. The classification of these systems is included in Table 1.3.[51]

Table 1.3 Classification of nano-imaging systems

• Nano-imaging systems
• Positron-emission tomography (PET)
• Single-photon-emission CT (SPECT)
• Fluorescence reflectance imaging
• Fluorescence-mediated tomography (FMT)
• Fibre-optic microscopy
• Optical frequency-domain imaging
• Bioluminescence imaging
• Laser-scanning confocal microscopy
• Multiphoton microscopy

Nano-imaging diagnosis is an indispensable diagnostic tool for numerous diseases such as cancer, cardiovascular diseases and neurological syndromes. The advantages of nano-imaging diagnosis are the early detection of the disease, as well as the monitoring of the stages of the disease. In the future, imaging-diagnosis could support the development of individualised medicine and the real-time assessment of therapeutic and surgical efficacy. These techniques should be non-invasive and sensitive, and provide objective information on cell survival, function and localisation.[51]

Gold nanoparticles are considered very powerful labels for sensors because of the variety of analytical techniques that can be applied to detect them. The size of gold nanoparticles ranges around 10–20 nm in diameter. Small pieces of DNA can be attached to gold particles and the nanoparticles assemble onto a sensor surface in the presence of a complementary target. By this technique, a number of different DNA sequences can be detected simultaneously, using multiple DNA strands in the surface.[52]

Quantum dots (QDs), having a wide range of applications, represent a very promising tool for molecular diagnostics and genotyping due to their unique properties. They are inorganic fluorophores with high sensitivity, broad excitation spectra, stable fluorescence with simple excitation, and no need for lasers. Among their numerous applications, are cancer and viral diagnosis.[52] In the first case, QDs are conjugated with cancer cells in living animals. The QDs are luminescent and stable so they can be combined with fluorescence microscopy to follow cells at high resolution in living animals (Fig. 1.10). In the second case, antibody-conjugated nanoparticles detect rapidly and sensitively the respiratory syncytial virus (RSV) and estimate the relative levels of surface protein expression.[74] A major development in the case of QDs is the use of dual-colour QDs or fluorescence energy transfer nanobeads that can be simultaneously excited with a single light source. A QD system can detect the presence of particles of the RSV in a matter of hours.[52]

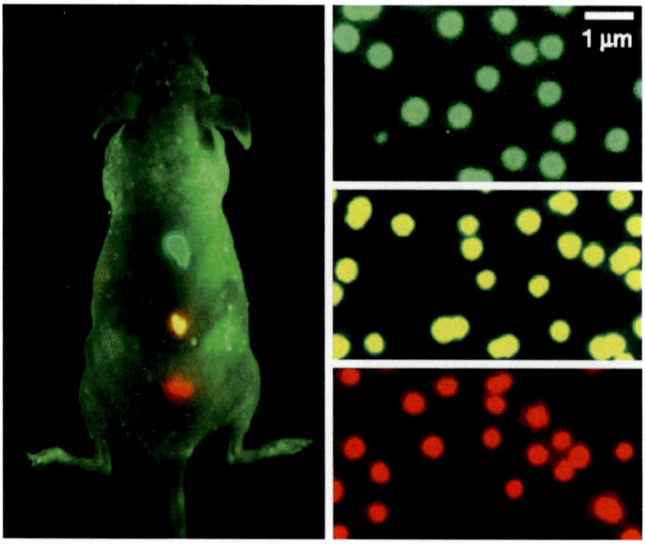

Figure 1.10 Quantum dot labelled cancer cells illustrating dots emitting green, yellow or red light.[75]

Magnetic nanoparticles are iron nanoparticles with a size that ranges from 15 to 20 nm. These nanoparticles are used for in vivo diagnosis as labelling molecules for bioscreening. For example, magnetic nanoparticles are very useful for cell-tracking cells and for calcium sensing. The use of magnetic nanoparticles is usually combined with MRI for in vivo imaging. Table 1.4 lists some applications of magnetic nanoparticles in combination with MRI.[52]

Table 1.4 Applications of magnetic nanoparticles—MRI

• Applications of magnetic nanoparticles—MRI
• Ferrofluids for calmodulin and its targets detection
• Superparamagnetic nanoparticles for lymph-node metastases
• Ultrasmall SPIO for ischemic lesions diagnosis
• Dextran-coated iron oxide nanoparticle for intracranial tumour detection

1.5 Targeted Drug Delivery

1.5.1 Nanocarrier-Based Drug Delivery Systems

A major focus of nanomedicine research—and an integral part of the overall process of drug development—has been the idea of delivering the drug precisely and safely to its target site, in a controlled manner. The concept of drug targeting to the desired cell or tissue aims to achieve the maximum therapeutic effect with the minimum risk-to-benefit ratio.[76] For example, drug targeting enables the delivery of chemotherapy agents directly to tumours reducing the systemic side effects which are related to the accumulation of the drug to other tissues/organs.[77] To this end, nanocarriers serve as a useful potential tool for achieving the main objective of drug delivery, offering important advantages (Fig. 1.11).

Nanocarriers, are engineered to (i) freely traverse through the body and penetrate tissues (i.e., tumours), (ii) be taken up by cells via endocytosis, and (iii) recognise specific biophysical characteristics and thus reach the target-cells, while minimising drug loss and toxicity to the non-desired sites.[78]

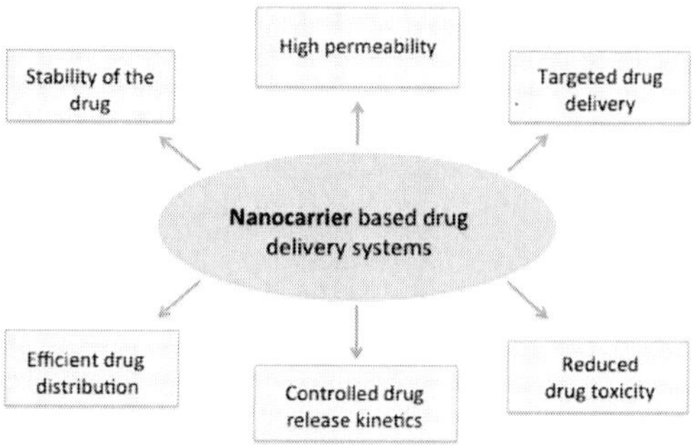

Figure 1.11 Advantages of nanocarrier-based drug delivery systems.

The ideal characteristics of a nanocarrier, which are summarised in Table 1.5, should include among others: in vivo efficacy, diminished undesirable side effects, stability, bio-availability, improved pharmacokinetic profiles, and controlled-release kinetics. Furthermore, the combination of various drugs, targeting different biological targets, as well as the combination of a drug and an imaging agent in one nanocarrier, represent only some of the future challenges. Thus, the design of *multifunctional nanocarriers* is a very promising future approach, which can lead to significant therapeutic results.

Table 1.5 Ideal characteristics and comments on the use of a nanocarrier[78,79]

Characteristic	Comments
Decreased toxicity	The use of biodegradable/biocompatible materials which are metabolised into non-toxic components and cleared through the circulation, reduces the risk of unwanted toxicities and adverse effects. Nanoformulations of drugs can also serve to decrease toxic side effects that limit their therapeutic efficacy (i.e., the nanoformulated version of doxorubicin).
Size (10–200 nm)	The range of the most effective particle diameters for drug delivery systems is, in most cases, 10–200 nm. The size must be large enough to avoid leakage into capillaries (20–200 nm), but not so large (>200) to be susceptible to macrophage -based clearance.
Stability	The nanocarrier should provide stability to the encapsulated drug, (i.e., be stable to the physiological environment of the body such as pH values, ionic strength, temperature etc.).
Clearance mechanism	Materials are cleared according to size. Small particles (0–30 nm) are rapidly cleared by renal excretion. Nanocarriers >30 nm are cleared by the mononuclear phagocytic system (MPS), consisting of macrophages located in the liver and the spleen.[80]
Long circulation times	Long circulation times to the circulatory system must be promoted for the nanocarriers, in order to find and sequester in the desired cells.
Ability to target the desired cell/tissue	The conjugated or bound drug-carrier complex should arrive and act preferentially at the selected target. Thus, it should be capable of surface bioconjugation to have the ability to recognise specific biophysical characteristics to target molecules within the body.[81]

1.5.2 Clinical Applications for Nanocarrier-Based Drug Delivery

The ability to apply nanomaterials as targeted delivery systems for drugs and other therapeutics holds promise for a wide variety of diseases (i.e., infectious diseases, metabolic and autoimmune diseases, central nervous system diseases, etc.) including many types of cancer.[75] For example, Fig. 1.12 depicts the variety of applications of a very challenging type of nanocarriers, *gold nanoparticles*, which can be chemically modified to deliver genes and drugs as well as target effectively the desired cell/tissue.

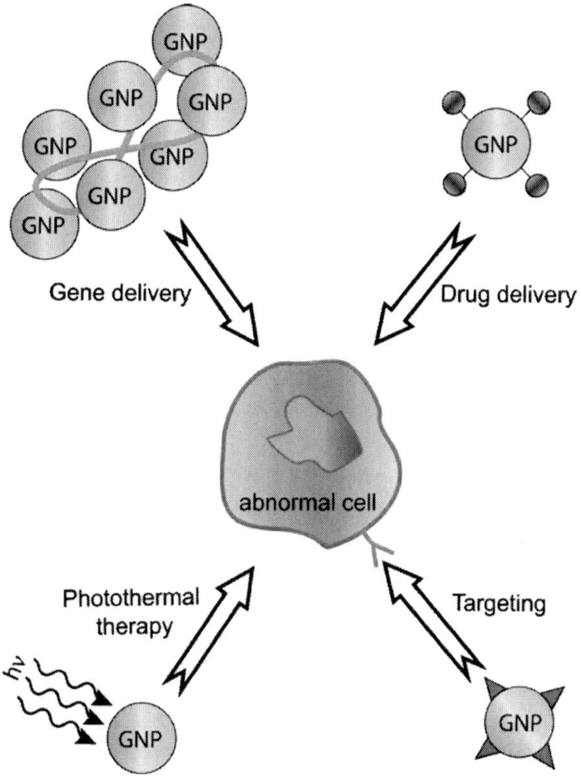

Figure 1.12 Various applications of gold nanoparticles in therapy.[82]

Some examples of different types of nanocarriers used for specific applications such as gene therapy, vaccination and cancer therapy are given in Table 1.6.

Table 1.6 Examples of nanocarrier-based drug delivery systems

System	Drug	Remark	Ref.
Polymeric NPs			
PLGA	OVA, TT	Suitable vaccine delivery system	83
	Paclitaxel	Suitable for various cancer therapies	84
Chitosan	*Toxoplasmi gondii*	Induction of immune response with decreased cytotoxicity	85
Hydrogel NPs	Antisense oligonucleotides	Efficient gene carrier-limited stability	86
Polymeric micelles	Taxol	Tumour targeting	87
Liposomes	gene	Efficient gene transporter	88
Dendrimers	DNA/genes	Targeting potential Ease of preparation	89

OVA: Ovalbumin; TT: Tetanus Toxoid.

Cancer: Chemotherapeutic agents are poor in tumour specificity and, therefore, exhibit dose-limiting toxicity. Nanocarrier-based drug delivery systems have the ability to target directly the cancer cell, release the drug in a controlled rate and maximise the therapeutic efficacy. Most tumours possess some unique pathophysiologic characteristics (i.e., high vascular permeability or extensive angiogenesis-EPR effect), which provide a passive target for nanocarriers.[74] Nanoparticles (i.e., gold nanoparticles), functionalised liposomes, albumin-based-particles, dendrimers and polymeric micelles are candidate carriers for the delivery of chemotherapeutic agents. For example, nanoformulations for the delivery of the chemotherapeutic drugs doxorubicin and paclitaxel are Food and Drug Administration (FDA) approved. PEGylated liposomal formulations of doxorubicin have been found to extend the half-life of the drug substantially and the albumin-based paclitaxel nanoformulation, *Abraxane*, efficiently targets tumour cells.[90] Another very important aspect is that nanocarriers have been found to enhance the delivery of chemotherapeutics across the blood brain barrier (BBB), while cationic nanospheres, polymers, lipids and hydrogels can be used to deliver selected gene therapy products.[75,91]

Infectious diseases: In the case of infectious diseases, biodistribution and penetration of cells including macrophages and dendritic cells should be enhanced. Nanoformulated drugs are being designed to specifically deliver therapeutics to sites of infection and in regions of the body that are often difficult to reach using traditional or available treatments (i.e., mucosal surfaces). Nanoparticles (i.e., PLGA, chitosan etc.), liposomes, dendrimers and micelles are candidate carriers for the delivery of antigens. Nanoparticles and liposomes may also be used in order to enhance the absorption through the mucosal tissues, thus facilitating, the delivery of the antigens to the mucosal surfaces of the body.[75,91]

Metabolic and Autoimmune diseases: Metabolic and autoimmune diseases (i.e., diabetes and osteoporosis) are generally treated with chronic and frequent drug injections. Thus, nanocarrier-mediated drug delivery system should provide a controlled and sustained delivery of the drug to achieve the optimum therapeutic effect with the minimum number of doses. A controlled drug biodistribution to provide greater targeting to the inflammatory cells is also needed. In this case, nanoparticles and liposomes are the most effective candidates.

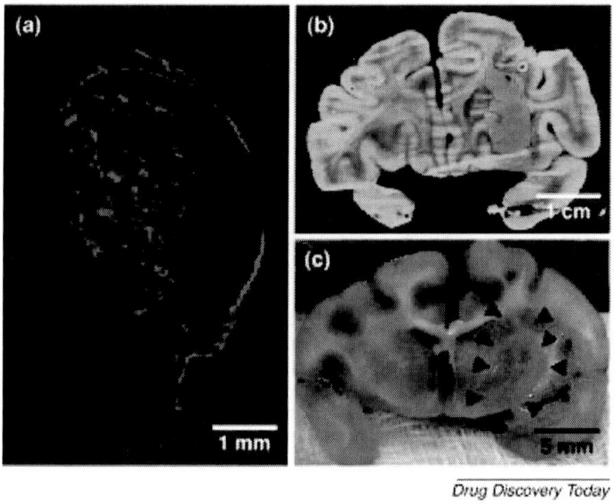

Figure 1.13 Extensive distribution of fluorescent PEG-modified liposomes with encapsulated Gadoteridol (GDL), as a contrast agent, in CNS. (a) Almost complete coverage of rodent striatum; (b) distribution visualised by UV light in primate putamen; (c) view of liposome distribution in putamen immediately after infusion procedure.[94]

Central nervous system diseases (CNS): In the case of the CNS diseases, the delivery methods used so far do not provide adequate delivery and distribution of the drugs across the blood brain barrier. Nanoparticles have the potential to overcome the barriers and effectively transport to the CNS. PEG-modified liposomes, as well as liposomes combined with viral components (virosomes), have also been employed to target specific sites in the brain.[91,92,93]

1.5.3 Drug Targeting Approaches

The targeting of drugs offers enormous advantages, but it is equally challenging. A better understanding of the physiological barriers, which a drug needs to overcome, will enable the development of the ideal targeted drug delivery systems. The main hurdles to drug targeting include the physiological barriers (i.e., the phagocytic cells of the RES system), and the biochemical and pharmaceutical challenges concerning the development of the appropriate techniques of conjugating targeting ligands to the nanocarriers. The targeting of drug to a specific site is not the only challenge to be faced. Retaining the drug for the desired duration to the specific site in order to elicit the pharmacological action also remains a crucial factor.[95] To this end, many methods have been developed, which can further be classified into two key approaches: active and passive. The main characteristics of each approach are summarised in Table 1.7 and the mechanisms of passive and active targeting of drugs are shown in Fig. 1.14.

Table 1.7 Drug targeting approaches[94]

Passive targeting	**Active targeting**
Pathophysiological factors of the tissue	*Biochemical targets*
• Infection/Inflammation • EPR effect	• Organs • Cellular • Intracellular
Physiochemical properties of the nanocarrier	*Physical/external stimuli*
• Size • Molecular weight	• Ultrasound • Magnetic field

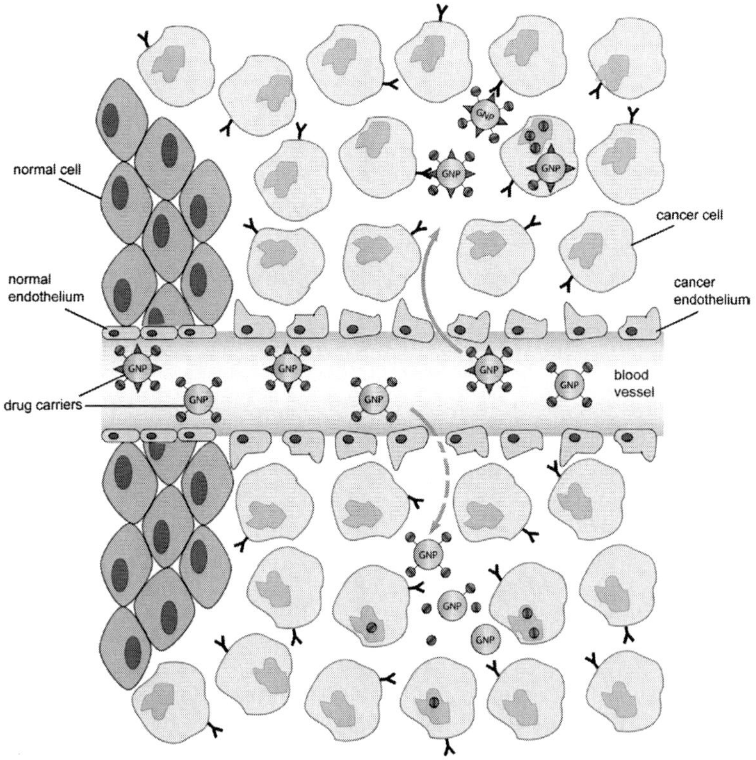

Figure 1.14 A schematic approach of drug delivery via 'active' and 'passive' targeting, solid and dotted lines, respectively.[82]

Passive targeting

Passive targeting depends on the characteristics of the nanocarrier as well as the pathophysiology of the disease in order to preferentially accumulate the drug at the site of interest and avoid non-specific distribution. It occurs due to extravasation of the nanoparticles at the diseased site where the microvasculature is leaky. The subsequent extravasation of complexes takes place either via *transcytosis*, whereby macromolecules are internalised from the blood at points of invagination of the cell membrane, or *paracellularly*, via diffusion through the tight junctions of endothelial cells.[81] For example, passive targeting occurs to tumour and inflamed tissues. Liposomes, polymeric nanoparticles and micelles are examples of nanocarriers that have efficiently targeted their cargo to tumour tissues in a passive manner.

Active targeting

As passive targeting approaches are limited, there has been great effort in the development of active approaches to drug targeting. The conjugation of active agents (targeting ligands) to the nanocarriers is the most frequently used method of directing them to their target sites, enabling their recognition by specific receptors on the surface of cancer cells or other target cells.[75,95] To this end, various techniques have been devised, including *covalent* and *non-covalent conjugation.*[96] It should be noted, therefore, that the coupling reactions must not affect the biological activity of the ligand and should not adversely affect the structure of the nanocarriers.[95,97]

1.6 Nanotoxicology and Safety Aspects

Nanotoxicology was coined as a term in 2004–2005 despite the fact that concerns about the adverse effects of nanomaterials on human health were voiced several years earlier. Nanotoxicology, as a field, aims to study the adverse effects of engineered nanomaterials on living organisms and ecosystems, in order to prevent and eliminate adverse responses and also evaluate the risk-to-benefit ratio of using nanoparticles in medical settings.[91, 98]

The chemical composition of the nanomaterials, along with their submicron size, appear to be the dominant indicators for their drastic and toxic effects. For example, nanomaterials, even when made of inert elements like gold, become highly active at nanometre dimensions. They can enter the host organism via the venous, dermal, subcutaneous, nasal, intraperitoneal and oral routes and can directly access the blood stream either via inhalation or via ingestion (Fig. 1.15).[99,100] The substantially small size of the nanomaterials, as mentioned in Section 1.2, allows them to enter human cells, cross the cellular membranes and move into the cytoplasm and reside for an uncertain period of time, possibly infecting other organs before being excreted out of the host organism.[101]

Thus, a complete understanding of the size, shape, composition and interactions of the nanostructures with the biological systems is necessary in order to clarify whether, and to what extent, the exposure of humans, animals and plants to engineered nanostructures could produce adverse responses.[103] To this end, the new subdiscipline of nanotechnology called nanotoxicology has emerged, in order to explore the backround of the interactions that occur between the host organism and the nanostructured constructs focusing on the crucial parameters of the above constructs (i.e., size and chemical properties) that dictate the toxic effects.

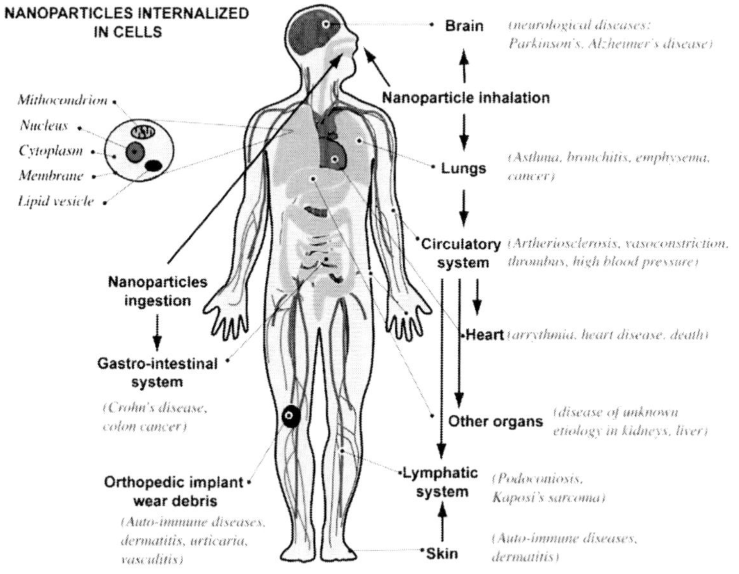

Figure 1.15 Schematics of human body with pathways of exposure to nanoparticles, affected organs, and associated diseases from epidemiological, in vivo and in vitro studies.[102]

For all these reasons, nanoparticle formulations should be tested on a case-by-case basis, depending on their portal of entry. A thorough understanding of the biological responses to the nanomaterials is also needed in order to develop effective testing procedures and protocols. Furthermore, a close collaboration between those working in drug delivery and particle toxicology is necessary for the exchange of concepts, methods and know-how to move the issue ahead.[99] A more extensive approach on the toxicology aspects of specific nanoparticles is given in Chapter 12.

1.7 Conclusions

Nanotechnology, undoubtedly, has the potential to revolutionise conventional medicine. Nanocarriers are entities that can contribute to therapy, imaging and early diagnosis of various diseases. For example, rapid and portable diagnostics, multifunctional drug delivery systems, and biodegradable polymeric scaffolds for tissue regeneration are some paradigms of the possibilities that nanomedicine offers. Nanomolecular imaging procedures enable the surface characterisation on a nanometer scale at predefined locations, while the opportunity to target specific surfaces (i.e., targeted drug delivery) is enabled by the chemical interactions in the nanoscale. Despite the rapid and constant development of nanomedicine tools and techniques, there is still insufficient knowledge regarding the toxic impact of the engineered nanostructures to human health. Thus, nanomedicine is a very promising field, with the potential to give solutions to many important health issues.

References

1. Feynman RP (1960) There's plenty of room at the bottom: an invitation to enter a new field of physics. *Eng Sci*: http://www.zyvex.com/nanotech/feynman.html (accessed February 2, 2005).
2. Whitesides GM. (2003) The 'right' size in nanobiotechnology. *Nat Biotechnol* (21), 1161–1165.
3. Tomalia DA (2005) Birth of a new macromolecular architecture: dendrimers as quantized building blocks for nanoscale synthetic polymer. *Chemi Prog Polym Sci* (30), 294–324.
4. Biswas A, Bayer IS, Biris AS, Wang T, Dervishi E, Faupel F (2012) Advances in top-down and bottom-up surface nanofabrication: techniques, applications & future prospects. *Adv Colloid Interface Sci* (170), 2–27.
5. Zhang S, Marini DM, Hwang W, Santoso S (2002) Design of nanostructured biological materials through self-assembly of peptides and proteins. *Curr Opin Chem Biol* (6) 865–871.
6. European Technology Platform on NanoMedicine (2005).
7. European Science Foundation (2005) Nanomedicine, an ESF–European Medical Research Councils (EMRC) forward look report.
8. Boisseau P, Loubaton B (2011) Nanomedicine, nanotechnology in medicine. *C R Physique* (12), 620–636.
9. Kawasaki ES, Audrey Player T (2005) Nanotechnology, nanomedicine, and the development of new, effective therapies for cancer. *Nanomed Nanotechnol Biol Med* (1) 101–109.

10. Wei Q, Mao K, Wu D, Dai Y, Yang J, Du B, Yang Y, Li H (2010) A novel label-free electrochemical immunosensor based on graphene and thionine nanocomposite. *Sensors Actuators B* (149) 314–318.
11. Seaton A, Donaldson K (2005) *Nanoscience, nanotoxicology, and the need to think small.* www.thelancet.com (365), 923–924.
12. Barnes LM, Phillips GJ, Davies JG, Lloyd AW, Cheek E, Tennison SR, AP Rawlinson, Kozynchenko OP, Mikhalovsky SV (2009) The cytotoxicity of highly porous medical carbon adsorbents. *Carbon* (47), 1887–1895.
13. Kunzmann A, Andersson B, Thurnherr T, Krug H, Scheynius A, Fadeel B (2011) Toxicology of engineered nanomaterials: Focus on biocompatibility, biodistribution and biodegradation. *Biochimica et Biophysica Acta* (1810) 361–373.
14. Nishiyama N, Kataoka K (2006) Current state, achievements, and future prospects of polymeric micelles as nanocarriers for drug and gene delivery. *Pharmacol Therap* (112), 630–648.
15. Vergaro V, Scarlino F, Bellomo C, Rinaldi R, Vergara D, Maffia M, Baldassarre F, Giannelli G, Zhang X, Lvov YM, Leporatti S (2011) Drug-loaded polyelectrolyte microcapsules for sustained targeting of cancer cells. *Adv Drug Deliv Rev* (63), 847–864.
16. Torchilin V (2009) Multifunctional and stimuli-sensitive pharmaceutical nanocarriers. *Eur J Pharm Biopharm* (71), 431–444.
17. Jabr-Milane L, van Vlerken L, Devalapally H, Shenoy D, Komareddy S, Bhavsar M, Amiji M (2008) Multi-functional nanocarriers for targeted delivery of drugs and genes. *J Control Release* (130), 121–128.
18. Veiseh O, Kievit FM, Ellenbogen RG, Zhang M (2011) Cancer cell invasion: treatment and monitoring opportunities in nanomedicine. *Adv Drug Deliv Rev* (63) 582–596.
19. Jabr-Milane L, van Vlerken L, Devalapally H, Shenoy D, Komareddy S, Bhavsar M, Amiji M (2008) Multi-functional nanocarriers for targeted delivery of drugs and genes. *J Control Release* (130), 121–128.
20. Logothetidis S, *Nanomedicine and Nanobiotechnology* (2012) Springer-Verlag Berlin Heielberg.
21. Kumari A, Yadav SK, Yadav SC (2010) Biodegradable polymeric nanoparticles based drug delivery systems. *Colloids Surf B Biointerfaces* (75), 1–18.
22. http://vindicopharma.com/wp-content/uploads/2009/06/Figure-1.jpg.

23. Hillaireau H, Couvreur P (2009) Nanocarriers entry into the cell: relevance to drug delivery. *Cell Mol Life Sci* (66), 2873–2896.

24. Danhier F, Feron O, Préat V (2010) To exploit the tumor microenvironment: Passive and active tumor targeting of nanocarriers for anti-cancer drug delivery. *J Control Release* (148), 135–146.

25. Torchilin VP (2006) Multifunctional nanocarriers. *Adv Drug Deliv Rev* (58), 1532–1555.

26. Koren E, Apte A, Jani J, Torchilin VP (2012) Multifunctional PEGylated 2C5-immunoliposomes containing pH-sensitive bonds and TAT peptide for enhanced tumor cell internalization and cytotoxicity, *J Control Release* (160), 264–273

27. Ma Y, Nolte RJM, Cornelissen JJLM (2012) Virus-based nanocarriers for drug delivery. *Adv Drug Deliv Rev* (64), 811–825*xxx*.

28. Arsawang U, Saengsawang O, Rungrotmongkol T, Sornmee P, Wittayanarakul W, Remsungnen T, Hannongbua S (2011) How do carbon nanotubes serve as carriers for gemcitabine transport in a drug delivery system? *J Mol Graph Model* (29), 591–596.

29. Prakash S, Malhotra M, Shao W, Tomaro-Duchesneau C, Abbasi S (2011) Polymeric nanohybrids and functionalized carbon nanotubes as drug delivery carriers for cancer therapy. *Adv Drug Deliv Rev* (63), 1340–1351.

30. Singh R, Lillard Jr. JW (2009) Nanoparticle-based targeted drug delivery. *Exp Mol Pathol* (86), 215–223.

31. Tokumasu F, Fairhurst RM, Ostera GR, Brittain NJ, Hwang J, Wellems TE, Dvorak JA (2005) Band 3 modifications in plasmodium falciparum-infected AA and CC erythrocytes assayed by autocorrelation analysis using quantum dots. *J Cell Sci* (118), 1091–1098.

32. Matsumura G, Hibino N, Ikada Y, Kurosawa H, Shin'oka T (2003) Successful application of tissue engineered vascular autografts: clinical experience. *Biomaterials* (24), 2303–2308.

33. Bouten CVC, Dankers PYW, Driessen-Mol A, Pedron S, Brizard AMA, Baaijens FPT (2011) Substrates for cardiovascular tissue engineering. *Adv Drug Deliv Rev* (63) 221–241.

34. Hutmacher DW, Garcia AJ (2005) Scaffold-based bone engineering by using genetically modified cells. *Gene* (347), 1–10.

35. Liu C, Xia Z, Czernuszka JT (2007) Design and development of three-dimensional scaffolds for tissue engineering. *Trans IChemE, Part A, Chem Eng Res Des*, 85(A7), 1051–10.

36. Chung C, Burdick JA (2008) Engineering cartilage tissue. *Adv Drug Deliv Rev* (60), 243–262.
37. Drosse I, Volkmer E, Capanna R, De Biase P, Mutschler W, Schieker M (2008) Tissue engineering for bone defect healing: An update on a multi-component approach. *Injury, Int J Care Injured* (39S2), S9–S20.
38. Csaki C, Schneider PRA, Shakibaei M (2008) Mesenchymal stem cells as a potential pool for cartilage tissue engineering. *Ann Anat* (190), 395—412.
39. Doblar M, Garcia JM, Gomez MJ (2004) Modelling bone tissue fracture and healing: a review. *Eng Fracture Mech* (71), 1809–1840.
40. Klein TJ, Schumacher BL, Schmidt TA, Li KW, Voegtline MS, Masuda K, Thonar EJ, Sah RL (2003) Tissue engineering of stratified articular cartilage from chondrocyte subpopulations. *OsteoArthritis Cartilage* (11), 595-602.
41. Klein TJ, Schumacher BL, Schmidt TA, Li KW, Voegtline MS, Masuda J, Thonar EJ, Sah RL (2003) Tissue engineering of stratified articular cartilage from chondrocyte subpopulations, *Osteoarthritis Cartilage* (11), 595-602.
42. Mesa JM, Zaporojan V, Weinand C, Johnson TS, Bonassar L, Randolph MA, Yaremchuk MJ, Butler PE (2006) Tissue engineering cartilage with aged articular chondrocytes in vivo, *Plast Reconstr Surg* (118), 41–49.
43. Li Y, Tew SR, Russell AM, Gonzalez KR, Hardingham TE, Hawkins RE (2004) Transduction of passaged human articular chondrocytes with adenoviral, retroviral, and lentiviral vectors and the effects of enhanced expression of SOX9, *Tissue Eng* (10) 575–584.
44. Chung C, Mesa J, Miller GJ, Randolph MA, Gill TJ, Burdick JA (2006) Effects of auricular chondrocyte expansion on neocartilage formation in photocrosslinked hyaluronic acid networks. *Tissue Eng* (12), 2665–2673.
45. van Osch GJVM, Mandl EW, Jahr H, Koevoet W, Nolst-Trenite G, Verhaar JA (2004) Considerations on the use of ear chondrocytes as donor chondrocytes for cartilage tissue engineering, *Biorheology* (41), 411–421.
46. Isogai N, Kusuhara H, Ikada Y, Ohtani H, Jacquet R, Hillyer J, Lowder E, Landis WJ (2006) Comparison of different chondrocytes for use in tissue engineering of cartilage model structures, *Tissue Eng* (12) 691–703.
47. Tay AG, Farhadi J, Suetterlin R, Pierer G, Heberer M, Martin I (2004) Cell yield, proliferation, and postexpansion differentiation capacity of human ear, nasal, and rib chondrocytes, *Tissue Eng* (10), 762–770.

48. Mauck RL, Soltz MA, Wang CCB, Wong DD, Chao PHG, Valhmu WB, Hung CT, Ateshian GA (2000) Functional tissue engineering of articular cartilage through dynamic loading of chondrocyte-seeded agarose gels, *J Biomech Eng-Trans ASME* (122), 252–260.

49. Miyanishi K, Trindade MCD, Lindsey DP, Beaupre GS, Carter DR, Goodman SB, Schurman DJ, Smith RL (2006) Dose- and time-dependent effects of cyclic hydrostatic pressure on transforming growth factor-beta 3-induced chondrogenesis by adult human mesenchymal stem cells in vitro, *Tissue Eng* (12), 2253–2262.

50. Berthiaume F, Yarmush ML, *Tissue Eng*, 817–842.

51. Houria Boulaiz et al. (2011) Nanomedicine: application areas and development prospects, *Int J Mol Sci* (12), 3303–3321.

52. Jain KK (2007) *Applications of Nanobiotechnology in Clinical Diagnostics, Clinical Chemistry* 53:11.

53. Jain KK (2003) Nanodiagnostics: application of nanotechnology in molecular diagnostics. *Expert Rev Mol Diagn* (3), 153–161.

54. Singh RP, Oh BK, Ch JW (2010) Application of peptide nucleic acid towards development of nanobiosensor arrays, *Bioelectrochemisty* (79), 153–161.

55. YeoHeung Yun, et.al (2007) Nanotube electrodes and biosensors, *Nanotoday* (2).

56. European Technology Platform on Nanomedicine. Nanotechnology for health. Vision paper and basis for a strategic research agenda for nanomedicine. *Eur Comm* (2005) (1), 1–39.

57. Caygill RL, Eric Blair G, Millner PA (2010) A review on viral biosensors to detect human pathogens, *Anal Chim Acta* (681) 8–15.

58. Perez JM, Simeone FJ, Saeki Y, Josephson L, Weissleder R. (2003) Viral-induced self-assembly of magnetic nanoparticles allows the detection of viral particles in biological media. *J Am Chem Soc* (125) 10192–10193.

59. Boisselier E, Astruc D (2009) Gold nanoparticles in nanomedicine: preparations, imaging, diagnostics, therapies and toxicity, *Chemical Society Reviews*.

60. McCaffrey Ta, Fu C, Du B, Eksinar S, Kent KC, Bush Jr. H, Kreiger K, Rosengart T, Cybulsky MI, Silverman ES, Collins T (2000) High-level expression of Egr-1 and Egr-1-inducible genes in mouse and human atherosclerosis. *J Clin Invest.* (105), 653–662.

61. Perou CM, Jeffrey SS, Van de Rijn M, Rees CA, Eisen MB, Ross DT, Pergamenschikov A, Williams CF, Zhu SX, Lee JC, Lashkari D, Shalon D,

Brown PO, Botstein D (1999) Distinctive gene expression patterns in human mammary epithelial cells and breast cancers. *Proc Natl Acad Sci USA* (96), 9212–9217.

62. Perou CM, Sørlie T, Eisen MB, Rijn M, Jeffrey SS, Rees CA, Pollack JR, Ross DT, Johnsen H, Akslen LA, Fluge O, Pergamenschikov A, Williams C, Zhu SX, Lønning PE, Børresen-Dale AL, Brown PO, Botstein D (2000) Molecular portraits of breast tumours. *Nature* (406), 747–752.
63. Alon U, Barkai N, Notterman DA, Gish K, Ybarra S, Mack D, Levine AJ (1999) Broad patterns of gene expression revealed by clustering analysis of tumor and normal colon tissues probed by oligonucleotide arrays. *Proc Natl Acad Sci USA* (96), 6745–6750.
64. Kaminski N, Allard JD, Pittet JF, Zuo F, Griffiths MJ, Morris D, Huang X, Sheppard D, Heller RA (2000) Global analysis of gene expression in pulmonary fibrosis reveals distinct programs regulating lung inflammation and fibrosis. *Proc Natl Acad Sci USA* (97), 1778–1783.
65. Telford M (2004) Nanoarrays for ultrasensitive biodetection, *Nanotoday*.
66. Ghallab YH, Badawy W (2010) *Lab-on-a-Chip: Techniques, Circuits, and Biomedical Applications*. Artech House: Norwood, MA, USA, pp. 1–220.
67. Kim J, Byun D, Mauk MG, Bau HH (2009) A disposable, self-contained PCR chip. *Lab Chip* (9), 606–612.
68. Oosterbroek E, Van den Berg A (eds) (2003) *Lab-on-a-Chip: Miniaturized Systems for (Bio)Chemical Analysis and Synthesis*, 2nd ed. *Elsevier Science: Amsterdam, The Netherlands,* pp. 1-402.
69. Engel E, Michiardi A, Navarro M, Lacroix D, Planell JA (2007) Nanotechnology in regenerative medicine: the materials side, *Trends Biotechnol (*26).
70. Park J-H, Gu L, von Maltzahn G, Ruoslahti E, Bhatia SN, Sailo MJ (2009) Biodegradable luminescent porous silicon nanoparticles for in vivo applications. *Nat Mater* (8), 331–336.
71. Gao XH, Cui YY, Levenson RM, Chung LWK, Nie SM (2004) In vivo cancer targeting and imaging with semiconductor quantum dots. *Nature Biotech.* (22), 969–976.
72. Liu Z, et al. (2008) Circulation and long-term fate of functionalized, biocompatible single-walled carbon nanotubes in mice probed by Raman spectroscopy. *Proc Natl Acad Sci USA* (105) 1410–1415 [PubMed: 18230737].
73. Kim D, Park S, Lee JH, Jeong YY, Jon S (2007) Antibiofouling polymer-coated gold nanoparticles as a contrast agent for in vivo X-ray computed

tomography imaging. *J Am Chem Soc* (129), 7661–7665 [PubMed: 17530850].

74. Agrawal A, Tripp RA, Anderson LJ, Nie S. (2005) Real-time detection of virus particles and viral protein expression with two-color nanoparticle probes. *J Virol* (79), 8625–8680.
75. Emerich DF, Thanos CG (2006) The pinpoint promise of nanoparticle-based drug delivery and molecular diagnosis, *Biomol Eng* (23) 171–184.
76. Mishra B, Patel BB, Tiwari S (2010) Colloidal nanocarriers: a review on formulation technology, types and applications toward targeted drug delivery. *Nanomed Nanotechnol Biol Med* (6) 9–24.
77. Hughes GA (2005) Nanostructure-mediated drug delivery. *Nanomed Nanotechnol Biol Med* (1), 22–30.
78. Adair JH, Parette MP, Altunoglu EI, Kester M (2010) *Nanoparticulate alternatives for drug delivery, ACS Nano* (4) 4967–4970.
79. Emerich DF, Thanos CG (2007) Targeted nanoparticle-based drug delivery and diagnosis. *J Drug Target* 15(3) 163–183
80. Gaumet M, et al. (2008) Nanoparticles for drug delivery: the need for precision in reporting particle size parameters. Eur J Pharm Biopharm (69) 1–9.
81. Malam Y, Loizidou M, Seifalian AM (2009) Liposomes and nanoparticles: nanosized vehicles for drug delivery in cancer. Trends Pharma*col Sci* (30) 592–599.
82. Ghosh P, et al. (2008) *Adv Drug Deliv Rev* 60 (2008) 1307–1315.
83. Rawat M, Singh D, Saraf S., Saraf S (2006) N*anocarriers: promising vehicle for bioactive drugs. Biol Pharm Bull* (29) 1790—1798.
84. Parveen S, Misra R, Sahoo SK (2012) Nanoparticles: a boon to drug delivery, therapeutics, diagnostics and imaging. Nanom*ed Nanotechnol (8), 147–166.*
85. Kim TH, Park IK, Nan JW, Choi YJ, Cho CS (2004) Bioma*terials (25) 3783–3792.*
86. Gupta M, Gupta AK(2004) J Co*ntrol Rel (99) 157–166.*
87. Torchilin VP, Lukyanov AN, Gao Z, Sternberg PB (2003) Proc *Natl Acad Sci USA (100) 6039–6044.*
88. Zhang Y, Calon F, Zhu C, Boado RJ, Pardridge WM (2003) Hum Gene Ther, 14, 1–12.
89. Choi Y, Thomas T, Kotlyar A, Islam MT, Baker JR (2005) Chem *Biol (12)* 35—43.

90. McMillan E, Batrakova E, Gendelman HE. Cell delivery of therapeutic nanoparticles. *Prog Mol Biol Translational Sci* (104).

91. Pardridge WM (2005) The blood brain barrier: bottleneck in brain drug development. *NeuroRx* (2) 3–14.

92. Shi N, Zhang Y, Zhu C, Boado RJ, Pardridge WM (2001) Brain-specific expression of an exogenous gene after i.v. administration. *Proc Natl Acad Sci USA* (98) 12754–12759.

93. Jiang C, Koyabu N, Yonemitsu Y, Shimazoe T, Watanabe S, Naito M, Tsuruo T, Ohtani H, Sawada Y (2003) In vivo delivery of glial cell-derived neurotrophic factor across the blood–brain barrier by gene transfer into brain capillary endothelial cells. *Hum Gene Ther* (14) 1181–1191.

94. Krauze MT, et al. (2006) Real-time imaging and quantification of brain delivery of liposomes. *Pharm Res* (23) 2493–2504.

95. Vasir JK, Reddy MK LabhasetwarVD (2005) *Nanosystems in drug targeting: opportunities and challenges. Curr Nanosci* (1) 47–64.

96. Nobs L, Buchegger F, Gurny R, Allemann E (2004) *J Pharm Sci* (93), 1980–1992.

97. Ganta S, Devalapally H, Shahiwala S, Amiji M (2008) A review of stimuli-responsive nanocarriers for drug and gene delivery. *J Control Release* (126) 187–204.

98. De Jong, WH, Born PJA (2008) *Drug delivery and nanoparticles: applications and hazards. Int J Nanomed* (3) 133–149.

99. Oberdorster G, Maynard A, et al. (2005) Principles for characterizing the potential human health effects from exposure to nanomaterials: elements of a screening strategy, *Particle Fibre Toxicol* (2) 1–35.

100. Ryman Rasmussen J, Jessica P, et al., Penetration of intact skin by quantum dots with diverse physicochemical properties, *Toxic Sci* (91), 159–165.

101. Pan Y, Leifert A, Ruau D, Neuss S, Bornemann J, Schmid G, Brandau W, Simon U, Jahnen-Deschent W (2009) Gold nanoparticles of diameter 1,4 nm trigger necrosis by oxidative stress and mitochondrial damage, *Small* (5), 2067–2076.

102. Buzea B, Pacheco Blandino, II, Robbie K (2007) Nanomaterials and nanoparticles: sources and toxicity. *Biointerphases* (2) MR17–MR172.

103. Fischer HC, Chan WCW (2007) Nanotoxicity: the growing need for in vivo study. *Curr Opin Biotechnol* (18) 565–571.

Chapter 2

Nanomedicine Combats Atherosclerosis

Varvara Karagkiozaki
Nanomedicine Group, Lab for Thin Films—Nanosystems and Nanometrology, Department of Physics, Aristotle University of Thessaloniki, 54124 Thessaloniki, Greece
vakaragk@physics.auth.gr

2.1 Introduction

Cardiovascular diseases (CVD), caused by atherosclerosis, are the most frequent causes of mortality and morbidity in developed societies. The pathogenesis of atherosclerosis involves a complex series of events occurring at the molecular or even the nanoscale level. A thorough understanding of the underlying mechanisms is essential for the development of strategies for the prevention of the disease, and for the development of new and effective therapies. This chapter reviews nanomedicine advances for better diagnosis, monitoring, and treatment of such a multifaceted disease and explores how nanoparticles can be leveraged to develop novel molecular tools for early detection of high risk lesions. Molecular targeting and site specific treatment with synthetic theranostic

Horizons in Clinical Nanomedicine
Edited by Varvara Karagkiozaki and Stergios Logothetidis

ISBN 978-981-4411-56-1 (Hardcover), 978-981-4411-57-8 (eBook)
www.panstanford.com

nanosystems represents a paradigm shift and makes the promise of combating atherosclerosis a realistic possibility.

The first part of the chapter highlights our current understanding of the pathophysiology of atherosclerosis, the vulnerability of plaque for rupture and the causes of plaque disruption that lead to the clinical manifestation of acute coronary syndrome. This knowledge basis enables the elucidation of nanomedicine utilities in identifying and treating the vulnerable plaques.

Emphasis is placed on the state–of-the-art stents that have been widely used to restore normal blood flow to ischemic myocardial sites and their drawbacks in the light of substantial nanotechnology based improvements. Finally, an outlook of the challenges and future perspectives of nanomedicine in cardiovascular field is presented.

Nanomedicine stands for the application of nanomaterials in medical technology. The term was first mentioned in the book *Unbounding the Future: The Nanotechnology Revolution* in 1991.[1] The European Commission has adopted cross-cutting definition of nanomaterials to be used for regulatory purposes as being 50 % or more of the particles with 1–100 nm size in at least one dimension. However, structures of several hundreds of nm or a few micrometres are also considered under nanotechnology applications.

Nanotechnology seeks to develop new materials by precisely engineering atoms and molecules to yield new molecular assemblies for favourable biological responses to translate into addressing medical challenges.[2] Owing to their nanoscale dimensions, such materials can interact with cell membrane receptors, peptides and cellular components to provide insights into the cell "nanoworld" and its function. Synthesis of nanomaterials may be achieved with a "top down" approach by miniaturising existing microscopic materials, or with a "bottom up" approach involving self-assembly of molecules into reproducible and well-defined nanoscale constructs.

Nanomedical pillars range from nanodiagnostics, to targeted drug delivery and tissue engineering strategies.[3] Nanoparticles offer exciting advantages and enormous potential to meet medical needs. The nanoparticle advantages include increased surface area, site specific targeting, delivery of insoluble drugs, enhanced rates of absorption, distribution, and metabolism of drugs in the body. Nanoparticles can circumvent biological barriers resulting in targeted delivery of therapeutic payloads at the diseased sites.[4] New drug-delivery nanomedicine approaches can improve the efficacy and safety of existing approved medicines and can lead to the development and approval of new drugs with inherent properties. Such advances in particle-based drug delivery will significantly affect the lives of patients afflicted with a variety of diseases such as CVD.

2.2 Understanding of Atherosclerosis

A better understanding of the atherosclerosis process and the factors that affect the plaque vulnerability enables the development of nanotechnology enabled practical tools that may aid patient management. To highlight potential imaging targets, it may be helpful to consider atherogenesis in terms of (a) a gradual progression of lesions over time, (b) a conversion of stable plaques to more advanced and vulnerable lesions, and (c) thrombotic complications.

To start with, atherosclerosis is mainly an inflammatory disease in which immune mechanisms interact with environmental and genetic risk factors to initiate, propagate and activate lesions in the arterial tree.[5]

The first event that initiates atherosclerosis is endothelial injury, leading through various steps to the eventual formation of atheromatous plaques (Fig. 2.1).

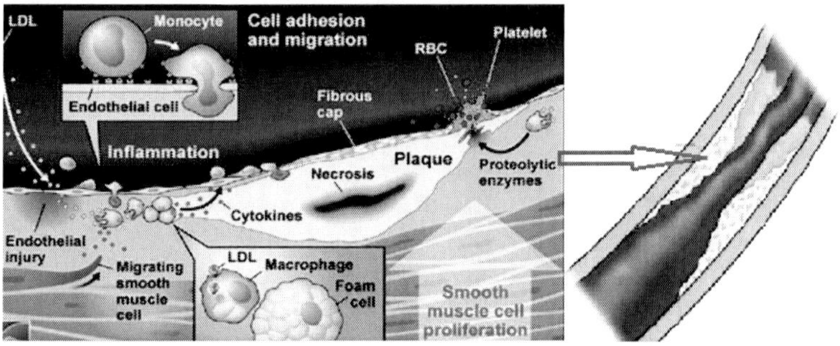

Figure 2.1 Schematic presentation of the cellular and molecular mechanisms of atherosclerosis that lead to the formation of plaques in the vessel wall and to arterial stenosis.

The endothelium plays a key role in vascular homeostasis through the release of a variety of autocrine and paracrine substances.[6] A healthy endothelium is also anti-atherogenic because of effects that encompass inhibition of platelet aggregation, smooth muscle cell (SMC) proliferation and leukocyte adhesion.

It is well established that nitric oxide (NO) is a crucial factor for the health and function of endothelial cells (ECs) due to its significant vasoprotective and cardioprotective effects, including inhibition of both platelet aggregation and inflammatory cell adhesion to ECs, disruption of proinflammatory cytokine – induced signalling pathways, inhibition of apoptosis, and regulation of tissue energy metabolism.[7]

During endothelial dysfunction, there is a reduced bio-availability of NO, causing the impairment of endothelium-dependent vasodilation and abnormalities in endothelial interactions with blood-borne cells such as leukocytes and platelets. In the dysregulated endothelium, intracellular granules like Wiebel-Palade bodies (storing von Willebrand- vWf and P-selectin) fuse with the cell membrane and consequently expose vWf and P-selectin on the endothelium surface. These factors can promote tethering and rolling of blood platelets at the vascular injury site via interaction with integrin GPIb/IX/V and P-selectin glycoprotein ligand-1 (PSGL-1), respectively, present on the platelet surface.[8] The P-selectin and other selectins (e.g., E- and L-) also mediate adhesion of inflammatory cells like leukocytes at the injury site. There is also an increased permeability of endothelium to low density lipoprotein (LDL) particles that enter the subendothelium. Within the arterial wall, these LDL particles undergo oxidation and endocytosis by the macrophages (via scavenger receptors pathways), leading to intracellular cholesterol accumulation and the formation of "foam" cells.[8]

The final outcome of this complex process is the development of atherosclerotic plaques that consist of lipid-rich necrotic cores made of foam cells and fibrous collagenous cap. The cap is formed by Extracellular Matrix (ECM) degradation and remodelling—processes mediated by macrophage and smooth muscle cell activity.[9] These dynamic interactions between the immune system, the endothelial barrier, and the lipoproteins form the mechanistic basis for our current understanding of atherosclerotic disease. Hence, the specific roles of all the cellular and molecular players in this dynamic interplay of atherogenesis still remain unknown.

Over time, mature lesions accumulate and form atherosclerotic plaques which may manifest clinically due to plaque related complications. Expansion of atherosclerotic lesions may lead to gradual stenosis of the lumen and, eventually, occlusion of the vessel initiating the clinical symptoms of angina and heart attack.

Atherosclerotic lesions (atheromata) are asymmetric focal thickenings of the innermost layer of the artery, the intima (Fig. 2.1). They consist of cells (such as blood-borne inflammatory and immune cells, vascular endothelial and smooth-muscle cells), connective-tissue elements, lipids, and debris. The atheroma is preceded by a fatty streak, an accumulation of lipid-laden cells beneath the endothelium that is prevalent in young people without causing any symptoms.[9]

Vulnerable Plaque Disruption: The mechanism responsible for the sudden conversion of a stable stenotic atherosclerotic disease to a life-threatening condition is usually plaque disruption with superimposed thrombosis. The risk of plaque disruption depends more on plaque composition and vulnerability (plaque type) than on the degree of stenosis.[10]

In cases where the matrix degradation is evident due to an increased activity of enzymes and of interferon gamma produced by inflammatory T-cells that downregulate the collagen production, the result is the progressive degradation and thinning of the fibrous plaque cap, making the plaque prone to rupture. Macrophages are capable of degrading ECM by phagocytosis or by secreting proteolytic enzymes such as plasminogen activators and a family of matrix metalloproteinases (MMPs: collagenases, gelatinases, and stromelysins) that may digest the fibrous cap. Besides macrophages, a wide variety of cells may produce MMPs. They are secreted in a latent zymogen form requiring extracellular activation, after which MMPs are capable of degrading all components of the extracellular matrix.[10]

MMP activity is controlled by: (i) inflammatory cytokines that induce the expression of MMP genes, (ii) plasmin that activates MMP proforms and, (iii) tissue inhibitors of metalloproteinase (TIMP-1 and TIMP-2) that suppress their action.

Often, due to stress and hemodynamic conditions, ruptures develop in such plaques and especially at the plaque shoulders causing intraplaque haemorrhage and exposure of the necrotic lipid core and tissue factor to flowing blood. As a result, a thrombus can be formed via platelet recruitment and activation of the coagulation cascade, leading to the occlusion of the artery.

The vulnerable high risk plaques for spontaneous rupture or erosion are characterised by the large lipid core size and thin fibrous cap, the loss of smooth muscle cells, the presence of inflammatory cells, and the proteolytic activity in the cap. Perhaps with the exception of thickness of the fibrous cap — <65 μm — the threshold values for any one of these characteristics to indicate plaque vulnerability are unknown.[10]

Histopathological examination of intact and disrupted plaques indicate that vulnerability to rupture depends on the (a) size of the atheromatous core, (b) inflammation within the plaque, (c) thickness and collagen content of the fibrous cap covering the core, (d) degree of inflammation within the cap and (e) cap's mechanical failure due to fatigue.[10]

The risk of plaque disruption is a function not only of plaque vulnerability but also of extrinsic forces. The circumferential wall tension (tensile stress) caused by the blood pressure is given by Laplace's law, which relates luminal pressure and radius to wall tension: the higher the blood pressure and the larger the luminal diameter, the more tension develops in the wall. If components within the wall (soft lipid core of the plaque) are unable to bear the imposed load, the stress is redistributed to adjacent structures (e.g., fibrous cap), where it may be concentrated at critical points (shoulder regions, for example). The stiffer the lipid core, the more stress it can bear, and the stress is correspondingly less redistributed to the adjacent fibrous cap. Interestingly, the thickness of the fibrous cap is most critical for the peak circumferential stress as the thinner the fibrous cap, the higher the stress that develops in it.[10] Thus, stable plaques that are rich in ECM and SMCs, and possess small lipid cores and thick fibrous caps, can withstand higher stresses without being ruptured.

According to Laplace's law, the tension created in fibrous caps of mildly or moderately stenotic plaques is greater than that created in caps of severely stenotic plaques, leading to a higher risk of rupture of the mildly and moderately stenotic plaques.

Besides the thickness of the fibrous cap, the focal activity of macrophages and blood pressure are strongly related to the plaque disruption. Hence, the plaque disruption may occur not only from the lumen into the plaque, but also in the opposite direction, because of an increase in intraplaque pressure caused by vasospasm, bleeding from vasa vasorum and plaque edema.[11]

2.3 Nanomedicine for Accurate Diagnosis of Atherosclerosis

The rapid growth of nanoscience and nanotechnology could greatly expand the clinical opportunities for molecular imaging in the fight against various diseases.[12] Especially, the application of nanotechnology to the diagnosis of atherosclerosis is a promising strategy because it can advance substantially the commonly used imaging modalities for plaque visualisation by the use of nanoscale contrast agents and diverse nanoparticles (NPs). These include fluorescent, radioactive, paramagnetic, superparamagnetic, electron dense and light scattering nanoscale contrast agents and diverse NPs. Based on the above description of molecular mechanisms of atherogenesis and its complications, nanoparticle-based therapeutic and diagnostic approaches have used injured endothelium, platelet hyperactivity, macrophage-mediated processes, matrix modelling events, coagulation factors and dysregulated metabolic activities as disease-specific targets.[9]

One of the key design considerations when fabricating synthetic NPs is achieving the optimal blood-circulating time necessary for biologic activity while avoiding non-specific cellular interactions. Several studies have demonstrated that nanoparticle size greatly affects the biodistribution and cellular uptake.[8] As small LDL particles have been shown to penetrate the endothelial barrier 1.7-fold more than large LDL particles,[13,14] causing atherogenesis analogous to lipoproteins, the size of nanoparticles that penetrates the endothelium is of great importance. Smaller NPs made of polylactide–polyglycolide acid (PLGA) with the size of 100nm demonstrated more than three-fold greater localisation in arterial tissue as compared to larger NPs (275 nm) in an ex vivo canine carotid artery model.[15]

There likely exists a subtle window of particle size range, reflecting the balance between the need for smaller dimensions for tissue diffusion and cell-specific internalisation and the larger dimensions necessary for optimal circulating times and avoidance of biological clearance mechanisms.[8]

Besides size, the surface properties of the nanoparticles determine their biological interactions with plasma proteins, the immune system, and filter organs like the lungs, liver, and kidneys.[16] These interactions affect the NPs' biodistribution, effectiveness and toxicity.

An abundance of specific cell types, such as macrophages and ECs, and the upregulation of cell surface receptors, such as vascular cell adhesion molecule-1 (VCAM-1), fibrin,[17] collagen III, and markers of angiogenesis are a few paradigms of NPs' targets, indicative of the vulnerable lesions.[18] Table 2.1 provides a summary of such targets in line with the diverse NPs used for their targeting and the appropriate ligands.

Table 2.1 Overview of nanoparticle-based systems widely applied in molecular imaging of vulnerable atherosclerotic plaques, in line with vascular targets, targeting ligands, payloads and imaging modalities.

Nanoparticle type	Vascular targets	Nanoparticle payload	Imaging method
Lipid-based	Macrophage scavenger receptor-B (CD36)	Gadolinium	MRI
Immunomicelles	Macrophage-scavenger receptor- A types I and II	Gd-DTPA	MRI
SPIO's	Macrophages	Epichlorohydrin-crosslinked dextran	MRI Fluorescence Nuclear
SPIO's	E-selectin receptor (injured endothelium)	IELLQAR	MRI
Liposomes	LDL- receptor LOX-1	Anti-LOX-1 Abs Indium	SPECT CT
Poly(sebacic acid)- PEG	VCAM-1	Anti-VCAM-1 Abs	Fluorescence
Magnetic	Integrin αvβ3	Peptidomimetic integrin αvβ3 antagonist	MRI
Dendrimers Poly(amidoamine)	E, P-selectin	Anti-E/P-selectin Abs	Fluorescence

Gd: Gadolinium; DTPA: Diethylene triamine pentaacetic acid; LDL: Low density lipoprotein; SPIO's: Small Superparamagnetic Iron-Oxide Nanoparticles; MRI: Magnetic Resonance Imaging; CT: Computed Tomography; PEG: Polyethylene Glycol; Abs:Antibodies; VCAM-1: Vascular Cell Adhesion Molecule 1; SPECT: Single-Photon Emission Computed Tomography.

A variety of NPs has been applied to enhance the imaging of inflammation and atherosclerotic plaques, including liposomes,[19] gold nanoparticles, quantum dots,[20] dextran coated ultra-small particles of iron oxide (USPIO), micelles, paramagnetic nanoparticles,[21] dendrimers,[22] etc. (Fig. 2.2).

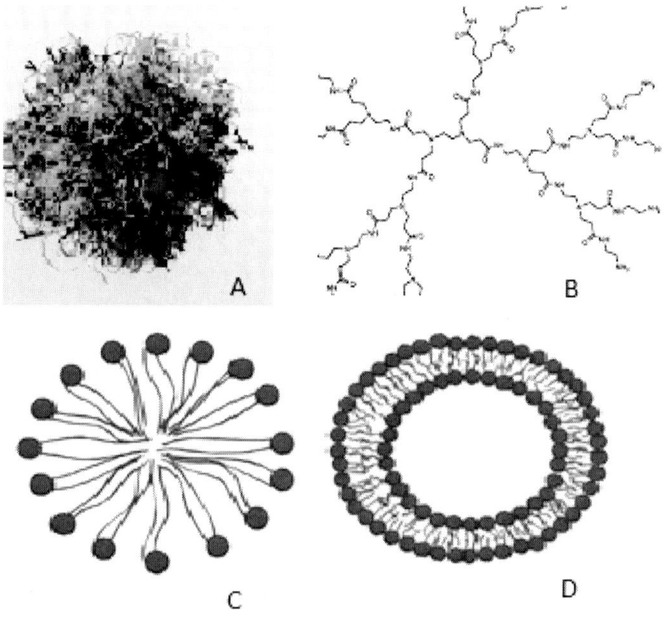

Figure 2.2 Schematic presentation of a: (A) nanoparticle, (B) dendrimer, (C) micelle, and (D) liposome.

As vulnerable plaques tend to exhibit high macrophage contents, many nanoparticulate contrast agents have been developed that are avidly recognised and uptaken by macrophages.

M. Lipinski et al.[23] developed gadolinium (Gd)-containing lipid-based NPs with a mean diameter of 125 nm that target the macrophage scavenger receptor-B (CD36) in order to improve cardiac magnetic resonance (CMR) detection and characterisation of human atherosclerosis. They demonstrated that these macrophage-specific (CD36) NPs bind human macrophages and improve CMR detection and characterisation of plaques and they could help identify high-risk human plaque before the development of an atherothrombotic event.

Another group developed Gd-DTPA diethylene triamine penta-acetic acid-carrying immunomicelles that were linked with a biotinylated monoclonal rat anti-mouse antibody to murine macrophage-scavenger receptor-A types I and II and infused into apoE−/− mice. A strong correlation between macrophage content in the plaques and signal-enhancement in MRI was observed.[23] Gadolinium-based contrast agents have been used to extend MRI capabilities to include smaller blood vessels, but because of concerns with renal toxicity, alternative materials such as superparamagnetic iron oxides have received much interest.[24,25]

SPIO's are small superparamagnetic iron-oxide nanoparticles with a crystalline magnetite structure, coated with dextran or dextran derivatives.[26] They are a promising group of imaging probes since MRI signal intensity is manipulated; they have a favorable toxicity, and a potential for cellular uptake.[27–29] Modification of the dextran coating by carboxylation leads to a shorter clearance half-life in blood.[30] Ferumoxytol, acarboxyalkylated polysaccharide coating is already described as a good first-pass contrast agent,[31] but uptake by macrophages is unspecific and too fast to enhance the uptake in macrophage-rich plaques.

In order to improve the limited uptake of SPIOs in macrophage cells of inflammatory atherosclerotic plaque, Nahrendorf et al.[32] synthesised epichlorohydrin-crosslinked dextran-coated iron oxide NPs for the targeted imaging of atherosclerotic lesions in-vivo. The authors were able to synthesise a multimodal nanoagent capable of detecting atherosclerotic plaques in apolipoprotein E–deficient (apoE-/-) mice by fluorescence, nuclear and MRI.[32] Hildebrandt et al.[33] used specific peptide labels such as the sequence IELLQAR for selective binding to the E-selectin receptor on the injured endothelium. The attachment of peptide oligomers to the dextran-coated iron oxide particles was achieved by electrostatic interactions.[33]

Although MRI provides excellent soft tissue contrast and does not expose the patient to ionising radiation, its low spatial resolution and inability to be applied in patients with metallic implantable devices such as stents, pacemakers etc., have hindered its use in coronary artery imaging.

As the limitations of MRI fall into the advantages of PET/SPECT, and vice versa, multimodal contrast agents that can be detected through multiple imaging platforms are of great interest.

For example, a multifunctional micelle coated magnetic iron oxide nanoparticle probe was developed from an iron oxide core coated with phospholipid for targeting. Different targeted ligands, radio-labelled ligands, and fluorescent dyes can be added to the coating for use with MRI, PET, or fluorescence imaging.[34] Li et al. have demonstrated targeted molecular imaging of atherosclerotic plaques in ApoE-/- knockout mice, by using liposomes surface-decorated with antibodies specific LDL receptor LOX-1 and loaded with Indium (111In) for SPECT-CT imaging.[35]

Deosarkar et al.,[36] by utilising the overexpression of VCAM-1 on ECs in atherosclerotic plaques as potential targets, demonstrated the efficacy of poly (sebacic acid)-co-PEG (PSAPEG) microparticles and NPs surface-modified with anti-VCAM-1 antibodies to undergo enhanced adhesion, binding and accumulation at atherosclerotic lesion sites in ApoE-/-mice.

Using a different ligand-receptor combination, another group of investigators generated NPs targeted to E- and P-selectin by covalently binding a ligand of E- and P-selectin to poly-dl-lactic acid polymer (PLA) nanoparticles and showed adhesion of these NPs to human umbilical vein endothelial cells.[37]

Plaque vascularity has been implicated in plaque growth and vulnerability. Angiogenesis of the plaques is associated with more advanced stages of human atherosclerosis and the extent of vasa vasorum correlates with atherosclerotic lesion size and lumen diameter in hypercholesterolemic animal models. The vasa vasorum, the microvasculature in the adventitial layer of large arteries are considered to be the conduit for nutrient supplies to atherosclerotic plaque.[38]

The cell adhesion molecule integrin αvβ3 is an important player in the process of angiogenesis occurring in atherosclerotic plaques and tumours. It is a heterodimeric adhesion molecule that is widely expressed by endothelial cells, monocytes, fibroblasts, and SMCs and plays a critical role in SMC migration and cellular adhesion,[39] both of which are required for the formation of new blood vessels. The utility of αvβ3 integrin–targeted NPs has been shown for the detection and characterisation of angiogenesis in correlation with tumour progression and atherosclerosis.[40,41]

Winter et al.[40,41] attached peptidomimetic integrin αvβ3 antagonist to magnetic nanoparticles for MRI imaging under a common clinical field strength of 1.5 Tesla. Two animal models were used: New Zealand white rabbits implanted with tumour, and an atherosclerosis model. In the atherosclerosis model, enhancement in the MRI signal was also observed among rabbits that received integrin-targeted nanoparticles. Histology and immunohistochemistry confirmed the proliferation of angiogenic vessels within the aortic adventitia of the atherosclerotic rabbits as compared with control animals.[41]

Dendrimers have successfully proved themselves as promising nanocarriers of drugs and antibodies with attractive application in pharmaceutical, nanotechnology and medicinal chemistry. Their structure is shaped like a tree or a star, with a central core, interior branches and terminal groups that decorate the surface. They are precisely defined, hyperbranched three-dimensional synthetic nanomaterials with 2–10 nm in diameter (Fig. 2.2B).[42]

Cavities inside the core and the interior branches can be modified to carry drugs, peptides and antibodies. The dendrimer surface contains many different sites to which drugs may be attached and also attachment sites for peptides, molecules such as polyethylene glycol (PEG) — which can be used to modify the way the dendrimer interacts with the body — and antibodies for site specific targeting.[42]

Aiming at atherosclerotic lesions, poly (amidoamine) (PAMAM) dendrimers have also been successfully targeted to E and P-selectin by attaching anti-E/P-selectin antibodies onto the PAMAM dendrimers.[43] These dendrimers increased gene delivery to endothelial cells expressing E-selectin in human saphenous vein segments ex vivo.

Recent studies have emphasised the role of "theranostic" NPs with dual application for both therapeutic and diagnostic purposes. They can be delivered to the specific targeted area for imaging while simultaneously acting as therapeutic moieties providing critical information about the efficacy of treatment. Preclinically, theranostics have been gradually applied to CVD with several interesting and encouraging findings. Due to their complexity, it is unlikely that theranostic NPs will be consistently applied clinically.[44]

Nanotheranostic approaches, however, may have significant benefits that can facilitate and speed up preclinical development. Advantages involve the ability to monitor biodistribution and investigate nanoparticle dynamics in the body over time.[45]

McCarthy et al.[46] developed theranostic cross-linked dextran-coated iron oxide (CLIO) nanoparticles targeted to macrophage-rich atherosclerotic lesions for local cell ablation. This agent is modified with a near infrared fluorophore in order to enable fluorescence imaging, as well as a potent chlorin-based photosensitiser, each at spectrally distinct wavelengths. After intravenous administration in apolipoprotein E deficient (apo$E^{-/-}$) mice, the irradiation of carotid atheroma with 650 nm light to activate the therapeutic component resulted in the eradication of inflammatory macrophages. The investigators conclude that this light-activated inflammatory cell ablation strategy is a novel paradigm for the treatment of atherosclerosis, and has several potential advantages over other theranostic techniques.[46]

2.4 Nanomedicine in Atherosclerosis Treatment

2.4.1 Therapeutic Nanoparticles

Despite the success of pharmacological and revascularisation therapies applied in clinical practice, CVD remains the most common cause of death in the world.[47] Thus, a clear need exists for additional strategies to combat atherosclerosis. The search for novel drug delivery technologies is essentially driven by the fact that many therapeutic agents failed due to their limited ability to reach the target tissue, poor selectivity of diseased cells, drug bioavailability and toxicity issues. Nanoparticulate systems are now available to improve bioavailability, drug loading capacity and specific tissue targeting efficiency, as well as to limit toxicity. Decorating nanoparticles with targeting ligands against biomarkers of a disease facilitates their specific delivery and accumulation within pathological tissue and their local drug delivery.

Morris et al.[48] used USPIO-enhanced MRI to evaluate the therapeutic effectiveness of a p38 kinase inhibitor, a critical molecule in the immune response, in an apolipoprotein E–deficient (ApoE KO) mouse model. The MRI studies revealed decreased accumulation of USPIO in the aortic root, and ex-vivo histology confirmed that the majority of USPIO were associated with macrophages.

In clinical studies, this method has been applied to assess statin treatment and the effects of high-dose and a low-dose of atorvastatin on carotid plaques. Continuation of a high-dose treatment for 12 weeks decreased USPIO uptake in carotid plaques, while the low-dose treatment had no significant effect. This study represents one of the first clinical examples of nanoparticle based imaging for the evaluation of drug treatment in atherosclerotic patients.[49]

Due to the inflammatory nature of atherosclerosis, anti-inflammatory drugs such as glucocorticoids have been proposed as a treatment, though the systemic side effects impede its application in clinical medicine. Lobatto et al.[50] synthesised paramagnetically labelled liposomes, with a size of 103 nm, to serve as drug carriers for glucocorticoids to be delivered in atherosclerotic lesions for local therapeutic efficacy. The MRI studies showed that the liposomes could be detected in atherosclerotic plaques.

In the work of Chono et al., dexamethasone-containing liposomes have been reported to enhance drug delivery and anti-atherosclerotic effects in atherogenic mice.[51] De Bittencourt et al.[52] have developed the LipoCardium, which is a unique liposome formulation that involve liposomes surface-decorated with VCAM-1-directed antibodies and loaded with cyclopentenone prostaglandins, which are anti-inflammatory agents. These liposomes were found to exhibit anti-inflammatory, and antiproliferative properties, selectively at atherosclerotic lesions in LDL-receptor knockout (LDLR-/-) mice.

Biodegradable nanoparticles formulated from PLGA have been extensively investigated for sustained and targeted/localised delivery of different agents. While such polymers have the advantage of sustaining release of the encapsulated therapeutic agent over a period of days to several weeks compared to natural polymers, they are in general limited by the use of organic solvents and hydrophobicity.[53]

Drug-loaded biodegradable nanoparticles can be locally delivered to the site of the atherosclerotic lesion.[54] In an in-vitro model using a microporous balloon catheter and testing particles of different sizes (120–230–1000 nm), it was shown that particles of 120 nm in size were present in all layers of the arterial wall, whereas particles of 230 and 1000 nm were mainly introduced to the intima of the arterial wall.[55] So by using biodegradable nanoparticles of different sizes, drug delivery to specific layers in the arterial wall can be achieved simultaneously.[56]

Dexamethasone- and rapamycin-loaded nanoparticles based on poly (ethylene oxide) and poly (DL-lactic-co-glycolic acid) block copolymers (PEO-PLGA) were prepared using the salting-out method by the group of Zweers et al.[55] They showed that these NPs inhibit smooth muscle cell proliferation and thus arterial restenosis.

2.4.2 Nanomedicine for Stents

In everyday clinical practice, there are many therapeutical options to deal with the complications of atherosclerosis, ranging from administration of drugs to atherosclerotic patients to invasive procedures such as percutaneous transluminal coronary angioplasty for re-opening of the stenotic or obstructed arteries. The intravascular stent, especially the drug eluting stent (DES), is certainly one of the most successful of cardiovascular devices that improves the efficacy of percutaneous coronary intervention.[57,58] The first generation of stents were metallic ones (including titanium, stainless steel, nitinol, etc.) since metals can sustain the mechanical stresses during stent expansion. Hence, such conventional metals are generally not compatible with tissues and may cause acute thrombosis and long-term restenosis.

The next generation of DES has already become a hot spot in the field of intervention therapy, as they suppress the arterial restenosis rate below 10% in selected cases.[59] This is due to the release of antiproliferative agents such as sirolimus (CYPHER stent, Cordis, Miami Lakes, Florida), paclitaxel (Taxus stent, Boston Scientific, Natick, Massachusetts), zotarolimus (Endeavor stent, Medtronic, Minneapolis, Minnesota), and everolimus (Xience stent, Abbott and Boston Scientific). These drugs inhibit smooth muscle proliferation and matrix production, reducing the neointimal restenosis.[60]

The drugs are loaded in the stent micro-reservoirs and a diversity of polymers is used as drug containers or diffusion barriers for controlled drug elution. In clinical cohorts, however, DES are found to cause late stent thrombosis after one year of stent implantation as compared to bare metal stents (BMS), because of the release of antiproliferative agents and the presence of polymers that disrupt the natural healing response of the endothelium (Fig. 2.3).[61] Such delay of stent endothelialisation and inhibition of vascular repair may cause the DES late stent thrombosis and the clinical onset of an acute coronary syndrome (ACS). The ACS covers the wide spectrum of unstable angina and Q/ or non-Q myocardial infarction that may be even fatal.

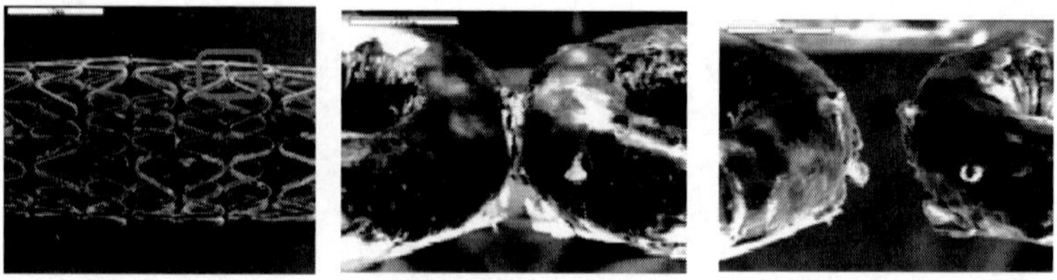

Figure 2.3 SEM images of (a) drug eluting stent, (b) drug eluting stent expanded using nominal pressure rating (3–4 atm) for inflation, and (c) drug eluting stent expanded at inflation pressure of 14 atm, where the polymer debris that may cause the disruption of endothelium are evident (From LTFN source).

The late stent thrombosis remains a significant clinical problem that needs to be addressed by advanced nanomedicine strategies that encompass novel nanomaterials and drug delivery nanosystems with superior biocompatibility and desirable drug release activities. A recent approach is to mimic the natural vascular tissue that exhibits nanostructured features due to the presence of proteins, e.g., collagen and elastin. Karagkiozaki et al. manufactured carbon-based nanocoatings with nanotopographical cues to repel platelets and to promote cell proliferation.[62] It is recognised that cells are sensitive to nanoscale characteristics of the biomaterial surface and that micro- and nanoscale grooves can be used as a means of orientating cells, guiding cell migration and proliferation. Other previous studies at our lab showed that the nanotopography, stoichiometry and electrical surface properties of the biomaterials are significant factors that determine the platelet behaviour and resultant thrombogenicity.[63–65]

The nanotopography modulates cell behaviour by changing the integrin clustering and focal adhesion assembly, leading to alterations in cytoskeletal organisation and mechanical profile. Titanium nitride nanocoatings with higher content of nitrogen, engineered by magnetron sputtering techniques, showed favorable stoichiometry and surface properties that repel platelets and enhance cell adhesion and proliferation.[65] A recent study gave evidence that nanostructured PLGA induced more fibronectin and vitronectin adsorption than the conventional PLGA, thus, leading to greater vascular cell responses on the nanostructured PLGA.[66]

Stampfl et al. developed a nanoscale coating with the highly biocompatible polymer polyzene-F (PZF), in combination with cobalt chromium and stainless steel stents. It was found in pig studies that the PZF-nanocoated cobalt chromium stents provided a significantly lower average loss in lumen diameter and the highest efficacy in reducing in-stent stenosis at long-term follow-up compared to PZF-nanocoated stainless steel stents. The PZF nanocoating proved to be biocompatible with respect to thromboresistance and inflammation and was suggested to be an interesting, passive stent platform.[67]

The most successful approaches to circumvent the issue of delayed stent endothelialisation involve the modification of the implant or scaffold surfaces with various functional molecules, such as anti-fouling polymers or cell growth factors. Especially, in tissue engineering, the biofunctionalisation of nanomaterials with bioactive peptides that have a distinct biological activity in promoting specific interactions with endothelial cell surface receptors enables the facilitation of cell attachment and proliferation. ECM-derived collagen, fibronectin, laminin and vitronectin have been shown to promote cell attachment since they possess intrinsic biological recognition sites for cells via specific integrin receptors.[68]

In the case of cardiovascular implants, various ECM peptide sequences, which have been determined to influence cell behaviour, have been grafted on materials to enhance biological properties (for example, REDV, PHSRN, RGD, etc.).[69] The covalent bonding of the peptides provides their higher stability under blood flow and towards cleaning or sterilisation conditions and changes in pH. The peptides are also completely stable against enzymatic degradation and should, therefore, exhibit excellent long-term stability. On the other hand, if the peptides are physically adsorbed onto the stent nanomaterials, they are susceptible to instability due to desorption over time, or denaturation caused by their conformational changes, or unfolding on the surface. Thus, a bioactivation strategy of stent nanocoatings can be applied to trigger the neo-endothelialisation process onto their surface.

Nanoporous stent coatings due to their high active surface area and high drug loading efficiency emerge as new drug delivery nanosystems for controlled drug release from stent. Previous studies at our lab showed that a multilayer biodegradable polymeric stent nanoplatform enables the delivery of multiplex drugs targeting to thrombosis, inflammation or restenosis that can be eluted at desirable time intervals from the stent surface.[70] Another group demonstrated that a nanoporous aluminium oxide coated stent delivers tracrolimus to rabbit carotid arteries in vivo.[71]

Nanoparticulate systems may also be used as drug delivery vehicles for DES. Bhargava et al.[72] evaluated a novel paclitaxel-eluting porous carbon-carbon nanoparticle coated stent in a porcine model. The stent surface consists of porous composite matrix made of amorphous carbon nanoparticles embedded in glassy polymeric carbon. This material was found to have good mechanical and chemical properties and offers the possibility of controlling drug release kinetics by adjusting the pore size.

It is worth mentioning that a favorable safety profile is a prerequisite for the nanomedicines that are previously described for clinical use. Nanotoxicity and regulatory requirements need to be addressed for their safety translation into clinical practice. Areas requiring evaluation include immunogenicity, biodistribution and accumulation of the nanoparticles into vital organs, radioactivity, chemical and physical toxicity, mutagenic and carcinogenic potential.

2.5 Conclusions and Future Perspectives

Atherosclerosis is an insidious disease and many decades may pass until the onset of clinical symptoms. There are clinical challenges to diagnose with accuracy the vulnerable atherosclerotic plaques that are susceptible to spontaneous rupture leading to acute obstruction of the coronary arteries and to myocardial infarction with its detrimental complications. Nanomedicine addresses this issue by the design and manufacturing of multimodal nanoparticulate systems for the early detection of the highly inflamed plaques, whereas the application of theranostic nanoparticles exhibiting both diagnostic and therapeutic capabilities enables their diagnosis and treatment simultaneously. A diversity of vascular targets within the plaques is used for the site specific targeted drug delivery of therapeutic agents to stabilize, regress or even eliminate the plaques.

Novel nanotechnology enabled systems present promising therapeutic possibilities and offer advantages that are lacking in current therapeutic strategies. Stents that are implanted into the stenotic arterial areas during coronary angioplasty can be advanced significantly by the use of (a) nanostructured and bioactive biomaterials that mimic the morphology of the native endothelium and induce re-endothelialisation onto stent surface and (b) nanoporous thin films and nanocontainers that deliver their therapeutic payloads at the diseased arterial sites and release them in a controllable manner. Via nanomedicine pathways, the drawbacks of drug eluting stents that may cause acute coronary syndrome can be overcome to advance the treatment of atherosclerosis in everyday clinical routine.

A future goal is the development of multifunctional nanoscale devices capable of providing clinical feedback by interfacing with biomarkers indicative of the multiple stages of atherosclerotic lesions and site specific triggered release of therapies. Clinical translation of such nanotechnologies will be facilitated by clinical trials to validate their safety profile and effectiveness.

Abbreviations

ACS: Acute Coronary Syndrome
BMS: Bare Metal Stents
CLIO: Cross-linked dextran-coated iron oxide nanoparticles
CVD: Cardiovascular Diseases
CMR: Cardiac Magnetic Resonance
CT: Computed Tomography
DES: Drug Eluting Stents
DTPA: Diethylene triamine pentaacetic acid
ECs: Endothelial cells
ECM: Extracellular Matrix
Gd: Gadolinium
ICAM-1: Intercellular adhesion molecule 1
LDL: Low Density Lipoprotein
MMPs: Matrix Metalloproteinases
MRI: Magnetic Resonance Imaging
NO: Nitric Oxide
NPs: Nanoparticles
PAMAM: Poly (amidoamine)
PEG: Polyethylene Glycol
PET: Positron Emission Tomography
PLA: Poly-l, dlactic acid
PLGA: Poly (DL-lactic-co-glycolic acid)
PZF: Polymer Polyzene-F
QDs: Quantum Dots

SPIO's: Small Superparamagnetic Iron-Oxide Nanoparticles
SMC: Smooth Muscle Cell
SPECT: Single Photon Emission Computed Tomography
TIMP: Tissue Inhibitor of Metalloproteinase
USPIOs: Ultra Small Superparamagnetic Iron-Oxide Nanoparticles
vWf: Von Willebrand
VCAM-1: Vascular Cell Adhesion Molecule 1

Acknowledgements

The author thanks Anna Maria Pappa for her help in the editing process of this book.

Disclosures and Conflict of Interest

The author declares that she has no affiliations or financial involvement with any organisation or entity discussed in this chapter. This includes employment, consultancies, honoraria, grants, stock ownership or options, expert testimony, patents (received or pending) or royalties. No writing assistance was utilised in the production of this manuscript and the author received no payment for the preparation of this chapter. The findings and conclusions here reflect the current views of the author. They should not be attributed, in whole or in part, to the organisations with which they are affiliated, nor should they be considered as expressing an opinion with regard to the merits of any particular company or product discussed herein. Nothing contained herein is to be considered as the rendering of legal advice.

References

1. Drexler, K.E., Peterson, C., and Pergamit, G. (eds) (1991) *Unbounding The Future: The Nanotechnology Revolution*, William Morrow, New York.
2. Karagkiozaki, V., Logothetidis, S., and Giannoglou, G. (2008) Advances in stent coating technology via nanotechnology tools and process. *J. Eur. Nanomed.* **1**(1): 24–28.
3. Karagkiozaki, V., Vavoulidis, E., and Logothetidis, S. (2012) Nanomedicine pillars and monitoring nanobiointeractions, in

Nanomedicine and Nanobiotechnology (Logothetidis, S., ed.), Springer, pp. 27–52.

4. Salata, O. (2004) Applications of nanoparticles in biology and medicine. *J. Nanobiotech.* **2**(3): 1–6.
5. Hansson, G.K. (2005) Mechanisms of disease Inflammation, atherosclerosis, and coronary artery disease. *N. Engl. J. Med.* **352**: 1685–1695.
6. Anderson, T.J. (1999) Assessment and treatment of endothelial dysfunction in humans. *J. Am. Coll. Cardiol.* **34** (3): 631–638.
7. Ungvari, Z., Kaley, G., De Cabo, R., Sonntag, W.E., and Csiszar, A. (2010) Mechanisms of vascular aging: new perspectives. *J. Gerontol. A. Biol. Sci. Med. Sci.* **65A** (10): 1028–1041.
8. Goonewardena, S.N. (2012) Approaching the asymptote: obstacles and opportunities for nanomedicine in cardiovascular disease. *Curr. Atheroscler. Rep.* **14** (3): 247–253.
9. Gupta, A.S. (2011) Nanomedicine approaches in vascular disease: a review. *J. Nanomed.* **7** (6): 763–779.
10. Falk, E., Shah, P.K., and Fuster V. (1995) Coronary plaque disruption. *Circulation* **92**(3): 657–671.
11. Barger, A.C., and Beeuwkes, R. (1990) Rupture of coronary vasa vasorum as a trigger of acute myocardial infarction. *Am. J. Cardiol.* **66**(suppl G): 41–43.
12. Wickline, S.A., Neubauer, A.M., Winter, P., Caruthers, S., Lanza, G., and Samuel A. (2006) Applications of nanotechnology to atherosclerosis, thrombosis, and vascular biology. *Arterioscler. Thromb. Vasc. Biol.* **26**: 435–441.
13. De Graaf, J., Hak-Lemmers, H.L., Hectors, M.P., Demacker, P.N., Hendriks, and J.C., Stalenhoef, A.F. (1991) Enhanced susceptibility to in vitro oxidation of the dense low density lipoprotein subtraction in healthy subjects. *Arterioscler. Thromb. Vasc. Biol.* **11**: 298–306.
14. Rizzo, M., and Berneis, K. (2006) Low-density lipoprotein size and cardiovascular risk assessment. *QJM* **99** (1): 1–14.
15. Panyama J., and Labhasetwar, V. (2003). Biodegradable nanoparticles for drug and gene delivery to cells and tissue. *Adv. Drug Deliv. Rev.* **55**: 329–347.
16. Logothetidis, S. (ed.) (2012) *Nanomedicine and Nanobiotechnology*, Springer.

17. Karagkiozaki, V., and Logothetidis, S., (2014) In *Nanomedicine Challenges in Thrombosis, Handbook of Clinical Nanomedicine: From Bench to Bedside* (Bawa, R., Audette, G.F., Rubinstein, I., ed.), Pan Stanford Publishing.
18. Nahrendorf, M., Jaffer, F.A., Kelly, K.A., Sosnovik, D.E., Aikawa, E., Libby, P., and Weissleder, R. (2006) Non-invasive vascular cell adhesion molecule-1 imaging identifies inflammatory activation of cells in atherosclerosis. *Circulation* **114**(14): 1504–1511.
19. Sumio Chono, S., Tauchi, Y., and Morimoto, K. (2006) Influence of particle size on the distributions of liposomes to atherosclerotic lesions in mice. *Drug Devel. Indust. Pharm.,* **32**: 125–135.
20. Jayagopal, A., Russ, P., and Haselton, F. (2007) Surface engineering of quantum dots for in vivo vascular imaging. *J. Bioconjug. Chem.* **18**(5): 1424–1433.
21. Maiseyeu, A., Mihai, G., Roy, S., Kherada, N., Simonetti, O., Sen, C., Sun, Q., Parthasarathy, S., and Rajagopalan, S. (2010) Detection of macrophages via paramagnetic vesicles incorporating oxidatively tailored cholesterol ester: an approach for atherosclerosis imaging. *J. Nanomed.* **5**(9): 1341–1356.
22. Karagkiozaki, V. (2013) Nanomedicine highlights in atherosclerosis. *J. Nanopart. Res.* **15**(3):1529–1546.
23. Lipinski, M.J., Frias, J.C., Amirbekian, V., Briley-Saebo, K.C., Mani, V., Samber, D., Abbate, A., Gilberto, J., Aguinaldo, S., Massey, D., Fuster, V., Vetrovec, G.W., and Fayad, Z.A. (2009) Macrophage-specific lipid-based nanoparticles improve cardiac magnetic resonance detection and characterization of human atherosclerosis. *JACC. Cardiovasc. Imag.* **2** (5): 637–647.
24. Briley-Saebo, K.C., Johansson, L.O., Hustvedt, S.O., Haldorsen, A.G., Bjornerud, A., Fayad, Z.A., and Ahlstrom, H.K. (2006) Clearance of iron oxide particles in rat liver—effect of hydrated particle size and coating material on liver metabolism. *Invest. Radiol.* **41**: 560–571.
25. Tang, T.Y., Muller, K.H., Graves, M.J., Li, Z.Y., Walsh, S.R., Young, V., Sadat, U., et al. (2009) Iron oxide particles for atheroma imaging. *Arterioscler. Thromb. Vasc. Biol.* **29:** 1001–1008.
26. Trivedi, R.A., Mallawarachi, C., UK-I, J.M., Graves, M.J., Horsley, J., Goddard, M.J., Brown, A., Wang, L., Kirkpatrick, P.J., Brown, J., et al. (2006) Identifying inflamed carotid plaques using in vivo USPIO-enhanced MR imaging to label plaque macrophages. *Arterioscler. Thromb. Vasc. Biol.* **26**: 1601–1606.

27. Litovsky, S., Madjid, M., Zarrabi, A., Casscells, W., Willerson, J.T., and Naghavi, M. (2003) Superparamagnetic iron oxide-based method for quantifying recruitment of monocytes to mouse atherosclerotic lesions in vivo. *Circulation* **107**: 1545–1549.
28. Yancy, A.D., Olzinski, A.R., Hu, T.C., Lenhard, S.C., Aravindhan, K., Gruver, S.M., Jacobs P.M., Willette, R.N., and Jucker, B.M. (2005) Differential uptake of ferumoxtran-10 and ferumoxytol, ultrasmall superparamagnetic iron oxide contrast agents in rabbit: critical determinants of atherosclerotic plaque labeling. *J. Magn. Reson. Imaging.* **21**: 432–442.
29. Wang, Y.X., Hussain, S.M., and Krestin, G.P. (2011) Superparamagnetic iron oxide contrast agents: physicochemical characteristics and applications in MR imaging. *Eur Radiol.* **11**(11): 2319–2331.
30. Tong, S., Hou, S., Zheng, Z., Zhou, J., and Bao, G. (2010) Coating optimization of superparamagnetic iron oxide nanoparticles for high T2 relaxivity. *Nano Lett.* **10**(11): 4607–4613.
31. Li, W., Tutton, S., Vu, A.T., Pierchala, L., Li, B.S.Y., Lewis, J.M., Prasad, P.V., and Edelman, R.R. (2005) First-pass contrast-enhanced magnetic resonance angiography in humans using ferumoxytol, a novel USPIO-based blood pool agent. *J. Magn. Reson. Imag.* **21**: 46–52.
32. Nahrendorf, M., Zhang, H., Hembrador, S., Panizzi, P., Sosnovik, D., Aikawa, E., Libby, P., Swirski, F., and Weissleder, R. (2008) Nanoparticle PET-CT imaging of macrophages in inflammatory atherosclerosis. *Circulation* **117**: 379–387.
33. Hildebrandt, N., Hermsdorf, D., Signorell, R., Schmitz, S.A., and Diederichsena, U. (2007) Superparamagnetic iron oxide nanoparticles functionalized with peptides by electrostatic interactions. *Arkivoc (v)*: 79–90.
34. Von zur Muhlen, C., Fink-Petri, A., Salaklang, J., Paul, D., Neudorfer, I., Berti, V., et al. (2010) Imaging monocytes with iron oxide nanoparticles targeted towards the monocyte integrin MAC-1 (CD11b/CD18) does not result in improved atherosclerotic plaque detection by in vivo MRI. *Contrast Media Mol. Imag.* **5**(5): 268–275.
35. Li, D., Patel, A.R., Klibanov, A.L., Kramer, C.M., Ruiz, M., Kang, B.Y, et al. (2010) Molecular imaging of atherosclerotic plaques targeted to oxidized LDL receptor LOX-1by SPECT/CT and Magnetic Resonance. *J. Circ. Cardiovasc. Imag.* **3**: 464–472.
36. Deosarkar, S.P., Malgor, R., Fu, J., Kohn, L.D., Hanes, J., and Goetz, DJ. (2008) Polymeric particles conjugated with a ligand to VCAM-1 exhibit selective avid and focal adhesion to sites of atherosclerosis. *J. Biotechnol. Bioeng.* **101**(2): 400–752.

37. Banquy, X., Leclair, G., Rabanel, J.M., Argaw, A., Bouchard, J-Fo, Hildgen, P., and Giasson, S. (2008) Selectins ligand decorated drug carriers for activated endothelial cell targeting. *J. Bioconjug. Chem.* **19**: 2030–2039.

38. Drinane, M., Mollmark, J., Zagorchev, L., Moodie, K., Sun, B., Hall, A., Shipman, S., Morganelli, P., Simons, M., and Mulligan-Kehoe, M.J. (2009) The anti-angiogenic activity of rPAI-123 inhibits vasa vasorum and growth of atherosclerotic plaque: rPAI-123 inhibits vasa vasorum and plaque growth. *J. Circ Res.* **104**(3): 337–345.

39. Antonov, A.S., Kolodgie, F.D., Munn, D.H., and Gerrity R.G. (2004) Regulation of macrophage foam cell formation by αVβ3 integrin: potential role in human atherosclerosis. *Am. J. Pathol.* **165**(1): 247–258.

40. Winter, P.M., Caruthers, S.D., Kassner, A., Harris, T.D., Chinen, L.K., Allen, J.S., Lacy, E.K., and Zhang, H. (2003) Molecular imaging of angiogenesis in nascent Vx-2 rabbit tumors using a novel αvβ3 -targeted nanoparticle and 1.5 tesla magnetic resonance imaging. *J. Cancer. Res.* **63**: 5838–5843.

41. Winter, P.M., Neubauer, A.M., Caruthers, S.D., Harris, T.D., Robertson, J.D., Williams, T.A., Schmieder, A., Hu, G., Allen, J., Lacy, E., Zhang, H., Wickline, S., and Lanza, G. (2006) Endothelial alpha(v)beta3 integrin-targeted fumagillin nanoparticles inhibit angiogenesis in atherosclerosis. *J. Arter. Thromb. Vasc. Biol.* **26**(9): 2103–2109.

42. Rajesh Babu, V., Mallikarjun, V., Nikhat, S.R., and Srikanth, G. (2010) Dendrimers: a new carrier system for drug delivery. *Int. J. Pharm. Ap.Sci.* **1** (1): 1- 10.

43. Babu, R.V., Mallikarjun, V., Nikhat, S.R., Srikanth, G., Theoharis, S., Krueger, U., Tan, P.H., Haskard, D.O., Weber, M., George, A.J.T. (2009) Targeting gene delivery to activated vascular endothelium using anti e/p-selectin antibody linked to PAMAM dendrimers. *J. Immunol.* Meth. **343**(2): 79–90.

44. Tang, J., Lobatto, M.E., Read, J.C., Mieszawska, A.J., Fayad, Z.A., and Mulder, W.J.M. (2012) Nanomedical theranostics in cardiovascular disease. *J. Curr. Cardiovasc Imag. Rep.* **5**(1): 19–25.

45. Sanhai, W.R., Sakamoto, J.H., Canady, R., and Ferrari, M. (2008*) Seven challenges for nanomedicine. Nat. Nanotechnol.* **3** (5): 242–244.

46. McCarthy, J.R, Korngold, E., Weissleder, R., and Jaffer, A.F. (2010) A light-activated theranostic nanoagent for targeted macrophage ablation in inflammatory atherosclerosis. *Small.* **6**(18): 2041–2049.

47. Leal, J., Luengo-Fernández, R., Gray, A., Petersen, S., and Rayner, M. (2006) Economic burden of cardiovascular diseases in the enlarged European Union. *Eur Heart J.* **27**: 1610–1619.
48. Morris, J.B., Olzinski, A.R., Bernard, R.E., Aravindhan, K., Mirabile, R.C., Boyce, R., et al. (2008) p38 MAPK inhibition reduces aortic ultrasmall superparamagnetic iron oxide uptake in a mouse model of atherosclerosis: MRI assessment. *Arterioscler. Thromb. Vasc. Biol.* **28**(2): 265–271.
49. Tang, T.Y., Howarth, S.P., Miller, S.R., Graves, M.J., Patterson, A.J., et al. (2009) The ATHEROMA (Atorvastatin Therapy: Effects on Reduction of Macrophage Activity) Study. Evaluation using ultrasmall superparamagnetic iron oxide-enhanced magnetic resonance imaging in carotid disease. *J. Am. Coll. Cardiol.* **53**(22): 2039–50.
50. Lobatto, M.E., Fayad, Z.A., Silvera, S., Vucic, E., Calcagno, C., Mani, V., et al. (2010) Multimodal clinical imaging to longitudinally assess a nanomedical anti-inflammatory treatment in experimental atherosclerosis. *J. Mol. Pharm.* **7**(6): 2020–2029.
51. Chono, S., Tauchi, Y., Deguchi, Y., and Morimoto, K. (2005) Efficient drug delivery to atherosclerotic lesions and the antiatherosclerotic effect by dexamethasone incorporated into liposomes in atherogenic mice. *J. Drug Target.* **13**(4): 267–276.
52. De Bittencourt, P.I.H., Lagranha, D.J., Maslinkiewicz, A., Senna, S.M., Tavares, A.M.V., Baldissera, L.P., et al. (2007) LipoCardium: Endothelium directed cyclopentanone prostaglandin-based liposome formulation that completely reverses atherosclerotic lesions. *Atheroscl.* **193**(2): 245–258.
53. Panyama, J., and Labhasetwara, V. (2003) Biodegradable nanoparticles for drug and gene delivery to cells and tissue. *J. Adv. Drug Del. Rev.* **55**: 329–347.
54. Song, C.X., Labhasetwar, V., Murphy, H., Qu, X., Humphrey, W.R., Shebuski, R.J., and Levy, R.J. (1997) Formulation and characterization of biodegradable nanoparticles for intravascular local drug delivery. *J. Control. Rel.* **43** (2–3): 197–212.
55. Zweers, M.L.T., Engbers, G.H.M, Grijpma, D.W., and Feijen, J. (2006) Release of anti-restenosis drugs from poly (ethylene oxide)-poly (dl-lactic-co-glycolic acid) nanoparticles. *J. Control. Rel.,* **114** (3): 317–324.
56. Guzman, L.A., Labhasetwar, V., Song, C., Jang, Y., Lincoff, M., Levy, R., and Topol, E.J., (1996) Local intraluminal infusion of biodegradable polymeric nanoparticles: a novel approach for prolonged drug delivery after balloon angioplasty. *Circulation* **94:** 1441–1448.

57. Fattori, R., & Piva, T. (2003). Drug-eluting stents in vascular intervention. *The Lancet*, **361**(9353), 247–249.

58. Stone, G., Moses, J., Ellis, S., Schofer, J., Dawkins, K., Morice, M., Colombo, A., Schampaert, E., Grube, E., Kirtane, A.J., Cutlip, D.E, Fahy, M., Pocock, S.J., Mehran, R., and Leon, M.B. (2007) Safety and efficacy of sirolimus & paclitaxel-eluting stents. *J. N. Engl. Med.* **356**: 998– 1008.

59. Nakazawa, G., Finn, A.V., Joner, M., Ladich, E., et al. (2008) Delayed arterial healing and increased late stent thrombosis at culprit sites after drug-eluting stent placement for acute myocardial infarction patients: an autopsy study. *J. Circ.* **118**(11): 1138–1145.

60. Luscher, T.F., Steffel, J., Eberli, F.R., Joner, M., Nakazawa, G., Tanner, F.C., and Virmani, R. (2007) Drug-eluting stent and coronary thrombosis: biological mechanisms and clinical implications. *J. Circ.* **115**(8): 1051–1058.

61. Bavry, A., Kumbhani, D., Helton, T., Borek, P., Mood, G., and Bhatt, D. (2006) Late thrombosis of DES: A meta-analysis of randomized clinical trials. *J. Am. Med.* **119**(12): 1056–1061.

62. Karagkiozaki, V., Karagiannidis, P., Kalfagiannis, N., Kavatzikidou, P., Patsalas, P., D.Georgiou, D., and Logothetidis S. (2012) Novel nanostructured biomaterials: implications for coronary stent thrombosis. *Int .J. Nanomedicine.* **7**: 6063–76.

63. Karagkiozaki, V., Logothetidis, S., Laskarakis, A., Giannoglou, G., and Lousinian, S. (2008) AFM study of the thrombogenicity of carbon-based coatings for cardiovascular applications. *J. Mater. Sci. Eng. B*, **152**(1–3): 16–21.

64. Karagkiozaki, V., Logothetidis, S., Lousinian, S., and Giannoglou, G. (2008) Impact of surface electric properties of carbon-based thin films on platelets activation for nano-medical and nano-sensing applications. *J. Int. Nanomed.* **3**(4): 461–469.

65. Karagkiozaki, V., Logothetidis, S., Kalfagiannis, N., Lousinian, S., and Giannoglou, G. (2009) AFM probing platelets activation behavior on titanium nitride nanocoatings for biomedical applications. *J. Nanomed.* **5**(1): 64–72.

66. Miller, D.C., Webster, T.J., and Haberstroh, K.M. (2005) Comparison of fibroblast and vascular cell adhesion to nano-structured poly (lactic-co-glycolic acid) films. *J. Appl. Bion. Biomech.* **2**(1): 1–8.

67. Stampfl, U., Sommer, C.M., Thierjung, H., Stampfl, S., Lopez-Benitez, R., Radeleff, B., Berger, I., and Richter, G.M. (2008) Reduction of late in-stent stenosis in a porcine coronary artery model by cobalt chromium

stents with a nanocoat of polyphosphazene (polyzene-F). *CardioVasc. Interv. Radiol.* **31**(6): 1184–1192.

68. Channon, K. (2002) Endothelium and pathogenesis of atherosclerosis. *J. Med.* **30**(4): 54–58.
69. De Mel, A., Jell, G., Stevens, M., and Seifalian A.M. (2008) Biofunctionalization of biomaterials for accelerated in situ endothelialization: a review. *J. Biomacrom.* **9**(11): 2969–2979.
70. Karagkiozaki, V., Vavoulidis, E., Karagiannidis, P., Gioti, M., Fatouros, D.G., Vizirianakis, I.S., and Logothetidis, S. (2012) Development of a nanoporous and multilayer drug-delivery platform for medical implants. *Int . J. Nanomedicine.* **7**: 5327–38.
71. Kang, H., Kim, D., Sung, J., Park, S., Yoo, J., and Ryu, Y. (2007) Controlled drug release using nanoporous anodic aluminum oxide on stent. *Thin Solid Films* **515**: 5184–5187.
72. Bhargava, B., Reddy, K., Karthikeyan, G., Raju, R., Mishra, S., and Singh, S. (2006) A novel paclitaxel-eluting porous carbon–carbon NP coated nonpolymeric cobalt–chromium stent: Evaluation in a porcine model. *J. Cath. Cardiovasc. Interv*. **67**: 698–702.

Chapter 3

Nanomedicine Advancements in Cancer Diagnosis and Treatment

Eric Michael Bratsolias Brown
University of Wisconsin-Whitewater, Department of Biological Sciences, 800 W. Main Street, Upham Hall 357, Whitewater, Wisconsin 53190, USA
browne@uww.edu

3.1 Introduction

The war on cancer was declared in the United States with the passing of the National Cancer Act in 1971.[1] The European Union (EU), through its Association of European Cancer Leagues, has recently developed the third version of its *European Code Against Cancer*: a list of eleven commandments on general lifestyle choices that can be adopted by all individuals to reduce the number of cancer associated deaths.[2] According to the recent World Health Organization's Global status report on noncommunicable diseases, cancers accounted for 7.6 million deaths worldwide (or 21% of noncommunicable deaths).[3] Considering these examples as testaments (and there are countless others), it is quite evident that the battle against cancer has been

Horizons in Clinical Nanomedicine
Edited by Varvara Karagkiozaki and Stergios Logothetidis

ISBN 978-981-4411-56-1 (Hardcover), 978-981-4411-57-8 (eBook)
www.panstanford.com

long and will endure for the foreseeable future. It is no surprise that many resources over the last 40 years have been dedicated toward finding new manners to diagnose and treat the various types of cancers. As the world's population continues to age, it is certain that increasing resources and tools will be required to address public health needs surrounding cancer.

Currently, cancer is typically treated with a combination of surgery, chemotherapy, and/or radiation. Although these approaches combined with earlier detection methods have resulted in greatly improved prognoses for many cancer patients, these techniques are often invasive or associated with fatigue, nausea, hair loss, sterilization, and many other undesired side effects.[4,5] Additionally, many cancers, such as pancreatic cancers,[6] are not particularly responsive to these current types of treatment, or are detected too late as is often the case with ovarian cancers.[7] With an ever-increasing and aging population, the number of cancer cases is anticipated to rise in the years ahead. For example, although the rate of mortality is expected to decline, the number of total cancer deaths in Europe is expected to climb to 1.3 million in 2012.[8] The American Cancer Society estimates that 1,638,910 new cancer cases will appear in 2012 and there will be 577,190 deaths.[9] It is predicted that there will be 234,727 new cancer cases and 73,313 cancer deaths in Korea in 2012 and this cancer burden is expected to continue to increase in future years.[10] These few examples highlight the large economic and societal impacts that cancer has around the world. For these reasons, it is essential that new types of diagnostics and therapeutics are explored in cancer research to enable earlier detection and improve patient treatment. Through employing newly developed nanotechnological tools, nanomedicine may offer such alternatives.

The general consensus of the EU Commission on Nanotechnology[11] and the United States Nanotechnology Initiative[12] is that nanotechnology is defined by two criteria: (1) study of materials that are between 1–100 nm in at least one dimension and (2) study of materials that have unique properties which differ from the same material in bulk form. The latter qualification is arguably deemed the more important attribute of nanomaterials, while the size restrictions imposed by the first qualification are not absolute. The use of interdisciplinary nanotechnology-based platforms and materials in medicine has gradually adopted the descriptor of nanomedicine (Fig. 3.1). Within the field of nanomedicine, much attention has been focused on the use of nanomaterials to improve cancer diagnostics and therapeutics, with the ultimate goal of creating single nanovehicles which can enhance both of these parameters simultaneously (theranostics).[13–17] Other reviews have discussed nanomedicine at large,[18] how nanoparticles can be used in cellular imaging,[19] or focused more heavily on how the electronic detail of nanomaterials leads to this nanotechnologic behavior.[20] This chapter will provide a representative cross-section of nanomedicine advancements in diagnosis and treatment of cancer, as well as nanotheranostics. Allowing for appropriate digressions, the focus herein will be heavily upon the use of gold, iron oxide, and titanium dioxide nanoparticles in nanomedicine, as the former two are some of the most widely studied and the last is a recently emerging platform in the cancer field.

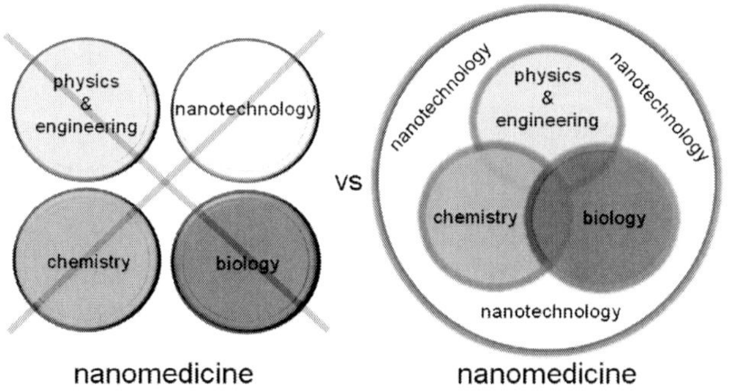

Figure 3.1 This schematic depicts the viewpoint that nanotechnology is not a separate field of study that complements distinctly traditional fields, leading to the creation of the newly emerging field of nanomedicine (left). Rather, nanotechnology is a new extension of existing fields within an interdisciplinary frame, which leads to the generation of novel tools within nanomedicine (right).

3.2 Unique Properties of Gold, Iron oxide, and Titanium Dioxide Nanoparticles That Benefit Cancer Applications

Although a diverse set of nanomaterials such as carbon nanostructures,[21–23] quantum dots,[24–26] and liposomes[27,28] are being investigated for cancer use in nanomedicine, the focus on this chapter will rest upon metal colloid gold, iron oxide, and titanium dioxide nanoparticles due to their varied attributes and unique promise. Each of these three types of nanoparticles has its own unique properties, which can be tailored to benefit different aspects of cancer diagnostics and therapeutics. As research in nanomedicine continues to progress, it is increasingly apparent that the goal of creating single theranostic devices (combining both diagnostic and therapeutic capabilities) is attainable with each gold, iron oxide, and titanium dioxide nanoplatforms (Fig. 3.2).

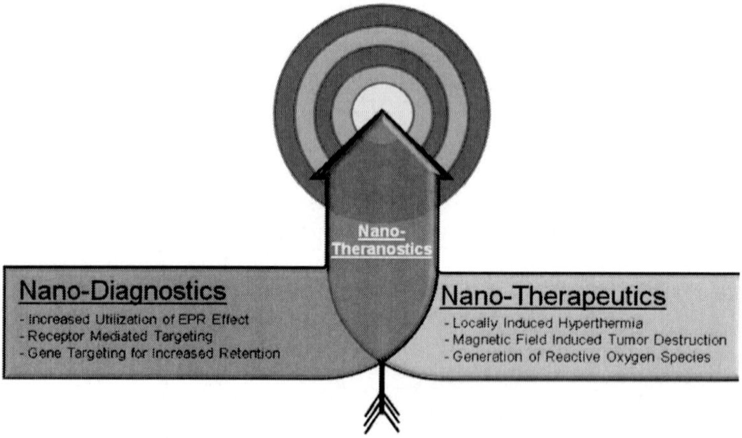

Figure 3.2 Nano-theranostics tools simultaneously detect and eliminate targeted cancer cells through rational and multifunctional design.

As reviewed in Hartman et al., metal collides, such as gold, nickel, silver, and platinum, are usually synthesized by coating a thin layer of metal over a dielectric core (such as silica).[29] Gold nanoparticles are generally considered to possess the highest level of promise for nanomedicine due to their low toxicity, physiological inertness, and anti-corrosive properties.[30] The ability to tailor the visible light emission from gold nanoparticles of different sizes and shapes arises from the ability of the surface electrons within the gold shell to oscillate in unison when exposed to an electromagnetic field (surface plasmon resonance, or SPR).[31,32] Through altering particle sizes to tune their absorption or scattering parameters, gold nanoparticles have been used in multiple imaging modalities, such as photoacoustic tomography, optical coherence tomography, confocal microscopy, and many others.[32–35] Later discussion will focus on the manner in which these characteristics are beneficial in diagnosis and therapy of cancer.

Metal oxides such as iron oxide and titanium dioxide have been the subject of much attention in cancer research over the last decade. Superparamagnetic iron oxide nanoparticles (SPIO) have attracted interest in the nanomedicine community for their unique properties. Under a magnetic field, these magnetic nanoparticles possess neighboring electrons which align in a single, magnetic domain, which results in a magnetic moment called superparamagnetism.[20] Although iron oxide nanoparticles hold much promise for MRI imaging, they are also intrinsically toxic and thus must be coated with polyethylene glycol (PEG) and polyethylene oxide (PEO),[36] dextran,[37] polysaccharides,[38] or other biocompatible materials. TiO_2 nanoparticles have been shown capable of ultra-violet (UV) light excitation (~3.2 eV), which results in the production of reactive oxygen species that can degrade chemical and biological agents.[39–41] In bulk, TiO_2, Ti atoms are hexacoordinated (octahedral). The Ti atoms in 4.5 nm TiO_2 nanoparticles, however, are pentacoordinated (square pyramidal) at the nanoparticle surface.[42] These undercoordinated "corner defects" of TiO_2 nanoparticles lead to high reactivity of the nanoparticles with ortho-substituted bidentate ligands (such as dopamine) which repair the undercoordination of the surface.[43] The ability to conjugate such ligands to the surface of TiO_2 nanoparticles (combined with their photocatalytic abilities) yields many benefits in targeting (cancer cells and genes) and treatment of cancer (localized production of reactive oxygen species), as will be illustrated.

3.3 Use of Gold, Iron Oxide, and Titanium Dioxide Nanoparticles in Cancer Diagnosis

Early detection enhances the prognosis of almost every type of cancer and relies upon imaging agent sensitivity and targeting. Nanotechnology promises to augment early cancer detection in both of these regards, by providing imaging agents that are more easily detected and targeted with higher precision to their desired location. This targeting can be achieved either through more efficient utilization of the enhanced permeability and retention (EPR) effect unique to tumors, or by rationally increasing specific targeting/

retention of nanoparticles through conjugation of peptides or nucleic acids.

Nanoparticles can be designed to create more sensitive imaging agents. Gold colloids have been studied with great interest for their human medicinal applications for thousands of years.[44] Due to their high electron density, gold nanoparticles have been used heavily in transmission electron microscopy (TEM) over the last four decades,[45] and this has enabled visualization of targets in cells and tissues. In addition to electron microscopy, gold nanoparticles are detectable by a diverse set of biomedical techniques such as surface plasmon resonance,[46–49] atomic force microscopy,[50–52] and fluorescence microscopy.[53] These examples also highlight the utility of gold nanoparticles in multimodal cancer imaging.

Magnetic resonance imaging (MRI) is a widely used imaging technique that detects water proton signals in the body. It is, however, often the case that contrast agents must be incorporated into the technique to achieve sufficient contrast. For the reasons mentioned previously, SPIOs possess a high relaxivity, image enhancement, and thus diagnostic potential in humans.[54–58] As reviewed in Hartman et al.,[29] the circulation half-lives of SPIOs in humans vary tremendously from anionic coated SPIOs (1 h) and starch coated SPIOs (two to three days). SPIOs are typically engulfed by macrophages in the spleen, liver, and bone marrow and metabolized over the course of a week. Iron-oxide nanoparticles are usually constructed of an Fe_3O_4 core with a polymer shell; they are useful as imaging agents, and can even be sensitized for enhanced ultrasound.[59–61] Xie et al. have encapsulated dopamine modified iron oxide nanoparticles with human serum albumin matrices to increase biocompatibility and labelled them with (64)Cu-DOTA and Cy5.5. Using a subcutaneous U87MG xenograft mouse model, these nanoparticles were detectable via a triple modality of positron emission tomography (PET), near-infrared fluorescence, and magnetic resonance imaging.[62]

TiO_2 nanoparticles also possess high utility in cancer imaging by enabling multimodal imaging capabilities. Bare TiO_2 nanoparticles can be detected with transmission[63–65] and scanning electron[66,67] microscopies, as well as atomic force microscopy.[40] Due to the aforementioned ability to conjugate enediol bidentate ligands through nanoparticle corner defects, dyes of interest can be coated on the nanoparticle surface to enable fluorescence detection of TiO_2 nanoparticles in cells and tissues.[68–71] Discussion in this chapter will demonstrate the manner in which multimodal imaging of TiO_2 nanoparticles aids cancer diagnostics.

Many nanoplatforms target tumors passively, but preferentially over normal tissues, via the EPR effect. The EPR effect was first described in the 1980s[72] and is an attractive feature for nanomedicine, given the size compatibility of nanoparticles and gaps resulting from angiogenesis.[73] It has been reported that these gaps resulting from EPR are 100–2000 nm, whereas normal tissues have smaller 2–6 nm junctions. The time of tumor development, however, is an important factor to consider, since the sizes of these gaps are always narrowing as the tumor develops.[74] For example, using this increased angiogenesis, radiolabelled long-circulating poly(ethylene glycol) (PEG)-coated hexadecylcyanoacrylate nanospheres have demonstrated enhanced targeting of brain tumors compared to their non-PEG-coated counterparts.[75]

Research using gold nanoparticles for cancer diagnostics has progressed steadily and some efforts to use the EPR effect to target vascularized tumors have entered into clinical trials, such as for refractory head and neck cancer (i.e., Auroshell).[76] Perrault et al. have demonstrated that the size of PEGylated gold nanoparticles highly affects their pharmacokinetic behavior, with large nanoparticles remaining near the vasculature and smaller counterparts penetrating into the tumor.[77] Despite these advances, targeting non-vascularized tumors has proven more difficult. To overcome this obstacle, Kennedy et al. have attempted to use T cells as "vehicles" to deliver gold nanoparticles to such tumors.[78] Using a human xenograft mouse model, they demonstrated that 45 nm gold nanoparticles can be loaded into T cells without adversely affecting the migration or cytokine production of the latter. Compared to free PEGylated gold nanoparticles, the T cell delivery system increases *in vivo* tumor targeting of gold nanoparticles by four times. Diagnostic approaches using iron oxide and TiO_2 nanoparticles are also attempting to hijack EPR to preferentially target tumor cells. For example, PAMAM dendrimers-doxorubicin (PAMAM-DOX) and superparamagnetic iron oxide (Fe_3O_4) nanoparticle conjugates possessed a hydrolytic release profile at a lysosomal pH of 5.0 and yielded increased DOX delivery to the tumor site via EPR.[79]

Almost all nanoplatforms also seek to enhance diagnosis of cancer through receptor-based targeting. Aurimune (CYT-6091), PEGylated gold nanoparticles decorated with tumor necrosis factor alpha for targeting of various types of solid tumors, have successfully completed phase I clinical trials for safety and are now entering phase II clinical trials and being combined with chemotherapeutics to treat pancreatic cancer, melanoma, soft tissue sarcoma, ovarian, and breast cancer patients.[80] Four to 16 nm gold nanoparticles that have been surface conjugated with a monocyclic RGDf, (RGD peptide analogue) have been shown to target the $\alpha(v)\beta_3$ integrin membrane protein in endothelial cells.[81] There have also been challenges associated with transitioning promising *in vitro* peptide-gold conjugates to demonstrate successful *in vivo* efficacy. While surface modified RGDfK gold nanorods were successfully bound and internalized by targeted endothelial cells *in vitro*, tumor targeting *in vivo* was substandard due to rapid clearing of the conjugates from the blood.[82] As with gold nanoparticles, it is possible to covalently conjugate various ligands (such as antibodies) to the surface of SPIOs to enhance targeting as is seen in the cases of breast[83] and rectal cancers.[84] Josephson et al. and Zhao et al. have conjugated the cell membrane penetrating HIV-1 tat peptide to iron oxide nanoparticles and demonstrated a 100-fold enhanced accumulation in lymphocytes.[85,86] SPIOs have also been conjugated with hormones to detect breast cancer metastasis[87] or folic acid to be used as a more broad spectrum method to target tumors.[88] Urokinase plasminogen activator (uPA)-targeted magnetic iron oxide nanoparticles can target breast cancer cells both *in vitro* and *in vivo*[89] Magnetic particles absorbed with epirubicin and targeted with a magnetic field for 45 minutes demonstrated a pharmakinetic advantage in approximately half of the patients in preliminary phase I clinical studies.[90] Detailed information on the factors affecting targeting, however, needs to be more fully elucidated in future studies. In the TiO_2 nanoparticle regime, the transmembrane protein HER2 (overexpressed on cancer cells) was successfully targeted *in vitro* by anti-HER2 antibody-TiO_2 conjugates.[91] Xu et al. demonstrated that TiO_2 nanoparticles conjugated with a specific antibody against the carcinoembryonic antigen were able to successfully discriminate LoVo cancer cells.[92]

An additional nanodiagnostic platform that is drawing attention is the attempt to influence gene expression of desired gene targets through conjugation of complementary nucleic acid sequences to the nanoparticle surface. Thiolated sense RNA sequences (with hybridized antisense RNA) have been conjugated to gold nanorods and irradiated with NIR to yield a 10% knockdown of HER2 protein breast carcinoma cells (BT474).[93] Using the same strategy, with the addition of folate receptor targeting, Lu et al. were able to demonstrate efficient downregulation of NF-κB p65.[94] It is important to note that this effect was only seen in those cells that received the nanoconjugate and NIR irradiation (no effect was seen in non-irradiated cells that were grown in the same mouse). Mok et al. have demonstrated that nanovectors composed of iron oxide magnetic nanoparticles, coated with siRNA, polyethyleneimine, and chlorotoxin, induced significant cytotoxic and gene silencing effects at acidic pH conditions in C6 glioma cells.[95] These effects were not seen at normal physiological pH, further supporting the use of this approach in the tumor microenvironment. Iron oxide nanoparticles have also been decorated with synthetic siRNA that targets the tumor-specific anti-apoptotic gene BIRC5, resulting in a significant downregulation of this gene of interest within subcutaneous mouse models of breast cancer.[96] Gene targeting has also been attempted in the TiO_2 nanoparticle regime. Zanardi et al. have developed a miniaturized fluorescence *in situ* hybridization (FISH) method using cluster-assembled nanoTiO_2 to be used on both fixed and fresh haematological samples and which requires >10 fold less probe usage.[97] This new technology will be beneficial in both genetic screening and diagnosis of onco-hematological malignancies. Numerous studies have demonstrated that ssDNA and DNA analogs can be conjugated to the surface of TiO_2 nanoparticles (creating nanoconjugates), using dopamine as a linker. Using this strategy, TiO_2 nanoconjugates have been synthesized that target select DNA sequences in a sequence specific manner *in vitro*[40,98] and within cells in culture.[40,99,100] Though promising results have been achieved using TiO_2 nanoconjugates to target genes *in vitro* and intracellular, it is too early to make predictions how this strategy will fare *in vivo* as this technology is still in its infancy.

3.4 Use of Gold, Iron Oxide, and Titanium Dioxide Nanoparticles in Cancer Therapy

In addition to furthering cancer diagnostics, nanoparticles offer many benefits to cancer therapeutics. Different nanoparticles absorb different, specified wavelengths of electromagnetic radiation, and this parameter is adjustable through altering either the nanoparticle composition or surface coating. This trait can be manipulated in cancer therapeutics to locally raise temperature or locally generate reactive oxygen species to treat cancers in a highly controlled fashion.

NIR light can penetrate relatively deep into human tissue and gold nanoparticles readily absorb light in this region, so gold nanoparticles are considered promising candidates in nanomedicine for thermal ablation of tumors.[101] It is estimated through both *in vivo* and theoretical experiments that the temperature of a tumor needs to be raised to 43°C to induce death.[102] As a practical demonstration of this promise, two recent studies have demonstrated that NIR activated gold nanoparticles are capable of eliminating human breast carcinoma tumors in culture, canine transmissible venereal tumors in mice,[101] mice carcinoma tumors,[33] as well as the aforementioned clinical trial using Auroshell gold nanoparticles for thermal ablation of head and neck cancers.[76]

Iron oxide nanoparticles that are directly injected into glioblastoma and exposed to an alternating current (Nano-Cancer) have recently completed a phase II clinical trial for efficacy and received EU regulatory approval.[103] Aminosilane-coated iron oxide nanoparticles were first used for tumor ablation through direct injection of the tumor tissue and application of an alternative magnetic field.[104,105] Following successful studies in rats, a pilot clinical study using iron oxide nanoparticles explored the treatment of prostate cancer.[106–109] There is some concern that iron toxicity with iron oxide nanoparticles may be an issue which needs to be addressed.[110]

TiO_2 nanoparticles are also being explored as a cancer therapy, since they readily absorb UV light and this wavelength can be shifted through multiple means (Fig. 3.3). Jonan-ku et al. have demonstrated that sonodynamic therapy (SDT), which uses ultrasound to activate chemical sensitizers, can be used in conjugation with TiO_2 nanoparticles to significantly decrease the growth of melanoma (C32) cells *in vitro* and inhibit the growth of subcutaneously implanted C32 solid tumors in mice.[111] It has also been shown that water soluble single-crystalline TiO_2 nanoparticles can be activated by UV light to induce killing of cancer cells *in vitro*.[112] This is a promising use of TiO_2 nanoparticles in cancer therapy. UV, light however, is a well-documented mutagen and may have negative effects on the tissue surrounding any area of treatment. For these reasons, and to broaden the applications of use for TiO_2 nanoparticles in cancer therapeutics, Blatnik et al. and Kamps et al. have demonstrated that dye surface-coated TiO_2 nanoconjugates can be activated with visible light to produce localized generation of reactive oxygen species which leads to the degradation of neighboring biological agents such as plasmid DNA *in vitro* and membrane associated proteins in HeLa cells.[68]

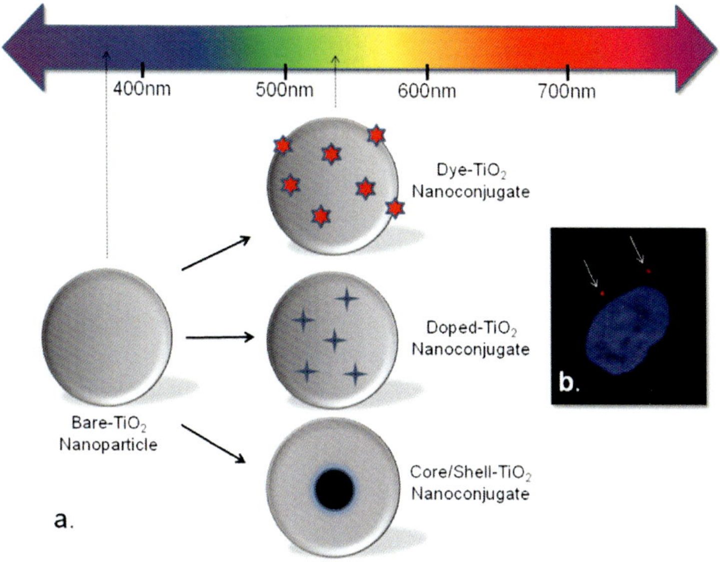

Figure 3.3 (a) The absorbance wavelength of bare TiO_2 nanoparticles can be red-shifted through numerous means such as surface conjugation of dyes (alizarin red s, ARS),[68,71] internal doping with metals (iron oxide),[113] or incorporation of an internal core (CdSe).[114] These and such other modifications broaden applications by allowing visible or IR light excitation. (b) Many of the aforementioned chemical modifications to TiO_2 nanoparticles can serve a dual purpose, such as ARS enabling intracellular fluorescence detection[68,71] of TiO_2 nanoconjugates in HeLa cells (white arrows) as demonstrated by this 1 μm slice confocal microscope image. (red = ARS-TiO_2 nanoconjugates; blue = Hoechst 33342).

Gold, iron oxide, and TiO_2 nanoparticles also benefit cancer therapeutics by acting as carriers and enhancing delivery of other compounds such as chemotherapeutics or other agents to their desired locations. Sershen et al. have demonstrated that gold nanoparticles can be combined with a temperature sensitive hydrogel in the form of a composite. When the gold is activated with 1064 nm light, the temperature of the composite (and thus the hydrogel) raises to induce significant drug release from the hydrogel.[115] Santra et al. demonstrated that Taxol and DiI co-encapsulating folate-functionalized iron oxide can enable both optical and magnetic resonance imaging as well as targeted cancer therapy.[116] TiO_2 nanoparticles have been used as carriers for chemotherapeutics such as daunorubicin (DNR) and doxorubicin which, otherwise, may be associated with serious side effects or suboptimal site delivery. DNR-TiO_2 nanocomposites, capable of higher rates of DNR release at pH 5-6 (compared to 7.4), yielded markedly enhanced anticancer activity in human leukemia K562 cells.[117] Iron oxide-titanium dioxide core-shell nanocomposites can be used as nanocarriers for doxorubicin to overcome common mechanisms of drug resistance in ovarian cancer cells.[69] Other studies have used TiO_2 nanomaterials to further enhance anticancer benefits observed from plant derivatives. When activated TiO_2 nanofibres are combined with celastrol (an active component in *T. wilfordii*) a synergistic effect of increased cytotoxicity was witnessed in HepG2 cells.[118] These results demonstrate the benefits of using TiO_2 nanoparticles in reducing required drug quantities and thus associated side effects.

3.5 Development of Gold, Iron Oxide, and Titanium Dioxide Nanoparticles for Cancer Theranostics

One of the greatest promises of nanomedicine in cancer research is the ability to combine diagnostics and therapeutics into a single theranostic tool. Ideally, this would allow real-time imaging and treatment of cancer. The result would be earlier detection, as well as more specific and less invasive treatments.

Recently, Fales et al. reported the synthesis of surface-enhanced Raman scattering (SERS) gold nanostars that were coated with a silica shell containing methylene blue. The incorporation of methylene blue photosensitizes the nanostar to allow 633 nm excitation and production of singlet oxygen, which leads to cytotoxic effect on BT549 breast cancer cells.[119] Such combination of diagnostics (SERS detection) and therapeutics (singlet-oxygen production) demonstrates the manner in which a single nanostructure can enhance theranostic capabilities. Iron oxide core-gold shell nanoparticles possessing multifunctional imaging and therapeutic capabilities have been developed by Melancon et al. These core-shell nanoparticles demonstrate a significant contrast enhancement in T2-weighted MRI as well as the ability to raise temperatures within targeted tumors *in vivo* by 21°C when irradiated by an 808 nm light. Additionally, a two fold increase in nanoparticle delivery to tumors *in vivo* was realized when nanoparticles were localized with an external magnet.[120] Quan et al. have shown the tumor targeting attributes of human serum albumin (HSA) coated iron oxide (HINP) nanoparticles through multiple imaging methods.[62] The additional incorporation of doxorubicin (Dox) in the HINP allowed for the release of Dox in a sustained manner, inducing cellular death in a 4T1 murine breast cancer xenograft model.[121] Others have used a combination of the fibrin binding, tumor-honing peptides (CREKA and CRKDKC) conjugated to elongated superparamagnetic iron oxide nanoparticles (nanoworms) to cause extensive clotting in the tumor vessels of mice possessing orthotopic human prostate cancer tumors. Reduced blood flow and necrosis were witnessed in these tumors, leading to a significant reduction in tumor growth. No such effects were evident in the vessels of normal tissues.[122] This technology demonstrates the benefits of using a single vehicle as both an imaging agent and therapy in the form of tumor clotting. It has been shown that SPIOs can induce hyperthermia upon exposure to cycling of a magnetic field.[123] Previous phase I clinical trials demonstrated that SPIO induced hyperthermia resulted in decreases in prostate-specific antigen (PSA), which further attests to the benefits of this approach.[124–127] Peptide-targeted TiO_2 nanoparticles, applied to a mouse xenograft model and irradiated with ultrasound for five sessions at 1.0 W/cm^2 for 60 s each, yielded tumor growth that was impeded for up to 28 days.[128]

Although theranostics in cancer nanomedicine is of great interest and rapid advances are being made, sufficient *in vivo* data required to make broad sweeping generalizations or claims are currently lacking (especially in the TiO_2 nanoparticle regime). Undoubtedly, some nanotheranostics will pass clinical trials while others will fail these tests. Though it is difficult to estimate the rate of these successes, it is clear that some nanotechnologies will realize their multifunctional promise.

3.6 Conclusion

In the over four decades since the war on cancer began, substantial progress has been made in both detection and therapy to improve the prognoses of cancer patients and to extend lives. Increasing life expectancies and an ever-aging population, however, almost guarantee that the societal cancer burden will continue to rise and new diagnostics and therapeutics will be required. Nanomedicine will likely serve as a major field to address this need and supply the new tools necessary for enhanced diagnosis and therapy of both currently "treatable" and "non-treatable" cancers. Directly stemming from the unique properties of nanomaterials, perhaps one of the most promising advancements of nanomedicine within cancer research is the creation of multifunctional nanotheranostics that are progressing toward the clinic.

References

1. The National Cancer Act of 1971 [Senate Bill 1828–Enacted December 23, 1971 (P.L. 92-218)]Journal of the National Cancer Institute Mar; 48(3), 577 (1972).
2. P. Boyle, P. Autier, H. Bartelink, J. Baselga, P. Boffetta, J. Burn, H. J. G. Burns, L. Christensen, L. Denis, M. Dicato, V. Diehl, R. Doll, S. Franceschi, C. R. Gillis, N. Gray, L. Griciute, A. Hackshaw, M. Kasler, M. Kogevinas, S. Kvinnsland, C. La Vecchia, F. Levi, J. G. McVie, P. Maisonneuve, J. M. Martin-Moreno, J. N. Bishop, F. Oleari, P. Perrin, M. Quinn, M. Richards, U. Ringborg, C. Scully, E. Siracka, H. Storm, M. Tubiana, T. Tursz, U. Veronesi, N. Wald, W. Weber, D. G. Zaridze, W. Zatonski and H. zur Hausen, European Code Against Cancer and scientific justification: third version (2003), *Ann Oncol* 14 (7), 973 (2003).
3. World Health Organization., *Global status report on noncommunicable diseases*. World Health Organization, Geneva, Switzerland, (2011), p v.
4. J. M. Binkley, S. R. Harris, P. K. Levangie, M. Pearl, J. Guglielmino, V. Kraus and D. Rowden, Patient perspectives on breast cancer treatment side effects and the prospective surveillance model for physical rehabilitation for women with breast cancer, *Cancer* 118 (8 Suppl), 2207 (2012).
5. M. D. Stubblefield, M. L. McNeely, C. M. Alfano and D. K. Mayer, A prospective surveillance model for physical rehabilitation of women with breast cancer: Chemotherapy-induced peripheral neuropathy, *Cancer* 118 (8 Suppl), 2250 (2012).
6. N. A. Hamilton, T. C. Liu, A. Cavatiao, K. Mawad, L. Chen, S. S. Strasberg, D. C. Linehan, D. Cao and W. G. Hawkins, Ki-67 predicts disease recurrence

and poor prognosis in pancreatic neuroendocrine neoplasms, *Surgery* 152 (1), 107–113 (2012).

7. C. Redman, S. Duffy, N. Bromham and K. Francis, Recognition and initial management of ovarian cancer: summary of NICE guidance, *BMJ* 342, d2073 (2011).
8. M. Malvezzi, P. Bertuccio, F. Levi, C. La Vecchia and E. Negri, European cancer mortality predictions for the year 2012, *Ann Oncol* 23 (4), 1044 (2012).
9. *American Cancer Society: Cancer Facts and Figures.* American Cancer Society, Inc. 2012.
10. K. W. Jung, S. Park, Y. J. Won, H. J. Kong, J. Y. Lee, H. G. Seo and J. S. Lee, Prediction of cancer incidence and mortality in Korea, 2012, *Cancer Res Treat* 44 (1), 25 (2012).
11. T. COMMUNICATION FROM THE COMMISSION TO THE COUNCIL, E. P. A. T. E. E. A. SOCIAL and COMMITTEE, Nanosciences and Nanotechnologies: An action plan for Europe 2005–2009. Second Implementation Report 2007–2009. SEC (2009) 1468.
12. NATIONAL NANOTECHNOLOGY INITIATIVE: Leading to the Next Industrial Revolution. A Report by the Interagency Working Group on Nanoscience, Engineering and Technology Committee on Technology. National Science and Technology Council. February 2000. Washington, D.C.
13. J. Xie and S. Jon, Magnetic nanoparticle-based theranostics, *Theranostics* 2 (1), 122 (2012).
14. F. Kiessling, S. Fokong, P. Koczera, W. Lederle and T. Lammers, Ultrasound microbubbles for molecular diagnosis, therapy, and theranostics, *J Nucl Med* 53 (3), 345 (2012).
15. N. Harrison, An update from the European Society of Pharmacogenomics and Theranostics, *Pharmacogenomics* 13 (2), 133 (2012).
16. X. Y. Chen, One Year after a Successful Start of Theranostics, *Theranostics* 2 (1), 1 (2012).
17. X. W. Cai, F. Yang and N. Gu, Applications of magnetic microbubbles for theranostics, *Theranostics* 2 (1), 103 (2012).
18. R. Duncan and R. Gaspar, Nanomedicine(s) under the microscope, *Mol Pharmaceut* 8 (6), 2101 (2011).
19. K. T. Thurn, E. M. B. Brown, A. Wu, S. Vogt, B. Lai, J. Maser, T. Paunesku and G. E. Woloschak, Nanoparticles for applications in cellular Imaging, *Nanoscale Res Lett* 2 (9), 430 (2007).

20. M. R. Fernández-Garcia, J. A., *Nanomaterials: Inorganic and Bioinorganic Perspectives: Metal Oxide Nanoparticles*. Brookhaven National Laboratory, (2007).
21. X. J. Wang and Z. Liu, Carbon nanotubes in biology and medicine: An overview, *Chinese Sci Bull* 57 (2–3), 167 (2012).
22. B. Rezaei, N. Majidi, S. Noori and Z. M. Hassan, Multiwalled carbon nanotubes effect on the bioavailability of artemisinin and its cytotoxity to cancerous cells, *J Nanopart Res* 13 (12), 6339 (2011).
23. D. L. Harris and R. Bawa, The carbon nanotube patent landscape in nanomedicine: an expert opinion, *Expert Opin Ther Pat* 17 (9), 1165 (2007).
24. C. Chen, J. Peng, S. R. Sun, C. W. Peng, Y. Li and D. W. Pang, Tapping the potential of quantum dots for personalised oncology: current status and future perspectives, *Nanomedicine-UK* 7 (3), 411 (2012).
25. H. E. Zhau, P. Z. Hu, G. D. Zhu, R. X. Wang, D. Berel, Y. Z. Wang, L. Fazli, D. Luthringer, A. Rogatko and L. W. K. Chung, Quantum dot multispectral imaging detects cell signaling for prostate cancer progression, *Eur Biophys J Biophys* 40, 65 (2011).
26. M. Adeli, M. Kalantari, M. Parsamanesh, E. Sadeghi and M. Mahmoudi, Synthesis of new hybrid nanomaterials: promising systems for cancer therapy, *Nanomed-Nanotechnol* 7 (6), 806 (2011).
27. W. T. Al-Jamal and K. Kostarelos, Liposomes: From a clinically established drug delivery system to a nanoparticle platform for theranostic nanomedicine accounts *Chem Res* 44 (10), 1094 (2011).
28. M. Q. Chu, S. Zhuo, J. Xu, Q. N. Sheng, S. K. Hou and R. F. Wang, Liposome-coated quantum dots targeting the sentinel lymph node, *J Nanopart Res* 12 (1), 187 (2010).
29. K. B. Hartman, L. J. Wilson and M. G. Rosenblum, Detecting and treating cancer with nanotechnology, *Mol Diagn Ther* 12 (1), 1 (2008).
30. L. R. Hirsch, A. M. Gobin, A. R. Lowery, F. Tam, R. A. Drezek, N. J. Halas and J. L. West, Metal nanoshells, *Ann Biomed Eng* 34 (1), 15 (2006).
31. H. Wang, D. W. Brandl, P. Nordlander and N. J. Halas, Plasmonic nanostructures: Artificial molecules, *Accounts Chem Res* 40 (1), 53 (2007).
32. J. A. Copland, M. Eghtedari, V. L. Popov, N. Kotov, N. Mamedova, M. Motamedi and A. A. Oraevsky, Bioconjugated gold nanoparticles as a molecular based contrast agent: Implications for imaging of deep tumors using optoacoustic tomography, *Mol Imaging Biol* 6 (5), 341 (2004).

33. A. M. Gobin, M. H. Lee, N. J. Halas, W. D. James, R. A. Drezek and J. L. West, Near-infrared resonant nanoshells for combined optical imaging and photothermal cancer therapy, *Nano Lett* 7 (7), 1929 (2007).
34. K. Fu, J. T. Sun, A. W. H. Lin, H. Wang, N. J. Halas and R. A. Drezek, Polarized angular dependent light scattering properties of bare and PEGylated gold nanoshells, *Curr Nanosci* 3 (2), 167 (2007).
35. C. Loo, L. Hirsch, M. H. Lee, E. Chang, J. West, N. J. Halas and R. Drezek, Gold nanoshell bioconjugates for molecular imaging in living cells, *Opt Lett* 30 (9), 1012 (2005).
36. C. C. Berry and A. S. G. Curtis, Functionalisation of magnetic nanoparticles for applications in biomedicine, *J Phys D Appl Phys* 36 (13), R198 (2003).
37. P. Reimer and T. Balzer, Ferucarbotran (Resovist): a new clinically approved RES-specific contrast agent for contrast-enhanced MRI of the liver: properties, clinical development, and applications, *Eur Radiol* 13 (6), 1266 (2003).
38. C. Gruttner and J. Teller, New types of silica-fortified magnetic nanoparticles as tools for molecular biology applications, *J Magn Magn Mater* 194 (1-3), 8 (1999).
39. Y. Sun, Y. Y. Bi and F. Shi, Synthesis and UV photosensitive properties of TiO_2-amphiphilic copolymer composite nanoparticles *Acta Chim Sinica* 65 (1), 67 (2007).
40. T. Paunesku, T. Rajh, G. Wiederrecht, J. Maser, S. Vogt, N. Stojicevic, M. Protic, B. Lai, J. Oryhon, M. Thurnauer and G. Woloschak, Biology of TiO_2-oligonucleotide nanocomposites, *Nat Mater* 2 (5), 343 (2003).
41. A. Michelmore, W. Q. Gong, P. Jenkins and J. Ralston, The interaction of linear polyphosphates with titanium dioxide surfaces, *Phys Chem Chem Phys* 2 (13), 2985 (2000).
42. T. Rajh, J. M. Nedeljkovic, L. X. Chen, O. Poluektov and M. C. Thurnauer, Improving optical and charge separation properties of nanocrystalline TiO_2 by surface modification with vitamin C, *J Phys Chem B* 103 (18), 3515 (1999).
43. T. Rajh, L. X. Chen, K. Lukas, T. Liu, M. C. Thurnauer and D. M. Tiede, Surface restructuring of nanoparticles: An efficient route for ligand-metal oxide crosstalk, *J Phys Chem B* 106 (41), 10543 (2002).
44. L. Dykman and N. Khlebtsov, Gold nanoparticles in biomedical applications: recent advances and perspectives, *Chem Soc Rev* 41 (6), 2256 (2012).

45. W. P. Faulk and G. M. Taylor, An immunocolloid method for the electron microscope, *Immunochemistry* 8 (11), 1081 (1971).
46. J. M. Zook, V. Rastogi, R. I. Maccuspie, A. M. Keene and J. Fagan, Measuring agglomerate size distribution and dependence of localized surface plasmon resonance absorbance on gold nanoparticle agglomerate size using analytical ultracentrifugation *ACS Nano* 5 (10), 8070 (2011).
47. X. Wang, Y. Xu, Y. Chen, L. Li, F. Liu and N. Li, The gold-nanoparticle-based surface plasmon resonance light scattering and visual DNA aptasensor for lysozyme, *Anal Bioanal Chem* 400 (7), 2085 (2011).
48. M. Inuta, R. Arakawa and H. Kawasaki, Use of thermally annealed multilayer gold nanoparticle films in combination analysis of localized surface plasmon resonance sensing and MALDI mass spectrometry, *Analyst* 136 (6), 1167 (2011).
49. W. C. Law, K. T. Yong, A. Baev, R. Hu and P. N. Prasad, Nanoparticle enhanced surface plasmon resonance biosensing: application of gold nanorods, *Opt Express* 17 (21), 19041 (2009).
50. T. H. Liu, S. H. Hsu, Y. T. Huang, S. M. Lin, T. W. Huang, T. H. Chuang, S. K. Fan, C. C. Fu, F. G. Tseng and R. L. Pan, The proximity between C-termini of dimeric vacuolar H+-pyrophosphatase determined using atomic force microscopy and a gold nanoparticle technique, *FEBS J* 276 (16), 4381 (2009).
51. R. Jin, X. He, K. Wang, L. Yang, H. Li, Y. Jin and W. Tan, Characterization of different sequences of DNA on si substrate by atomic force microscopy and gold nanoparticle labeling, *J Nanosci Nanotechnol* 7 (2), 418 (2007).
52. M. P. Bui, T. J. Baek and G. H. Seong, Gold nanoparticle aggregation-based highly sensitive DNA detection using atomic force microscopy, *Anal Bioanal Chem* 388 (5–6), 1185 (2007).
53. U. Taylor, S. Petersen, S. Barcikowski, D. Rath and S. Klein, Verification of gold nanoparticle uptake by bovine immortalised cells using laser scanning confocal microscopy, *Cytom Part A* 75A (8), 714 (2009).
54. W. Li, S. Tutton, A. T. Vu, L. Pierchala, B. S. Y. Li, J. M. Lewis, P. V. Prasad and R. R. Edelman, First-pass contrast-enhanced magnetic resonance angiography in humans using ferumoxytol, a novel ultrasmall superparamagnetic iron oxide (USPIO)-based blood pool agent, *J Magn Reson Imaging* 21 (1), 46 (2005).
55. M. Taupitz, S. Wagner, J. Schnorr, I. Kravec, H. Pilgrimm, H. Bergmann-Fritsch and B. Hamm, Phase I clinical evaluation of citrate-coated monocrystalline very small superparamagnetic iron oxide particles as

a new contrast medium for magnetic resonance imaging, *Invest Radiol* 39 (7), 394 (2004).

56. O. Clement, N. Siauve, M. Lewin, E. de Kerviler, C. A. Cuenod and G. Frija, Contrast agents in magnetic resonance imaging of the liver: present and future, *Biomed Pharmacother* 52 (2), 51 (1998).
57. P. Reimer, C. Marx, E. J. Rummeny, M. Muller, M. Lentschig, T. Balzer, K. H. Dietl, U. Sulkowski, T. Berns, K. Shamsi and P. E. Peters, SPIO-enhanced 2D-TOF MR angiography of the portal venous system: Results of an intra-individual comparison, *J Magn Reson Imaging* 7 (6), 945 (1997).
58. S. J. Mclachlan, M. R. Morris, M. A. Lucas, R. A. Fisco, M. N. Eakins, D. R. Fowler, R. B. Scheetz and A. Y. Olukotun, Phase-I Clinical-evaluation of a new iron-oxide MR contrast agent, *J Magn Reson Imaging* 4 (3), 301 (1994).
59. F. H. Xu, C. M. Cheng, F. J. Xu, C. F. Zhang, H. Xu, X. Xie, D. Z. Yin and H. C. Gu, Superparamagnetic magnetite nanocrystal clusters: A sensitive tool for MR cellular imaging, *Nanotechnology* 20 (40), (2009).
60. C. J. Xu and S. H. Sun, Superparamagnetic nanoparticles as targeted probes for diagnostic and therapeutic applications, *Dalton Trans* (29), 5583 (2009).
61. B. Wang, F. Zhang, J. H. Qiu, X. H. Zhang, H. Chen, Y. Du and P. Xu, Preparation of Fe_3O_4 superparamagnetic nanocrystals by coprecipitation with ultrasonic enhancement and their characterization, *Acta Chim Sinica* 67 (11), 1211 (2009).
62. J. Xie, K. Chen, J. Huang, S. Lee, J. H. Wang, J. Gao, X. G. Li and X. Y. Chen, PET/NIRF/MRI triple functional iron oxide nanoparticles, *Biomaterials* 31 (11), 3016 (2010).
63. I. Djerdj, A. M. Tonejc, M. Bijelic, V. Vranesa and A. Turkovic, Transmission electron microscopy studies of nanostructured TiO_2 films on various substrates, *Vacuum* 80 (4), 371 (2005).
64. T. Akita, M. Okumura, K. Tanaka, K. Ohkuma, M. Kohyama, T. Koyanagi, M. Date, S. Tsubota and M. Haruta, Transmission electron microscopy observation of the structure of TiO_2 nanotube and Au/TiO_2 nanotube catalyst, *Surf Interface Anal* 37 (2), 265 (2005).
65. A. M. Tonejc, M. Goti, B. Grzeta, S. Music, S. Popovi, R. Trojko, A. Turkovi and I. MuSevic, Transmission electron microscopy studies of nanophase TiO_2, *Mat Sci Eng B-Solid* 40 (2–3), 177 (1996).
66. A. B. Moghaddam, M. R. Ganjali, R. Dinarvand, A. Mohammadi and P. Norouzi, Electrochemical and scanning electron microscopic studies of the influence of anatase TiO_2 nanoparticles on the electropolymerization of aniline, *Mendeleev Commun* 18 (2), 90 (2008).

67. K. Mustafa, J. Wroblewski, K. Hultenby, B. S. Lopez and K. Arvidson, Effects of titanium surfaces blasted with TiO_2 particles on the initial attachment of cells derived from human mandibular bone: A scanning electron microscopic and histomorphometric analysis, *Clin Oral Implan Res* 11 (2), 116 (2000).

68. J. Blatnik, L. Luebke, S. Simonet, M. Nelson, R. Price, R. Leek, L. Zeng, A. Wu and E. M. B. Brown, Dye surface coating enables visible light activation of TiO_2 nanoparticles leading to degradation of neighboring biological structures, *Microsc Microanal* v18, 134 (2012).

69. H. C. Arora, M. P. Jensen, Y. Yuan, A. G. Wu, S. Vogt, T. Paunesku and G. E. Woloschak, Nanocarriers enhance doxorubicin uptake in drug-resistant ovarian cancer cells, *Cancer Res* 72 (3), 769 (2012).

70. K. T. Thurn, H. Arora, T. Paunesku, A. Wu, E. M. Brown, C. Doty, J. Kremer and G. Woloschak, Endocytosis of titanium dioxide nanoparticles in prostate cancer PC-3M cells, *Nanomedicine-UK* 7 (2), 123 (2011).

71. K. T. Thurn, T. Paunesku, A. G. Wu, E. M. B. Brown, B. Lai, S. Vogt, J. Maser, M. Aslam, V. Dravid, R. Bergan and G. E. Woloschak, Labeling TiO_2 nanoparticles with dyes for optical fluorescence microscopy and determination of TiO_2-DNA nanoconjugate stability, *Small* 5 (11), 1318 (2009).

72. Y. Matsumura and H. Maeda, A new concept for macromolecular therapies in cancer chemotherapy: Mechanism of tumouritrophic accumulation of proteins and the antitumour agent SMANCS, *Cancer Res* 6, 6387 (1986).

73. I. Brigger, C. Dubernet and P. Couvreur, Nanoparticles in cancer therapy and diagnosis, *Adv Drug Deliver Rev* 54 (5), 631 (2002).

74. S. K. Hobbs, W. L. Monsky, F. Yuan, W. G. Roberts, L. Griffith, V. P. Torchilin and R. K. Jain, Regulation of transport pathways in tumor vessels: Role of tumor type and microenvironment, *Proc Natl Acad Sci USA* 95 (8), 4607 (1998).

75. I. Brigger, J. Morizet, G. Aubert, H. Chacun, M. J. Terrier-Lacombe, P. Couvreur and G. Vassal, Poly(ethylene glycol)-coated hexadecylcyanoacrylate nanospheres display a combined effect for brain tumor targeting, *J Pharmacol Exp Ther* 303 (3), 928 (2002).

76. Nanospectra [http://www.nanospectra.com/index.html]. (April 23, 2012).

77. S. D. Perrault, C. Walkey, T. Jennings, H. C. Fischer and W. C. W. Chan, Mediating tumor targeting efficiency of nanoparticles through design, *Nano Lett* 9 (5), 1909 (2009).

78. L. C. Kennedy, A. S. Bear, J. K. Young, N. A. Lewinski, J. Kim, A. E. Foster and R. A. Drezek, T cells enhance gold nanoparticle delivery to tumors *in vivo*, *Nanoscale Res Lett* 6, (2011).

79. Y. L. Chang, X. L. Meng, Y. L. Zhao, K. Li, B. Zhao, M. Zhu, Y. P. Li, X. S. Chen and J. Y. Wang, Novel water-soluble and pH-responsive anticancer drug nanocarriers: Doxorubicin-PAMAM dendrimer conjugates attached to superparamagnetic iron oxide nanoparticles (IONPs), *J Colloid Interf Sci* 363 (1), 403 (2011).

80. Cytimmune Sciences Inc., http://www.cytimmune.com/go.cfm?do=page.view&pid=26, (April 24, 2012).

81. M. B. Dowling, L. J. Li, J. Park, G. Kumi, A. Nan, H. Ghandehari, J. T. Fourkas and P. DeShong, Multiphoton-absorption-induced-luminescence (MAIL) imaging of tumor-targeted gold nanoparticles, *Bioconjugate Chem* 21 (11), 1968 (2010).

82. A. J. Gormley, A. Malugin, A. Ray, R. Robinson and H. Ghandehari, Biological evaluation of RGDfK-gold nanorod conjugates for prostate cancer treatment, *J Drug Target* 19 (10), 915 (2011).

83. M. A. Funovics, B. Kapeller, C. Hoeller, H. S. Su, R. Kunstfeld, S. Puig and K. Macfelda, MR imaging of the her2/neu and 9.2.27 tumor antigens using immunospecific contrast agents, *Magn Reson Imaging* 22 (6), 843 (2004).

84. A. Toma, E. Otsuji, Y. Kuriu, K. Okamoto, D. Ichikawa, A. Hagiwara, H. Ito, T. Nishimura and H. Yamagishi, Monoclonal antibody A7-superparamagnetic iron oxide as contrast agent of MR imaging of rectal carcinomaBrit, *J Cancer* 93 (1), 131 (2005).

85. L. Josephson, C. H. Tung, A. Moore and R. Weissleder, High-efficiency intracellular magnetic labelling with novel superparamagnetic-tat peptide conjugates, *Bioconjugate Chem* 10 (2), 186 (1999).

86. M. Zhao, M. F. Kircher, L. Josephson and R. Weissleder, Differential conjugation of tat peptide to superparamagnetic nanoparticles and its effect on cellular uptake, *Bioconjugate Chem* 13 (4), 840 (2002).

87. C. Leuschner, C. S. S. R. Kumar, W. Hansel, W. Soboyejo, J. K. Zhou and J. Hormes, LHRH-conjugated magnetic iron oxide nanoparticles for detection of breast cancer metastases, *Breast Cancer Res Trans* 99 (2), 163 (2006).

88. C. Sun, R. Sze and M. Q. Zhang, Folic acid-PEG conjugated superparamagnetic nanoparticles for targeted cellular uptake and detection by MRI, *J Biomed Mater Res A* 78A (3), 550 (2006).

89. L. L. Yang, X. H. Peng, Y. A. Wang, X. X. Wang, Z. H. Cao, C. C. Ni, P. Karna, X. J. Zhang, W. C. Wood, X. H. Gao, S. M. Nie and H. Mao, Receptor-targeted

nanoparticles for *In vivo* imaging of breast cancer, *Clin Cancer Res* 15 (14), 4722 (2009).

90. A. S. Lubbe, C. Bergemann, H. Riess, F. Schriever, P. Reichardt, K. Possinger, M. Matthias, B. Dorken, F. Herrmann, R. Gurtler, P. Hohenberger, N. Haas, R. Sohr, B. Sander, A. J. Lemke, D. Ohlendorf, W. Huhnt and D. Huhn, Clinical experiences with magnetic drag targeting: A phase I study with 4′-epidoxorubicin in 14 patients with advanced solid tumors, *Cancer Res* 56 (20), 4686 (1996).
91. M. C. Tsai, T. L. Tsai, D. B. Shieh, H. T. Chiu and C. Y. Lee, Detecting HER2 on cancer cells by TiO_2 spheres mile scattering, *Anal Chem* 81 (18), 7590 (2009).
92. J. Xu, Y. Sun, J. J. Huang, C. M. Chen, G. Y. Liu, Y. Jiang, Y. M. Zhao and Z. Y. Jiang, Photokilling cancer cells using highly cell-specific antibody-TiO_2 bioconjugates and electroporation, *Bioelectrochemistry* 71 (2), 217 (2007).
93. S. E. Lee, G. L. Liu, F. Kim and L. P. Lee, Remote optical switch for localized and selective control of gene interference, *Nano Lett* 9 (2), 562 (2009).
94. W. Lu, G. D. Zhang, R. Zhang, L. G. Flores, Q. Huang, J. G. Gelovani and C. Li, Tumor site-specific silencing of NF-kappa B p65 by targeted hollow gold nanosphere-mediated photothermal transfection, *Cancer Res* 70 (8), 3177 (2010).
95. H. Mok, O. Veiseh, C. Fang, F. M. Kievit, F. Y. Wang, J. O. Park and M. Q. Zhang, pH-Sensitive siRNA nanovector for targeted gene silencing and cytotoxic effect in cancer cells, *Mol Pharmaceut* 7 (6), 1930 (2010).
96. M. Kumar, M. Yigit, G. P. Dai, A. Moore and Z. Medarova, Image-guided breast tumor therapy using a small interfering RNA nanodrug, *Cancer Res* 70 (19), 7553 (2010).
97. A. Zanardi, D. Bandiera, F. Bertolini, C. A. Corsini, G. Gregato, P. Milani, E. Barborini and R. Carbone, Miniaturized FISH for screening of onco-hematological malignancies, *Biotechniques* 49 (1), 497 (2010).
98. E. M. B. Brown, T. Paunesku, A. Wu, K. T. Thurn, B. Haley, J. Clark, T. Priester and G. E. Woloschak, Methods for assessing DNA hybridization of peptide nucleic acid-titanium dioxide nanoconjugates, *Anal Biochem* 383 (2), 226 (2008).
99. T. Paunesku, T. Ke, R. Dharmakumar, N. Mascheri, A. Wu, B. Lai, S. Vogt, J. Maser, K. Thurn, B. Szolc-Kowalska, A. Larson, R. C. Bergan, R. Omary, D. Li, Z. R. Lu and G. E. Woloschak, Gadolinium-conjugated TiO_2-DNA oligonucleotide nanoconjugates show prolonged intracellular

retention period and T1-weighted contrast enhancement in magnetic resonance images, *Nanomedicine-UK* 4 (3), 201 (2008).

100. T. Paunesku, S. Vogt, B. Lai, J. Maser, N. Stojicevic, K. T. Thurn, C. Osipo, H. Liu, D. Legnini, Z. Wang, C. Lee and G. E. Woloschak, Intracellular distribution of TiO_2-DNA oligonucleotide nanoconjugates directed to nucleolus and mitochondria indicates sequence specificity, *Nano Lett* 7 (3), 596 (2007).

101. L. R. Hirsch, R. J. Stafford, J. A. Bankson, S. R. Sershen, B. Rivera, R. E. Price, J. D. Hazle, N. J. Halas and J. L. West, Nanoshell-mediated near-infrared thermal therapy of tumors under magnetic resonance guidance, *Proc Natl Acad Sci U S A* 100 (23), 13549 (2003).

102. Q. A. Pankhurst, J. Connolly, S. K. Jones and J. Dobson, Applications of magnetic nanoparticles in biomedicine, *J Phys D Appl Phys* 36 (13), R167 (2003).

103. Magforce Nanotechnologies AG, http://www.cytimmune.com/go.cfm?do=page.view&pid=26, (April 24, 2012).

104. K. Maier-Hauff, F. Ulrich, D. Nestler, H. Niehoff, P. Wust, B. Thiesen, H. Orawa, V. Budach and A. Jordan, Efficacy and safety of intratumoral thermotherapy using magnetic iron-oxide nanoparticles combined with external beam radiotherapy on patients with recurrent glioblastoma multiforme, *J Neuro-Oncol* 103 (2), 317 (2011).

105. A. Jordan, Hyperthermia classic commentary: 'Inductive heating of ferrimagnetic particles and magnetic fluids: Physical evaluation of their potential for hyperthermia' by Andreas Jordan et al., *Int J Hyperthermia*, 1993;9:51-68, *Int J Hyperthermia* 25 (7), 512 (2009).

106. K. Maier-Hauff, A. Jordan, D. Nestler, R. Scholz, R. Rothe, A. Feussner, U. Gneveckow, P. Wust and R. Felix, Magnetic Fluid Hyperthermia (MFH) as an alternative treatment of malignant gliomas, *Strahlenther Onkol* 181, 44 (2005).

107. M. Johannsen, B. Thiesen, A. Jordan, K. Taymoorian, U. Gneveckow, N. Waldofner, R. Scholz, M. Koch, M. Lein, K. Jung and S. A. Loening, Magnetic fluid hyperthermia (MFH) reduces prostate cancer growth in the orthotopic dunning R3327 rat model, *Prostate* 64 (3), 283 (2005).

108. M. Johannsen, U. Gneveckow, L. Eckelt, A. Feussner, N. Waldofner, R. Scholz, S. Deger, P. Wust, S. A. Loening and A. Jordan, Clinical hyperthermia of prostate cancer using magnetic nanoparticles: Presentation of a new interstitial technique, *Int J Hyperther* 21 (7), 637 (2005).

109. A. Jordan, R. Scholz, K. Maier-Hauff, F. K. H. van Landeghem, N. Waldoefner, U. Teichgraeber, J. Pinkernelle, H. Bruhn, F. Neumann, B.

Thiesen, A. von Deimling and R. Felix, The effect of thermotherapy using magnetic nanoparticles on rat malignant glioma, *J Neuro-Oncol* 78 (1), 7 (2006).

110. M. Mahmoudi, A. Simchi, M. Imani, M. A. Shokrgozard, A. S. Milani, U. O. Hafeli and P. Stroeve, A new approach for the *in vitro* identification of the cytotoxicity of superparamagnetic iron oxide nanoparticles, *Colloid Surface B* 75 (1), 300 (2010).

111. Y. Harada, K. Ogawa, Y. Irie, H. Endo, L. B. Feril, T. Uemura and K. Tachibana, Ultrasound activation of TiO_2 in melanoma tumors, *J Control Release* 149 (2), 190 (2011).

112. J. W. Seo, H. Chung, M. Y. Kim, J. Lee, I. H. Choi and J. Cheon, Development of water-soluble single-crystalline TiO_2 nanoparticles for photocatalytic cancer-cell treatment, *Small* 3 (5), 850 (2007).

113. R. Alexandrescu, I. Morjan, M. Scarisoreanu, R. Birjega, C. Fleaca, I. Soare, L. Gavrila, V. Ciupina, W. Kylberg and E. Figgemeier, Development of the IR laser pyrolysis for the synthesis of iron-doped TiO_2 nanoparticles: Structural properties and photoactivity, *Infrared Phys Techn* 53 (2), 94 (2010).

114. A. M. F. Garcia, M. S. F. Fernandes and P. J. G. Coutinho, CdSe/TiO_2 core-shell nanoparticles produced in AOT reverse micelles: applications in pollutant photodegradation using visible light, *Nanoscale Res Lett* 6, (2011).

115. S. R. Sershen, S. L. Westcott, N. J. Halas and J. L. West, Temperature-sensitive polymer-nanoshell composites for photothermally modulated drug delivery, *J Biomed Mater Res* 51 (3), 293 (2000).

116. S. Santra, C. Kaittanis, J. Grimm and J. M. Perez, Drug/dye-loaded, multifunctional iron oxide nanoparticles for combined targeted cancer therapy and dual optical/magnetic resonance imaging, *Small* 5 (16), 1862 (2009).

117. H. J. Zhang and B. A. Chen, Daunorubicin -TiO_2 nanocomposites as 'smart' pH-responsive DRUG delivery system, *Blood* 118 (21), 1485 (2011).

118. J. Y. Li, X. M. Wang, H. Jiang, X. H. Lu, Y. D. Zhu and B. A. Chen, New strategy of photodynamic treatment of TiO_2 nanofibers combined with celastrol for HepG2 proliferation *in vitro*, *Nanoscale* 3 (8), 3115 (2011).

119. A. M. Fales, H. Yuan and T. Vo-Dinh, Silica-coated gold nanostars for combined surface-enhanced raman scattering (SERS) detection and singlet-oxygen generation: A potential nanoplatform for theranostics, Langmuir 27 (19), 12186 (2011).

120. M. P. Melancon, A. Elliott, X. J. Ji, A. Shetty, Z. Yang, M. Tian, B. Taylor, R. J. Stafford and C. Li, Theranostics with multifunctional magnetic gold nanoshells photothermal therapy and T2* magnetic resonance imaging, *Invest Radiol* 46 (2), 132 (2011).

121. Q. M. Quan, J. Xie, H. K. Gao, M. Yang, F. Zhang, G. Liu, X. Lin, A. Wang, H. S. Eden, S. Lee, G. X. Zhang and X. Y. Chen, HSA coated iron oxide nanoparticles as drug delivery vehicles for cancer therapy, *Mol Pharmaceut* 8 (5), 1669 (2011).

122. L. Agemy, K. N. Sugahara, V. R. Kotamraju, K. Gujraty, O. M. Girard, Y. Kono, R. F. Mattrey, J. H. Park, M. J. Sailor, A. I. Jimenez, C. Cativiela, D. Zanuy, F. J. Sayago, C. Aleman, R. Nussinov and E. Ruoslahti, Nanoparticle-induced vascular blockade in human prostate cancer, *Blood* 116 (15), 2847 (2010).

123. R. E. Rosensweig, Heating magnetic fluid with alternating magnetic field, *J Magn Magn Mater* 252 (1-3), 370 (2002).

124. A. Jordan, K. Maier-Hauff, R. Wust, B. Rau and M. Johannsen, Thermotherapy using magnetic nanoparticles, *Onkologie* 13 (10), 896 (2007).

125. M. Johannsen, U. Gneveckow, K. Taymoorian, B. Thiesen, N. Waldofner, R. Scholz, K. Jung, A. Jordan, P. Wust and S. A. Loening, Morbidity and quality of life during thermotherapy using magnetic nanoparticles in locally recurrent prostate cancer: Results of a prospective phase I trial, *Int J Hyperther* 23 (3), 315 (2007).

126. M. Johannsen, U. Gneveckow, K. Taymoonan, B. Thiesen, N. Waldofener, R. Scholz, C. H. Cho, A. Jordan, P. Wust and S. A. Loening, Thermotherapy using magnetic nanoparticles in patients with locally recurrent prostate cancer: Results of a prospective phase I trial, *Eur Urol Suppl* 6 (2), 201 (2007).

127. M. Johannsen, U. Gneueckow, B. Thiesen, K. Taymoorian, C. H. Cho, N. Waldofner, R. Scholz, A. Jordan, S. A. Loening and P. Wust, Thermotherapy of prostate cancer using magnetic nanoparticles: Feasibility, imaging, and three-dimensional temperature distribution, *Eur Urol* 52 (6), 1653 (2007).

128. K. Ninomiya, C. Ogino, S. Oshima, S. Sonoke, S. Kuroda and N. Shimizu, Targeted sonodynamic therapy using protein-modified TiO_2 nanoparticles, *Ultrason Sonochem* 19 (3), 607 (2012).

Chapter 4

Nanomedicine and Blood Diseases

Emmanouil Nikolousis
Heart of England Teaching Hospital, NHS Trust, Bordesley Green, Birmingham, B9 5SS, UK
manos.nikolousis@heartofengland.nhs.uk

4.1 Introduction: Haematological Malignancy Outcomes

Haematological malignancies can be life-threatening conditions which impose a significant morbidity and mortality in different human populations, putting tremendous financial and labour burden on health systems across the world.[1] Despite the dramatic improvements in the management of these patients, thanks to the use of novel agents and the ongoing success in supportive care, the overall survival and disease-free survival figures remain below a widely accepted level.[2] The main causes for the suboptimal outcomes remain the complexity of the diagnosis for some of these disorders, the toxicity of the available treatment regimen—especially chemotherapy—and the resistance of these tumours to treatment usually attributed to the overexpression of the multi-drug resistance (MDR) gene.[3]

Horizons in Clinical Nanomedicine
Edited by Varvara Karagkiozaki and Stergios Logothetidis

ISBN 978-981-4411-56-1 (Hardcover), 978-981-4411-57-8 (eBook)
www.panstanford.com

Moreover, most of the patients diagnosed with a haematological disorder are above the age of 60 with other co-morbidities, which render the use of chemotherapy even more controversial. Undoubtedly, this increases the non-disease related mortality and puts more pressure on the adverse disease related outcomes.[4] But even for the younger patients, there are tremendous variations in outcomes that highlight the unpredictable responses to the ordinary chemotherapy regimen and the unexpected toxicities. One of the most striking examples of this is the differences in outcomes for the different young age groups in acute lymphoblastic leukaemia. Patients in the age group 2–10 years old have a five-year overall survival of 85%, compared to 50% for the age group 10–25 years. Although molecular genetics and modern diagnostics have aided in the direction of disease risk assessment, the use of a more toxic regimen for the high risk patients ended in more deaths as a result of the treatment toxicities.[5]

The impact of both disease and treatment toxicities on the quality of life of these patients cannot be underestimated either. Most of them will spend a significant proportion of their everyday life in a hospital, either to receive treatment or to have the side effects of the treatment managed by specialist services. For the health services themselves, it translates into increased acute bed occupancy, which essentially means less space for other patients in need of treatment and also a sharp rise in the utilisation of the taxpayer's money.[6]

The complexities of management for haematological patients are highlighted below where outcomes of acute leukaemias are discussed[7–9]:

4.1.1 Acute Myeloid Leukaemia Outcomes

Figure 4.1 demonstrates the dramatic improvement in the outcomes of AML in different age groups between 1980–1984 and 2000–2004.

Another difference, however, is the sharp decline in survival in the age groups of 55 and above, which becomes even more profound above the age of 65. In this age group, it is shown that the outcomes, irrespective of the year being treated, remain equally poor with no major difference. The reason might be that these patients (especially above the age of 65) have less chances of undergoing intensification of their treatment with an allogeneic bone marrow transplant and also that they experience more deaths as a direct consequence of the chemotherapy toxicities

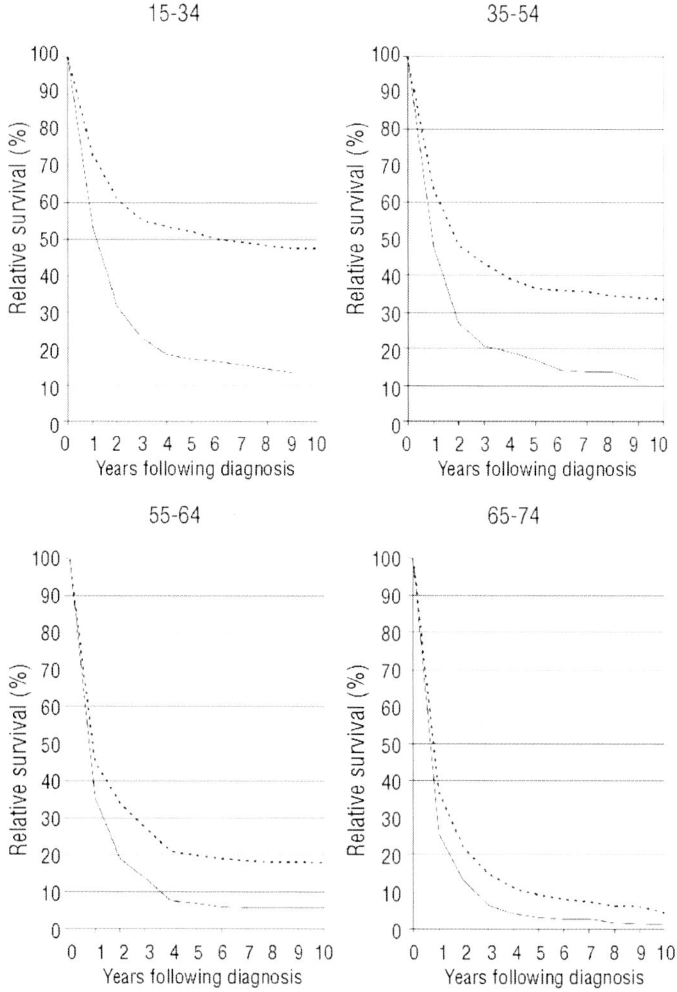

Figure 4.1 improvement in the outcomes of AML in different age groups between 1980–1984 (straight line) and 2000–2004 (dashed line).

Even for patients who are able to undergo a bone marrow transplant, the five-year overall survival does not exceed 40% for patients in their first or second complete remission (see Figure 4.2). For patients with advanced disease, however, the five-year overall survival falls to less than 20%, indicating that even the most intensive treatment gives a suboptimal response.

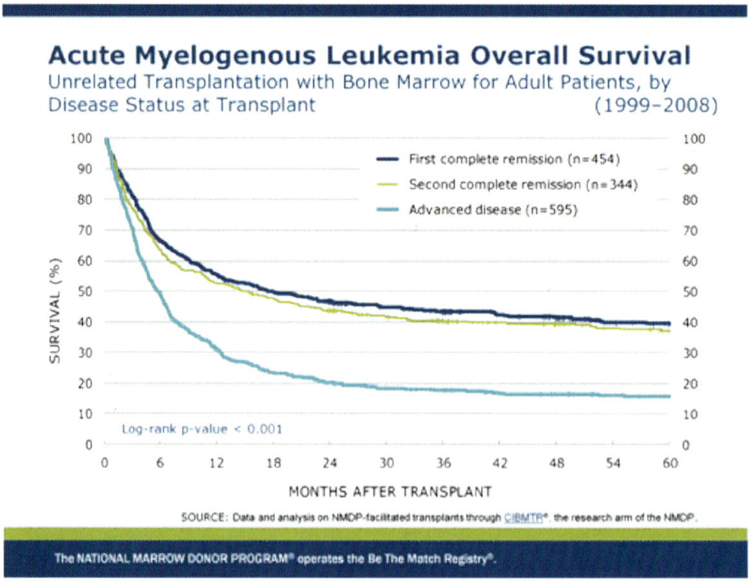

Figure 4.2 Allogeneic transplantation for acute myeloid leukaemia.

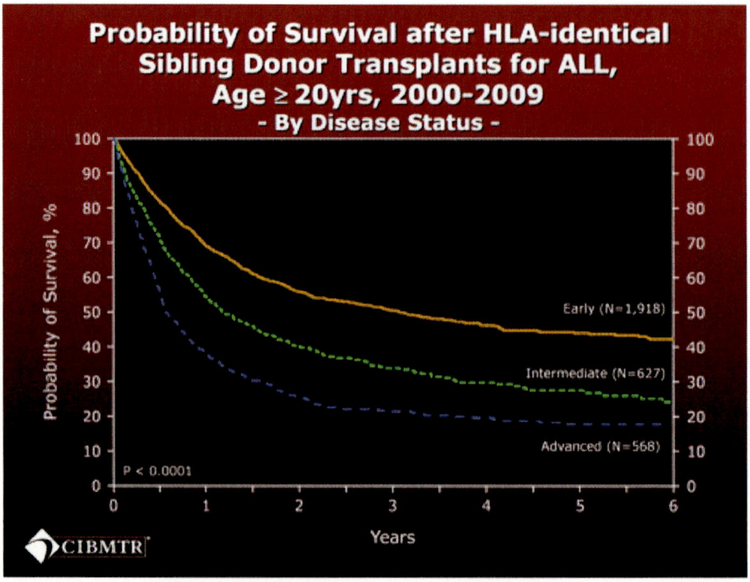

Figure 4.3 Acute lymphoblastic leukaemia.

The results are also disappointing in patients who received a sibling bone marrow transplant for acute lymphoblastic leukaemia, where the transplant related mortality is less than an unrelated donor transplant (see Figure 4.3).

As with the patients who underwent a transplant for acute myeloid leukaemia, the more advanced the disease, the worse the outcomes of the transplant.

4.2 Major Achievements in Haematology over the Last Decade

4.2.1 Targeted Therapy for Chronic Myeloid Leukaemia

In the last 20 years, many advances in the treatment of cancer can be traced to the cure of leukaemia. The landmark of these developments has been undoubtedly the development of a "targeted" specific therapy for chronic myeloid leukaemia (CML). This is an example of precise understanding of the molecular cause of a cancer which can lead to effective and non-toxic treatment options.

CML is one of the four main types of leukaemia, affecting approximately 5,000 people per year in the United States. The disease can occur at any age, but is primarily a disease of adults. Historically, patients with CML lived no more than three to five years and the mainstay of treatment has been hydroxycarbamide, with the only effect on the disease being a successful haematologic response for a specific period of time, or interferon—with all its related side effects—which reduced significantly the quality of life of these patients. Although these two different forms of therapy could result in a temporary benefit in most cases, the disease would quickly transform from a chronic leukaemia to an aggressive and fatal acute leukaemia.

CML has become the glowing example for targeted cancer therapy, demonstrating that a precise understanding of the cause of a disease allows an effective therapy to be developed. To achieve this, however, required decades of scientific discovery to unravel the cause and develop a targeted therapy. In 1960, Peter Nowell and David Hungerford, working in Philadelphia, described a shortened chromosome in the blood and bone marrow of patients with CML, which was the first consistent chromosomal abnormality associated with a human cancer. Later on, the identification of translocation of chromosomes 9 and 22 resulting in the Philadelphia chromosome was discovered. It was demonstrated that the consequence of this chromosome exchange was the production of an abnormal gene called *BCR-ABL*. This gene resulted in the excess growth of white blood cells in CML.

As the first task was accomplished and the target identified, the next step was to develop a drug which would inhibit the activity of *BCR-ABL* leading to normal haemopoesis. The compound that became known as Imatinib (Gleevec) was developed in 1992, and studies showed that this drug directly targeted the malignant cells containing the transcript, rather than both tumour and normal cells as with the conventional chemotherapy drugs. In 1998, the drug was tested in phase II trials in patients with end stage CML, who did not have an option for an allogeneic transplant and had exhausted standard treatment options. Within six months of starting the clinical trials of Imatinib, all the patients had an impressive haematological response, with their blood counts returning to normal. Later on, these findings were confirmed in much larger Phase III clinical trials, and Imatinib was approved by the U.S. Food and Drug Administration (FDA) in 2001, less than three years from the start of the clinical trials. Although imatinib is a major achievement, about 25% of patients will demonstrate resistance to Imatinib, even with the evolution of Imatinib to second line tyrosine kinase inhibitors (dasatinib,Nilotinib).These patients would certainly benefit from novel approaches.[10,11]

4.2.2 Immunotherapy for Non-Hodgkin Lymphoma

In the last 15 years, there has been a major improvement from large clinical trials of targeted therapies, which have been shown to be effective as a single agent or to boost the effectiveness of established treatments with combination chemotherapy. These novel drugs attack cancer via its biology, targeting specific molecular characteristics to impede malignant proliferation. Targeted therapies have been studied in most types of non-Hodgkin lymphoma and at all disease stages. Most of these agents are monoclonal antibodies, which are laboratory-developed proteins that mainly bind to surface proteins or receptors expressed by cancer cells; or small molecules that interfere with key cell survival machinery. The incidence of non-

Hodgkin lymphoma has been increasing since the 1970s, making it the fifth most common cancer in the U.S. Different studies have shown that the five-year survival of the lymphoma patients 30 years ago was only 30%, compared to the most recent studies which suggest that this number has more than doubled to 64%. This is largely due to the decreased number of relapsed patients who are treated with monoclonal anti-CD20 antibody targeted against the specific lymphoma cells without damaging the surrounding tissues of the body. The other major advantage of these targeted therapies was related to the relapsed lymphomas. With patients treated with conventional chemotherapy, once they relapse, it is more difficult to get them back to a remission status. Targeted treatment, however, can achieve this goal, although again this is not applicable to all relapsed patients.[12,13]

4.3 Nanotechnology and Its Use in Haematology: Background

The combination of physics, chemistry and biology at the nanometric scale of nanotechnology is a powerful technology, which is predicted to have a large impact on life sciences—particularly cancer treatment and diagnostics. The exploration of the nanoscale opens concrete opportunities for revealing new properties and undiscovered cell-particle interactions. Although multiple new treatments have been identified, cancer mortality is an ever-increasing menace that needs to be curbed soon.[14] In 90% of cancer cases, chemotherapy is the backbone of treatment. However, as mentioned in the introduction, the main drawbacks of chemotherapy is the high doses required to access the tumour sites with all the accompanying toxicity, the ability of the tumour to develop multiple drug resistance and the non-specific targeting of the chemotherapy drugs.[15] Although nanotechnology is not widely used at present in the clinical setting of haematooncology, the growing number of Phase I and II trials points towards the direction of a potential increase in use .

According to different studies, the main areas of interest for application are cancer imaging, molecular diagnosis and targeted therapy. The basic concept behind the use of nanotechnology in haematology is that nanoparticles, such as biodegradable micelles, semiconductor quantum dots and iron oxide nanocrystals, have functional or structural properties that are not available from the conventional chemotherapeutic agents. They can overcome issues in haematooncology related to drug delivery and retention and are considered potential candidates to carry drugs in a safe manner with the least possible side effects to the desired site of therapeutic action Their use could be in conjunction to other biotargeting ligands, such as monoclonal antibodies, peptides or small molecules to target malignant cells with high affinity and specificity, providing an effective non-toxic treatment.[16] Because of their small size, nanoscale devices could work synergistically with biomolecules on both the surface and inside cells. By gaining access to so many areas of the body, they have the potential to detect disease and deliver treatment in ways unimagined before. Moreover, in the field of diagnostics, "mesoscopic" size range of nanoparticles (5–100 nm in diameter) have large surface areas and functional groups for conjugating to multiple diagnostics (e.g., optical, radioisotopic or magnetic) and therapeutic (e.g., anticancer) agents. In cancer diagnostics, the combination of nanoparticles with imaging contrast agents provides a highly specific diagnostic system. The formation and use of multifunctional nanoparticle probes for molecular and cellular imaging, nanoparticle drugs for targeted therapy, and integrated nanodevices for early cancer detection and screening, has attracted great scientific interest in many countries. The exciting opportunities for personalised oncology in which cancer detection, diagnosis and therapy are tailored to each individual's molecular profile can be linked to the evolution of nanotechnology. Thus it can play a critical role at the site of cancer prevention in which genetic/molecular information is used to predict tumour development, progression and clinical outcome.[17]

One of the most difficult issues in the therapy of haematological malignancies is multidrug resistance (MDR). By the different pathways the nanoparticles enter cells, MDR can be minimised. There are two broad targeting modalities used by the anti-cancer polymer-drug conjugates: passive and active. It is well known that tumour tissues have anatomic characteristics that differ from normal tissues. Of great interest in nanotechnology has been the penetration and accumulation of the nanomolecules in tumours relative to normal tissues, leading to extended pharmacological effects.[18,19]

The different ways nanoparticles perform against cancer is very well summarised by Alliance Healthcare in the three major different pathophysiologic mechanisms below[20]:

4.3.1 Passive Targeting

There are now several nanocarrier-based drugs in the market, which rely on passive targeting through a process known as "enhanced permeability and retention". Because of their size and surface properties, certain nanoparticles can escape through blood vessel walls into tissues. In addition, tumours tend to have leaky blood vessels and defective lymphatic drainage, causing nanoparticles to accumulate in them, thereby concentrating the attached cytotoxic drug where it is needed, protecting healthy tissue and greatly reducing adverse side effects.

4.3.2 Active Targeting

On the horizon are nanoparticles that will actively target drugs to cancerous cells, based on the molecules that they express on their cell surface. Molecules that bind particular cellular receptors can be attached to a nanoparticle to actively target cells expressing the receptor. Active targeting can even be used to bring drugs into the cancerous cell, by inducing the cell to absorb the nanocarrier. Active targeting can be combined with passive targeting to further reduce the interaction of carried drugs with healthy tissue. Nanotechnology-enabled active and passive targeting can also increase the efficacy of a chemotherapeutic, achieving greater tumour reduction with lower doses of the drug.

4.3.3 Destruction from Within

Moving away from conventional chemotherapeutic agents that activate normal molecular mechanisms to induce cell death, researchers are exploring ways to physically destroy cancerous cells from within. One such technology—nanoshells—is being used in the laboratory to thermally destroy tumours from the inside. Nanoshells can be designed to absorb light of different frequencies, generating heat (hyperthermia). Once the cancer cells take up the nanoshells (via active targeting), scientists apply near-infrared light that is absorbed by the nanoshells, creating intense heat inside the tumour that selectively kills tumour cells without disturbing neighbouring healthy cells. Similarly, new targeted magnetic nanoparticles are in development that will both be visible through magnetic resonance imaging (MRI) and can also destroy cells by hyperthermia.

4.4 Nanotechnology and Diagnosis for Haematological Diseases

Part of the improvement in the haematological diseases over the last 20 years has been the improvement in diagnosis and the early detection of diseases, which enabled patients to have a better option for a curative treatment. A few examples in haematology are flow cytometry, use of positron emission tomography (PET) scan and molecular diagnostics. There is, however, still space for improvement and it appears that nanotechnology could contribute towards further development in this field.

Current imaging methods can only readily detect cancers once they have made a visible change to a tissue, by which time thousands of cells will have proliferated and, perhaps, metastasised. And even when visible, the nature of the tumour—malignant or benign—and the characteristics that might make it responsive to a particular treatment must be assessed through biopsies. Multifunctional nanomedicine is emerging as a highly integrated platform that allows for molecular diagnosis, targeted drug delivery, and simultaneous monitoring and treatment of cancer. Advances in polymer and materials science are critical for the successful development of these multi-component nanocomposites in one particulate system with such a small size confinement (<200 nm).[21] In a recent study, protein stabilised gold nanoclusters (Au-NCs) were examined in the flow cytometric diagnosis of leukaemia. Novel nano-bioprobes can be developed using protein protected fluorescent nanoclusters of Au for the molecular receptor targeted flow cytometry based detection and imaging of leukaemic cancer cells. The development of a gold nanoclustered based targeted fluorescent nano-bioprobe for the flow-cytometric detection of acute myeloid leukaemia (AML) cells were conjugated with monoclonal antibody against CD33 myeloid antigen, which is overexpressed in about 95% of the primitive population of AML cells. Au-NC-CD33 conjugates having an average size of ~12 nm retained bright fluorescence over an extended duration of a year, as the albumin protein protects Au-NCs against degradation. Nanotoxicity studies revealed excellent biocompatibility of Au-NC conjugates, as they showed no adverse effect on the cell viability and inflammatory response. There was a great difference in target specificity of the conjugates for detecting CD33 expressing AML KG1a leukaemic cells compared to human peripheral blood cells (PBMCs) which are CD33 (low): 95% vs. 8.4% within 1–2 h. The confocal imaging also demonstrated the targeted uptake of CD33 conjugated Au-NCs by leukaemia cells, thus confirming the flow cytometry results.[22]

Another critical area in diagnostics for haematology malignancies is the screening for biomarkers in tissues and fluids for diagnosis. This could also be enhanced and potentially upgraded by the use of nanotechnology. The concept behind this relies on the fact that individual cancers differ from each other and from normal cells by changes in the expression and distribution of tens to hundreds of molecules. As therapeutics advance, it may require the simultaneous detection of several biomarkers to identify a cancer for treatment selection. Towards this direction nanoparticles such as quantum dots, which, depending on their size, are able to emit light of different colours, would be able to identify different tumour biomarkers. The discrimination between cancer and normal cells can be enhanced by photoluminescence signals from antibody-coated quantum dots by the spectrum of light seen. With this method, different coloured quantum dots would be attached to antibodies for the detection of healthy and malignant cell lines.[23]

4.5 Nanotechnology in Specific Haematological Cancers

4.5.1 Mantle Cell Lymphoma

Mantle cell lymphoma (MCL) is a pre-germinal centre heterogeneous neoplasm characterised by cyclin D1 overexpression resulting from t(11;14)(q13;q32). Just with the use of conventional chemotherapy, MCL is incurable .The only curative option remains the allogeneic stem cell transplantation with the major disadvantage of a high morbidity and mortality due to the actual transplant. Therefore new treatment approaches are needed that target specific biologic pathways. Few papers exist in the literature on the use of nanomedicine in mantle cell lymphoma. The most interesting study by Singh et al. suggests a novel drug delivery nanovehicle enriched with the bioactive polyphenol, curcumin (curcumin nanodisks, or curcumin-ND). The biological explanation for this was based on the fact that cells treated with curcumin showed a dose-dependent increase in apoptosis

Curcumin (diferuloylmethane) is the active ingredient of the dietary spice turmeric (*Curcuma longa*) which is quite popular mainly in the Asian continent. In the past, there were many different groups which investigated its apoptotic effect and tried to translate this into a potential therapeutic manoeuvre in malignant disorders. Its main function is as a histone deacetylase inhibitor (HDAC) and it could be regarded as a new member of this class of drugs already used in haematological malignancies like azacytidine and decatibine in myelodysplastic syndromes and acute myeloid leukaemias. In a previous study at the Mayo Clinic in the US, curcumin has been shown to have a synergistic effect with green tea (*Camellia sinensis*), as described from studies against B-chronic lymphocytic leukemia (CLL) cells. This agent appears to suppress activation of nuclear factor κB, downregulates Syc activity, and also has an effect on transcription factors, growth factors, and cytokines, which are all specific molecular targets. Pathogenesis of mantle cell lymphoma as mentioned above is based on altered apoptosis pathways and activation of Akt .As a result, it is indeed feasible that curcumin via its many effects on these and similar pathways could be effective in mantle cells, as also to extend its use to other lymphoproliferative disorders if utilised correctly and in optimal fashion.[24]

Despite the obvious advantages of curcumin, its main disadvantage is related to the fact that it is very insoluble and has poor bioavailability, which makes its potential use in haemato-oncological malignancies very restricted. The same report by Singh et al describes the development of a novel drug delivery system enriched for this herb, incorporating the bioactive polyphenol into nanodisks (nanovehicle) containing a disk-shaped phospholipid bilayer whose edge is stabilised by a scaffold protein recombinant human apolipoprotein. By using this nanoparticle the biological activity of nano-bound curcumin, compared to free curcumin, is greatly enhanced, resulting in the delivery of an otherwise insoluble compound more effectively. This will interfere with the cell cycle causing increased growth arrest of the G1 phase of the cell cycle and decreased cyclin D1 levels within the cell. In the same study, the authors concluded that the curcumin nanodisks cause and induce apoptosis via reactive oxygen species (ROS) generation and activation of the caspase-3 pathway in cells of two MCL cell lines.[25]

4.5.2 CNS Lymphomas

Primary central nervous system lymphomas are aggressive lymphoid tumours which appear to be incurable with the use of chemotherapy or radiotherapy. Furthermore, chemotherapy drugs which are able to penetrate the blood-brain barrier (BBB) have debilitating side effects that are enhanced by the concomitant use of radiotherapy for these patients. One of the major problems is the limited availability of drugs that penetrate the blood brain barrier and the existing ones cannot be used with the maximum safety. But even with the use of these treatments, the outlook of these patients is really poor and rarely exceeds two years of survival.[26,27] The use of nanoparticles to deliver drugs to the brain across the BBB may provide a step forward and could offer a significant advantage compared to current strategies. The primary advantage of NP carrier technology is that NPs mask the BBB-limiting characteristics of the therapeutic drug molecule and in a similar way the nanoparticles would release the drug slower into the brain tissue, minimising the toxic effects to the brain. Further, influencing manufacturing factors (type of polymers and surfactants, NP size, and the drug molecule) are detailed in relation to movement of the drug delivery agent across the BBB. In the literature, there are currently Phase I/Phase II studies assessing nanoparticles for brain delivery. So far, there are different anaesthetic and chemotherapeutic agents. One of the drawbacks identified which could influence the efficacy and delivery of drugs, however, are physiological factors such as phagocytic activity of the reticuloendothelial system and protein opsonisation, resulting in a reduced amount of brain delivered drug. Both natural and synthetic polymers have been used as drug carriers, and several bioconjugates have been clinically approved or are in human clinical trials. Passive macromolecular drug delivery systems have achieved significant anti-lymphoma activity which could also be enhanced using further selectivity by active targeting. Attachment of targeting moieties to the polymer structure can selectively highlight the differences between cancer and normal brain cells through selective receptor-mediated endocytosis.[28]

4.5.3 Acute Leukaemias

Although treatment options for acute leukaemias have increased during the last decade, the results of these interventions are suboptimal, especially in older patients where they appear to be disappointing. One of the key chemotherapy drugs is daunorubicin, an anthracycline containing drug, which can induce remissions in about 60–70% of patients. These remissions, however, are certainly not long lasting mainly due to the presence of MDR in the leukaemic cells and the price paid for this is sometimes excessive toxicity, with cardiotoxicity being one of the most severe side effects.[29] One of the most important studies in the literature focused on daunorubicin-loaded magnetic nanoparticles in order to minimise serious side effects of daunorubicin containing chemotherapy regimen for acute leukaemias. This specific nanoparticle formulation possesses sustained drug-release and favourable antitumour properties, which may be used as a side effect-free treatment option for acute leukaemia therapy

The physical properties of Daunorubicin nanoparticles were investigated and their efficacy was evaluated in vitro on leukaemic cells by a standard WST-1 cell proliferation assay. Furthermore, cell apoptosis and intracellular accumulation of DNR were determined by FACSCalibur flow cytometry. The in vitro release data showed that the nanoparticles had excellent sustained release property and its use on K562 leukaemic cells resulted in inhibition of K562 leukaemic cell proliferation. In order to detect the concentration of daunorubicin in the leukaemic cells, fluorescent light techniques were used which demonstrated that daunorubicin nanoparticles could be taken up by K562 cells and persistently released daunorubicin in these cells.

In another study, which looked into the dose-limiting side effects of conventional chemotherapeutic agents and the therapeutic failure resulting from multidrug resistance (MDR) in acute leukaemia, a novel chemotherapy formulation of magnetic nanoparticles co-loaded with daunorubicin and 5-bromotetrandrin (DNR/BrTet-MNPs) was developed, and its effect on MDR K562 leukemic cells was explored. The results showed that the self-prepared DNR/BrTet-MNPs formulation demonstrated a sustained release of drug and displayed a dose-dependent antiproliferative activity on MDR leukemia K562/A02 cells. The other significant effect of this nanoparticle was the enhanced accumulation of intracellular daunorubicin in K562/A02 cells, and the downregulation of the transcription of the mdr1 gene and the expression of P-gp overcoming the resistance of leukaemic cells to treatment. This study confirmed the impressive effect of the novel DNR/BrTet-MNPs formulation, acting as a drug depot system for the sustained release of the loaded DNR and BrTet on multidrug resistance leukemia K562/A02 cells, and as a promising substrate for the development of a Phase II trial for overcoming MDR in patients with acute leukaemia.[30]

4.5.4 Chronic Myeloid Leukaemia

In 25-30% of the patients with CML there will be resistant to Imatinib mesylte leading to dose escalation or switching to a second line TKI. Despite this a significant proportion will remain resistant to treatment and an allogeneic stem cell transplant will be the treatment of choice with a curtive intent albeit the significant morbidity and mortality associated with the procedure. The most interesting study in the literature suggests packaging of imatinib into a biodegradable carrier, based on polyelectrolyte microcapsules. This manoeuvre increases drug retention and antitumor activity in CML stem cells. It can also be used for the ex vivo purging of malignant progenitors from patient autologous stem cells if an autograft is indicated, or if the patient does not have a donor for an allogeneic transplant . In this study, microparticles/capsules were obtained by the layer-by-layer (LbL) self-assembly of oppositely charged polyelectrolyte multilayers on removable calcium carbonate (CaCO(3)) templates and loaded with or without Imatinib mesylate. A leukemic cell line (KU812) and CD34(+) cells, freshly isolated from healthy donors or CML patients, were studied.[31]

Quantification of cell uptake of these particles was performed in both KU812, leukemic and normal CD34(+) stem cells. Imatinib-loaded polyelectrolyte microcapsules selectively targeted CML cells, by promoting apoptosis at doses that exert only cytostatic effects by Imatinib alone. The other impressive finding of this study, though, was that residual CML cells from patients' stem cell harvest products were reduced or eliminated more efficiently by using Imatinib-loaded PMCs, in comparison to the freely soluble Imatinib, with a purging efficiency of several logs. Along with the tumouricide activity, an impressive observation was the normal survival of the healthy haematopoetic CD34(+) stem-cell survival. This could potentially give a favourable long term outcome for patients who do not have the option of an allogeneic transplant, either because of different comorbidities—which make the allogeneic transplant procedure very risky—or due to the lack of a suitable donor.[32]

4.5.5 Multiple Myeloma

Multiple myeloma is an incurable plasma cell tumour which affects about 20000 patients in the U.S. yearly. The usual initial presentation can include bone fractures, acute renal failure, hypercalcaemia and anaemia separately, or in different combinations.[33] In any case, most of the patients had a median overall survival of 3–5 years before the introduction of the novel agents Lenalidomide and Bortezomib. In the most recent myeloma IX study, Zoledronic acid has been found to prevent further bony fractures due to myeloma. The study also found that zoledronic acid had a potential antitumour activity with a prolongation of overall survival for those patients who received it. The disadvantage of Zoledronic acid, however, is the short plasma half-life and rapid accumulation in bone limits that limits its potential use as an antitumor agent in extraskeletal tissues. In a recent Italian study, stealth liposomes encapsulating zoledronic acid to increase extraskeletal drug availability were used. In this study, Liposomal zoledronic acid induced inhibition of tumour growth and increased the overall survival in murine models of multiple myeloma significantly in comparison with zoledronic acid alone. According to the authors, treatment failure with zoledronic acid is likely based on the almost exclusive localisation and distribution of this agent in bone tissues, especially in the "osteoclastic niche".[34]

In Fig. 4.4, it is clearly shown that liposomal zoledronic acid 10 or 20 total μg induced a 58–68% of tumour growth inhibition (Table on the left) while PC3 xenografts showed a resistance to zoledronic acid treatment having only about 16–22% of tumour growth inhibition (Table on the right) after an equal dose of zoledronic acid. Therefore, the encapsulation of zoledronic acid in PEGylated liposomes that escape from reticuloendothelial system sensitised PC3 cells to the antitumour activity of zoledronic acid. The delivery of zoledronic acid in liposomes overcomes the bioavailability limitations of ZOL, allowing it to reach the "tumour niche".

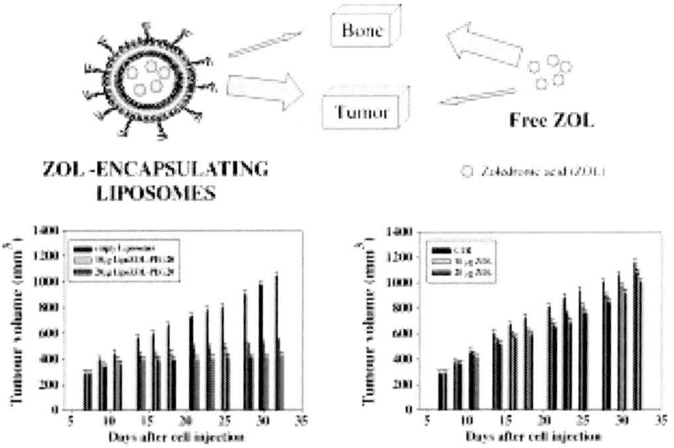

Figure 4.4 Liposomal zoledronic acid.

4.6 Major Challenges Facing the Use of Nanotechnology in Haematology

The main challenge for the use of nanotechnology is the lack of phase III clinical trials. Although most of the nanoparticles have been efficiently and safely used in a number of patients with haematological malignancies, the lack of a larger cohort of patients delays the use of nanoparticles in modern and wealthier health systems. At first glance, pushing for financial savings in nanomedicine has to be proven cost efficient in the current era of fiscal austerity with some of the Health Systems across the world. . Especially in haematology, the use of expensive antitumour medication (Rituximab, Imatinib, Dasatinib, Lenalidomide, etc.)[35] has added extra cost, although the use of these drugs result in prolonged survival, and therefore extend the follow up of these patients.

Thus, a change in the patients' management has to be strongly supported by well-designed randomised controlled trials which would directly compare the use of nanomedicine with the existing treatment regimen. On the other hand, even if the nanoparticles appear to be more expensive in the first place, they can still provide savings to haematology departments by preventing unnecessary hospitalisation due to treatment toxicities, and also by enabling patients to have a better quality of life. According to some recent data from the U.S., the cost of cancer treatment is taken over by the cost of the inpatient bed utilisation from the haematology patients, rather than the drugs themselves.

4.7 Conclusion

Nanotechnology will certainly be one of the medical innovations that will attract further attention in the field of haematooncology in the next ten years. The preliminary results look promising, but its use in large Phase III trials of haematological malignancies is essential. We should not forget the great promise given by gene therapy to different haematological cancers and other haematological diseases (i.e., Haemophilia, Haemoglobinopathies) ten years ago, but with very little use in the actual haematology practice, due to different obstacles and the side effects of treatment compared to the actual benefits. Health systems across the world invest in innovation technologies with a view to become more cost efficient, and also to improve the quality of life for these patients. In the U.K., there is a government Quality, Innovation, Productivity and Prevention (QIPP) incentive for the National Health Scheme (NHS) and the use of nanotechnology in solid cancers and haematological malignancies could well be part of this incentive. It is well known that blood cancers are very difficult to treat, let alone cure, and they impose a negative impact on the patient's life and a major burden on the health system.

References

1. Howell DA, Roman E, Cox H, Smith AG, Patmore R, Garry AC, Howard MR, Destined to die in hospital? Systematic review and meta-analysis of place of death in haematological malignancy, *BMC Palliat Care*. 2010; 9:9.
2. Alexander SC, Sullivan AM, Back AL, Tulsky JA, Goldman RE, Block SD, Stewart SK, Wilson-Genderson M, Lee SJ. Information giving and receiving in hematological malignancy consultations. *Psychooncology*. 2012; 21(3):297–306
3. Ireland R, Haematological malignancies: the rationale for integrated haematopathology services, key elements of organization and wider contribution to patient care. *Histopathology*. 2011; 58(1):145–154.

4. Sorror ML, Comorbidities and hematopoietic cell transplantation outcomes. *Hematol Am Soc Hematol Educ Program.* 2010; 2010:237–247.
5. Kobayashi Y. Molecular target therapy in hematological malignancy: front-runners and prototypes of small molecule and antibody therapy. *Jpn J Clin Oncol.* 2011; 41(2):157–164.
6. Sun CL, Francisco L, Baker KS, Weisdorf DJ, Forman SJ, Bhatia S, Adverse psychological outcomes in long-term survivors of hematopoietic cell transplantation: a report from the Bone Marrow Transplant Survivor Study (BMTSS). *Blood.* 2011; 118(17):4723–4731. Epub 5 Aug 2011.
7. National Marrow Donor Program (NMDP) UK website. *Transplant outcomes by disease and disease stage.*
8. Center for International Blood and Marrow Transplant Research website. *Transplant outcomes by disease and disease stage.*
9. Pulte D, Gondos A, Brenner H, Improvements in survival of adults diagnosed with acute myeloblastic leukemia in the early 21st century. *Haematologica.* 2008; 93(4):594–600.
10. Drucker B, ASH anniversary brochure 2008. *Targeted therapy for chronic myeloid leukemia.*
11. Tauchi T, Kizaki M, Okamoto S, Tanaka H, Tanimoto M, Inokuchi K, Murayama T, Saburi Y, Hino M, Tsudo M, Shimomura T, Isobe Y, Oshimi K, Dan K, Ohyashiki K, Ikeda Y; TARGET Investigators Seven-year follow-up of patients receiving imatinib for the treatment of newly diagnosed chronic myelogenous leukemia by the TARGET system. *Leuk Res.* 2011; 35(5):585–590.
12. Dotan E, Aggarwal C, Smith MR, Impact of rituximab (Rituxan) on the treatment of B-cell non-Hodgkin's lymphoma. *P T.* 2010; 35(3):148–157.
13. Chamarthy MR, Williams SC, Moadel RM, Radioimmunotherapy of non-Hodgkin's lymphoma: from the 'magic bullets' to 'radioactive magic bullets'. *Yale J Biol Med.* 2011; 84(4):391–407.
14. Amiji M. *Nanotechnology for Cancer Therapy.* Florida: CRC Press, 2007. p. 3.
15. Yeung DT, Parker WT, Branford S, Molecular methods in diagnosis and monitoring of haematological malignancies. *Pathology.* 2011; 43(6):566–579.
16. Hall JB, Dobrovolskaia MA, Patri AK, McNeil SE. Characterization of nanoparticles for therapeutics. *Nanomedicine.* 2007; 2(6):789–803.

17. Van Vlerken LE, Amiji MM. Multi-functional polymeric nanoparticles for tumour-targeted drug delivery. *Expert Opin Drug Deliv.* 2006; 3(2):205–216.

18. Parveen S, Sahoo SK. Polymeric nanoparticles for cancer therapy, *J Drug Target.* 2008; 16 (2):108–123.

19. Wang MD, Shin DM, Simons JW, Nie S. Nanotechnology for targeted cancer therapy. Expert Rev Anticancer Ther. 2007; 7(6): 833–837.

20. NCI Alliance Healthcare website. *Nanotechnology and cancer.*

21. Huang HC, Barua S, Sharma G, Dey SK, Rege K. Inorganic nanoparticles for cancer imaging and therapy. *J Control Release.* 2011; 155(3):344–357.

22. Retnakumari A, Jayasimhan J, Chandran P, Menon D, Nair S, Mony U, Koyakutty MC, *D33 monoclonal antibody conjugated Au cluster* nano-bioprobe for targeted flow-cytometric detection of acute myeloid leukaemia. *Nanotechnology.* 2011; 22(28):285102.

23. Khemtong C, Kessinger CW, Gao J, Polymeric nanomedicine for cancer MR imaging and drug delivery. *Chem. Commun.*, 2009, 3497–3510.

24. Singh ATK, Ghosh m, Forte TM, Ryan RO, Gordon LI. Curcumin nanodisk-induced apoptosis in mantle cell lymphoma. *Leuk Lymphoma.* 2011; 52(8): 1537–1543.

25. Tadmor T, Polliack A, Mantle cell lymphoma: curcumin nanodisks and possible new concepts on drug delivery for an incurable lymphoma. *Leuk Lymphoma.* 2011; 52(8): 1418–1420.

26. Ahluwalia MS. American Society of Clinical Oncology, 2011 CNS tumors update. *Expert Rev Anticancer Ther.* 2011; 11(10):1495–1497.

27. Kiefer T, Hirt C, Späth C, Schüler F, Al-Ali HK, Wolf HH, Herbst R, Maschmeyer G, Helke K, Kessler C, Niederwieser D, Busemann C, Schroeder H, Vogelgesang S, Kirsch M, Montemurro M, Krüger WH, Dölken G; Ostdeutsche Studiengruppe Hämatologie und Onkologie Long-term follow-up of high-dose chemotherapy with autologous stem-cell transplantation and response-adapted whole-brain radiotherapy for newly diagnosed primary CNS lymphoma: results of the multicenter Ostdeutsche Studiengruppe Hamatologie und Onkologie OSHO-53 phase II study. *Ann Oncol.* 2012; 23(7): 1809–12.

28. Lockman PR, Mumper RJ, Khan MA, Allen DD. Nanoparticle technology for drug delivery across the blood–brain barrier. *Drug Dev Ind Pharm.* 2002; 28(1): 1–13.

29. Duggan ST, Keating GM. Pegylated liposomal doxorubicin: A review of its use in metastatic breast cancer, ovarian cancer, multiple myeloma and AIDS-related Kaposi's sarcoma. *Drugs.* 2011; 71(18): 2531–2558.

30. Cheng J, Wang J, Chen B, Xia G, Cai X, Liu R, Ren Y, Bao W, Wang X, A promising strategy for overcoming MDR in tumor by magnetic iron oxide nanoparticles co-loaded with daunorubicin and 5-bromotetrandrin. *Int J Nanomed.* 2011; 6:2123–2131.

31. Palamà IE, Leporatti S, de Luca E, Di Renzo N, Maffia M, Gambacorti-Passerini C, Rinaldi R, Gigli G, Cingolani R, Coluccia AM. Imatinib-loaded polyelectrolyte microcapsules for sustained targeting of BCR-ABL+ leukemia stem cells. *Nanomedicine (Lond).* 2010; 5(3):419–431.

32. Singh A, Dilnawaz F, Sahoo SK, Long circulating lectin conjugated paclitaxel loaded magnetic nanoparticles: A new theranostic avenue for leukemia therapy. *PLoS One.* 2011; 6(11):e26803. Epub 16 November 2011.

33. Kumar A, Galeb S, Djulbegovic B, Treatment of patients with multiple myeloma: an overview of systematic reviews. *Acta Haematol.* 2011;125(1–2):8–22. Epub 8 December 2010. Review.

34. Marra M, Salzano G, Leonetti C, Tassone P, Scarsella M, Zappavigna S, Calimeri T, Franco R, Liguori G, Cigliana G, Ascani R, Immacolata La Rotonda M, Abbruzzese A, Tagliaferri P, Caraglia M, De Rosa G, Nanotechnologies to use bisphosphonates as potent anticancer agents: the effects of zoledronic acid encapsulated into liposomes *Nanomed: Nanotechnol Biol Med.* 2011; 7(6): 955–964.

35. Saloura V, Grivas PD, Lenalidomide: a synthetic compound with an evolving role in cancer management, *Hematology.* 2010; 15(5):318–331.

Chapter 5

Nanomedicine and Orthopaedics

Fares E. Sayegh
Aristotle University of Thessaloniki, Thessaloniki, Greece
fsayegh@auth.gr

5.1 Introduction

Musculoskeletal disorders (MSD) mainly include all types of bone and joint diseases and are considered to be the leading cause of disability in most of the world's population.[1,2] It is estimated that in most developed countries, more than one-half of all chronic conditions with MSD occur in people over 50 years of age.[3] The burden of these bone and joint diseases is enormous and the economic impact on the health care costs is also very high.[4,5] Moreover, the increase in the aging population and sedentary lifestyles are expected to increase the number of people that suffer from different musculoskeletal conditions and who require medical care, leading to further increase in health care costs.[6–8]

In the framework of general orthopaedic practice, bone and joint disorders may include all forms of arthritis, different spinal deformi-

Horizons in Clinical Nanomedicine
Edited by Varvara Karagkiozaki and Stergios Logothetidis

ISBN 978-981-4411-56-1 (Hardcover), 978-981-4411-57-8 (eBook)
www.panstanford.com

ties and disorders, trauma injuries, bone tumours, bone infections, and other different metabolic bone and joint diseases.[9–13]

Osteoarthritis (OA) is the most common form of arthritis. It is known as a chronic, progressive and degenerative disease of the joints that results in irreversible damage and loss of the articular cartilage.[14,16] This pathological condition is characterised by focal areas of loss of articular cartilage within the synovial joints, associated with hypertrophy of the bone (osteophytes and subchondral bone sclerosis) and thickening of the capsule.[15–19]

OA affects more than 70% of adults between 55 and 78 years of age. Women are more affected than men. It is a disease that commonly affects the weight-bearing joints, such as knees, hips, ankle and spine. It also affects the non-weight bearing joints such as the shoulder, toes and fingers. Patients with OA suffer from chronic musculoskeletal pain in the affected joint that causes stiffness, swelling and deformity of the insulted joints that usually leads to the reduction of their health-related quality of life.[15–18] A large number of patients with symptoms of advanced OA usually are disabled and cannot cope with their daily activities because of increased depression and social isolation. According to the World Health Organisation, osteoarthritis is one of the leading causes of disability in the elderly population in most of the developing countries.[18–21] It affects millions across Europe, and the direct cost of this disease is considerable and very high.[22–27]

Osteoarthritis is considered to be one of the challenging issues to both orthopaedic surgeons and researchers. It is one of the unsolved problems in orthopaedics that require the development of new methods that could early diagnose, prevent and offer effective and long lasting therapies for this disease. One of the major concerns and challenges of an orthopaedic surgeon is the early diagnosis and detection of OA. The ability to monitor the progression of OA early contributes to the relief and exemption of the patient from his/her symptoms. There are many useful diagnostic tools that help the surgeon to diagnose, prevent and even treat patients with different musculoskeletal disorders. The question is: could nanotechnology have a role in aiding the orthopaedic surgeon to make an early diagnosis of OA, or, could nanotechnology contribute to the treatment and repair of different musculoskeletal diseases?

5.2 The Role of Nanomedicine in Early Screening of Orthopaedic Diseases

Many different recent research works that involve nanobiosensor detectors and are based on carbon nanotubes technology are introduced. This technology allows rapid screening of bone quality, could identify individuals that may be at increased risk of injury for certain occupations, and could detect cancer and neurodegenerative skeletal disorders. A group of researchers developed the Indentation-type Atomic Force Microscopy for early detection and for micro- and nanomechanical analysis of aging articular cartilage and osteoarthritis. Their ultimate goal was to develop an arthroscopic atomic force microscopy, which can be directly introduced into a knee or hip joint for early diagnosis and detection of articular joint diseases.[28–30]

The Magnetic resonance spectroscopy is also a new technology tool that could evaluate bone development, intervertebral disc degeneration, and even different spinal cord diseases. The functional magnetic resonance (f-MRI) and positron emission tomography (PET) are also useful tools for detecting tumours, and those tumours that generally have a high metabolic activity.

5.3 Nano Biologically Active materials and Conservative Treatment of Early OA

Patients with early or advanced OA are usually afflicted with severe functional impairment and continuously seek different types of medications. The main purpose for treatment of OA is the relief of the patient from the unpleasant and disabling symptoms. As far as the conservative treatment of early OA is concerned, different types of drugs such as analgesics and anti-inflammatory, and symptoms and structure modifying drugs such as glucosamine C. sulfate,[31] chondroitin sulphate, hyaluronic acid and diacerein are mainly used. Physiotherapy, rest, exercises, and maintaining an acceptable body weight may improve joint mobility and also contribute to the partial relief of symptoms.[32] It is well known that long-term use of anti-inflammatory drugs is not advised as it may cause serious gastro-intestinal, renal, blood or other complications. It could be of tremendous help to use nanotechnology in target-drug delivery. It could be of great help if anti-inflammatory nanocapsules are used as they provide prolonged treatment to all forms of arthritis. Moreover, it would be beneficial if those nano biologically active materials could be sent directly to the diseased site. It is believed that the targeted drug delivery therapies to specific tissues in a body could be more effective than a drug that is delivered to all body tissues. There is already work on targeted drug delivery therapies for the treatment of cancer. Nanotechnology was able to discover new types of polymers which were shown to physically detect and destroy antibiotic-resistant bacteria and infectious diseases like Methicillin-resistant Staphylococcus aureus, known as MRSA. Therefore, the primary goals for research of nano-biotechnologies in drug targeting and delivery should include a more specific, safe, less toxic and biocompatible drug.[33, 34]

5.4 Nanotechnology and Nanobiomaterials in the Treatment of Advanced OA

In advanced OA of the joints (Fig. 5.1), surgery is the treatment of choice. Surgical debridement or corrective osteotomies of the pathologic joint are occasionally performed. The most common treatment in patients with severe damage of the articular joints is the total joint replacement, or joint arthroplasty. There are many different types of joint replacements. In hip arthroplasties, the orthopaedic prosthetic implant could be metal-on-polyethylene, metal-on-metal, ceramic-on-ceramic, ceramic-on-polyethylene, etc. The polyethylene component may be either cross-linked or highly cross-linked. Acrylic bone cement is also used to fix the prosthetic implants to the bone of the patient. In this case, two interfaces are created at the site of implant insertion; the bone-cement interface and the cement-implant interface. Most orthopaedic surgeons prefer the use of cementless press fit orthopaedic prosthetic implants. The press fit prosthetic implants are, therefore, in direct contact with the bone and only one bone-implant interface exists. Most of the orthopaedic prosthetic implants used are made of cobalt-chromium or titanium.

Most of the epidemiological studies show that total joint arthroplasties are continuously increasing in number and more than 50% of all joint arthroplasties will be expected to be performed in patients less than 65 years of age. At the current rate of increase in the number of joint replacements, researchers believe that younger patients will account for 52% of all total hip arthroplasties, and 55–62% of all primary and revision total knee arthroplasties, by 2030.[35]

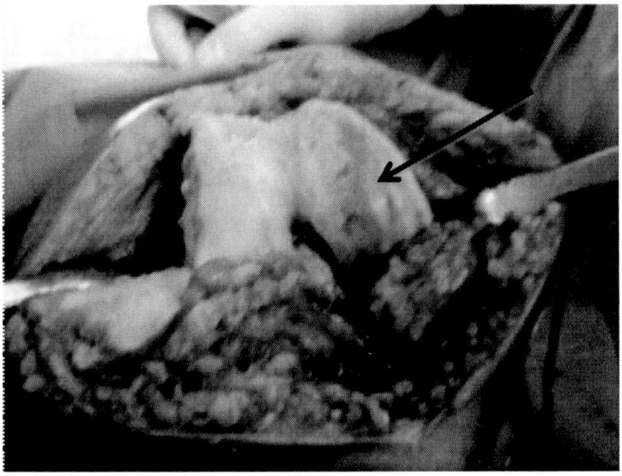

Figure 5.1 Intraoperative picture of an advanced osteoarthritic knee with the damaged articular surfaces.

Joint replacements, are, therefore not always successful; there is a failure rate that may reach 1–2% (Fig. 5.2). The prosthetic implant could fail, loosen, or get infected. Patients with failed joint arthroplasty should be re-operated and this type of operation is called revision arthroplasty. In some patients, more than one revision surgery in their lifetime could be necessary. Metal ions and debris that are released from the different types of prosthetic implants constitute an issue of concern to orthopaedic surgeons, as it is the source of toxicity to different body organs.

An ideal orthopaedic prosthetic implant is one that lasts long, does not loosen, and does not produce debris or metal wear. Implants that are less toxic to the body organs and resistant to different types of microorganisms are needed. An ideal implant is the type of prosthesis that adheres and bonds well to bone with minimal or even no fibroblast formation between the interfaces. To achieve all the above, researchers should first understand the relation between the implant and the bone, and second, study the natural cell behaviour of the bone. In other words, we need prosthetic implants that are more biocompatible.

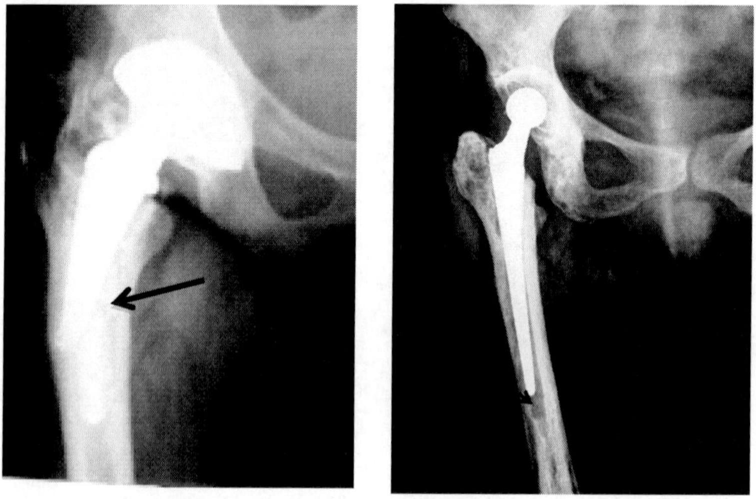

Figure 5.2 Failed right total hip replacement. Long arrow shows the hardware breakage of femoral prosthesis which may be the result of loosening of the femoral prosthetic implant. Short arrow shows osteolysis sites between the prosthetic implant and bone.

The natural bone tissue is considered to be a nanocomposite material; the bone cells are accustomed to interact with nanostructures. The extracellular matrix (ECM) in natural bone tissue is principally composed of helical chains of collagen fibrils that are 10–500 nm in length, of hydroxyapatite that exists as plate-like nanocrystals measuring 20–80 nm in length, and of the proteoglycans. The question is, what kind of nanobiomaterials do we need to obtain an ideal implant surface that is biocompatible with the bone tissues and could adhere and bond naturally to the bone?

We need nanobiomaterials with physicochemical properties that could make the metal surface more acceptable to bone cells and properly recognise and manipulate those cells. In other words, we need nanobiomaterials that allow the increase of osteoblast adhesion and inhibit or decrease the fibroblast function. A nanobiocompatible orthopaedic prosthesis should, therefore, be made of suitable surface roughness, suitable surface chemistry, and suitable wettability.

Nanometre surface roughness of the implant increases the efficacy of orthopaedic implants. A study by Price,[36] demonstrated increased osteoblast adhesion on polymer casts of nanophase carbon fibres when compared with polymer casts of conventional carbon fibres. A recent study by Webster[37] found that nanophase roughness on nanophase ceramics improved both osteoblastic and osteoclastic responses, and parallel inhibited the fibroblast function. Similarly, Mustafa[38] suggested that ceramics with increasing surface roughness (from 60 nm to 300 nm) decreased initial fibroblast adhesion compared with other smooth surfaces. Other studies showed that the surface chemistry of an implant plays an important role in protein adsorption and subsequent cell adhesion and the anodisation of the titanium orthopaedic implants' surface reported to increase bone growth.[39,40]

Studies also showed that implants with greater wettability surfaces that are more hydrophilic (Titanium and hydroxyapatite), the cell adhesion and osteoblastic activity are better.[41,42] Other studies showed that the surface wettability of alumina could be improved by reducing the alumina grain size from 167 to 24 nm[43] and the use of the helical rosette nanotubes as a biomimetic coating for orthopaedics.[44]

5.5 Tissue Engineering in Orthopaedics

Nowadays, the incidence of trauma is consistently increasing and the impact of such external forces varies according to the severity and type of injuries exercised on the musculoskeletal system. Degloving soft tissue injuries, loss of bone or soft tissues of a limb, open or closed comminuted fractures, articular joint destructions, ligamentous ruptures, nerve and vessel injuries, post-traumatic osteomyelitis and many other types of musculoskeletal damages are major problems that the orthopaedic surgeon should be aware of and ready to confront and treat.

Moreover, the increase in the incidence of bone and soft tissue tumours, either as primary or as a result of metastases from different sites of the body, may lead to pathological fractures or it may cause severe bone and soft tissue defects. Lately, a great effort is made to use both bone and soft tissue engineering and regeneration for restoring, maintaining, and improving the status of a damaged tissue. A lot of important research has been undertaken in the last several decades with the promise of developing biological substitutes that may help in restoring the function of a damaged hard or soft tissue.[45–49] Lately, a group of researchers managed to manufacture cytocompatible biomimetic nanomaterial scaffolds with encapsulating cells (such as stem cells, chondrocytes and osteoblasts, etc.) that can be used for bone, cartilage, vascular, and neural tissue engineering. The authors were able to explore the influence of the novel biomimetic nanofibrous or nanotubular scaffolds on regenerative medicine by following a bottom-up self-assembly process. Hartgerink[50] self-assembled peptide-amphiphile with the cell-adhesive ligand RGD (Arg-Gly-Asp) into supramolecular nanofibres. Hosseinkhani[51] studied the mesenchymal stem cell behaviour on self-assembled peptide- amphiphile nanofibre scaffolds. The authors observed increased osteogenic and chondrogenic differentiation of the mesenchymal stem cells in the 3-D peptide-amphiphile nanofibre scaffolds when compared with the 2-D static tissue culture. Kisiday[52] self-assembled a peptide hydrogel scaffold for cartilage repair (the peptide KLD-12, Lys-Leu-Asp). After four weeks culture in vitro, the encapsulated chondrocytes within the peptide hydrogel retained their morphology and developed a cartilage-like ECM matrix that is rich in proteoglycans and type II collagen.[52]

Other promising new nanostructured self-assembled chemistries that can be used in bone and cartilage tissue engineering include the osteogenic helical rosette nanotubes obtained through the self-assembly of DNA base pairs (Guanine-Cytosine) in aqueous solutions. Such tissue engineering promotes osteoblast adhesion and inhibits fibroblast adhesion and forms excellent mineralisation templates to assemble biomimetic nanotube-Hydroxyapatite structures. The synthetic and natural polymers, which are biodegradable and easy to fabricate, are considered excellent candidates for bone/cartilage tissue engineering. The nanoporous-nanofibrous polymer matrices can be fabricated through different methods such as electrospinning, chemical etching, phase separation, particulate leaching, and 3-D printing techniques.[53–57] Moreover, Xin[58] was able to study the viability, growth, and differentiation of the human mesenchymal stem cells (hMSCs) when seeded 1–4 weeks in PLGA electrospun nanofibre scaffolds. He noticed the continuous differentiation of the hMSCs into chondrogenic and osteogenic cells.

5.6 Conclusion

Nanomedicine and nanotechnology in orthopaedics represent an important growing area of research that may improve bonding between an implant and surrounding bone, which could play a great role in bringing further benefits to the orthopaedic patient.

In order to generate better scaffold and implant performance, it is critical to understand the molecular mechanisms governing cells and cell-material interaction.

Recent research showed that new biomaterials could be fabricated into nanostructures that simulate the native hierarchical structure of the bone and thus better adherence to the bone.

Further steps have been taken in bone/cartilage tissue engineering. In the future, with the help of nanomedicine and nanotechnology, orthopaedic surgeons would be able to repair bone losses or/and cartilage damage.

We should remember that for the last decades, the release of debris from the conventional orthopaedic implants and its adverse effects on the health of the patients has constituted a major issue to the entire orthopaedic society. A better understanding of the behaviour of nanoparticulate wear debris release and its behaviour on the human system is, therefore, of great importance and further studies concerning the potential risk and toxicity of the nanophase materials need to be conducted.

References

1. Picavet, HS, Hazes, JM (2003) Prevalence of self-reported musculoskeletal diseases is high. *Ann Rheum Dis.* 62:644–650.
2. Woolf, AD (2000) The bone and joint decade 2000–2010. *Ann Rheum Dis.* 59:81–82.
3. Sprangers, MA, De Regt, EB., Andries, F, et al. (2000) Which chronic conditions are associated with better or poorer quality of life? *J Clin Epidemiol.* 53:895–907.
4. World Health Organization (WHO) (2003) The Burden of Musculoskeletal Diseases at the Start of the New Millenium. Report of a WHO Scientific Group. *TRS 919*. Geneva, Switzerland.
5. Woolf, AD, Zeidler, H, Haglund, U, et al. (2004) Musculoskeletal pain in Europe: Its impact and a comparison of population and medical

perceptions of treatment in eight European countries. *Ann Rheum Dis.* 63:342–347.

6. Badley, EM, Rasooly, I, Webster, GK (1994) Relative importance of musculoskeletal disorders as a cause of chronic health problems, disability, and health care utilization: findings from the 1990 Ontario Health Survey. *J Rheumatol.* 21:505–514.
7. Bergman, S, Herrstrom, P, Hogstrom, K, et al. (2001) Chronic musculoskeletal pain, prevalence rates, and sociodemographic associations in a Swedish population study. *J Rheumatol.* 28:1369–1377.
8. Oliveria, SA, Felson, DT, Reed JI, et al. (1995) Incidence of symptomatic hand, hip, and knee osteoarthritis among patients in a health maintenance organization. *Arthritis Rheum.* 38:1134–1141.
9. Johnell, O, Kanis, JA, Oden, A, et al. (2004) Mortality after osteoporotic fractures. *Osteoporos Int.* 15:38–42.
10. Cooper, C (1997) The crippling consequences of fractures and their impact on quality of life. *Am J Med.* 103:12S–17S.
11. European Prospective Osteoporosis Study Group (2002) Incidence of vertebral fractures in Europe: results from the European prospective osteoporosis study (EPOS). *J Bone Miner Res.* 17:716–724.
12. Kanis, JA, Johnell, O, Oden, A, et al. (2000) Risk of hip fracture according to the World Health Organization criteria for osteopenia and osteoporosis. *Bone.* 27:585–590.
13. Kanis, JA, Johnell, O, Oden, A, De Laet, C, et al. (2002) Ten-year risk of osteoporotic fracture and the effect of risk factors on screening strategies. *Bone.* 30:251–258.
14. Lawrence, RC, Hochberg, MC, Kelsey, JL, et al. (1989) Estimates of the prevalence of selected arthritic and musculoskeletal diseases in the United States. *J Rheumatol.* 16:427–431.
15. Guccione, AA, Felson, DT, Anderson, JJ, et al. 1994 The effects of specific medical conditions on the functional limitations of elders in the Framingham Study. *Am J Public Health.* 84(3):351–358.
16. Dahaghin, S, Bierma-Zeinstra, SM, et al. (2005) Prevalence and pattern of radiographic hand osteoarthritis and association with pain and disability (the Rotterdam study). *Ann Rheum Dis.* 64: 682–687.
17. Felson, DT, Zhang, Y (1998) An update on the epidemiology of knee and hip osteoarthritis with a view to prevention. *Arthritis Rheum.* 4, 1343–1355.
18. Altman, RD (1991) Criteria for classification of clinical osteoarthritis. *J Rheumatol Suppl.* 27: 10–12.

19. Leyland, KM, Hart, DJ, Javaid, MK, et al. (2012) The natural history of radiographic knee osteoarthritis: A fourteen-year population-based cohort study. *Arthritis Rheum.* 64(7):2243–2251.

20. Peterson, IF (1996) Occurrence of osteoarthritis of the peripheral joints in European populations. *Ann Rheum Dis.* 55(9):659–661.

21. Bedson, J, Jordan, K, Croft, P (2005) The prevalence and history of knee osteoarthritis in general practice: a case-control study. *Fam Pract.* 22(1):103–108. Epub 2005 Jan 7.

22. Le Pen, C, Reygrobellet, C, Gerentes, I (2005) Financial cost of osteoarthritis in France: the "COART" France study. *Joint Bone Spine.* 72:567–570.

23. Gupta, S, Hawker GA, Laporte, A, et al. (2005) The economic burden of disabling hip and knee osteoarthritis (OA) from the perspective of individuals living with this condition. *Rheumatology (Oxford).* 44:1531–7.

24. Leardini, G, Salaffi, F, Caporali, R, et al. (2004) For the Italian Group for Study of the Costs of Arthritis. Direct and indirect costs of osteoarthritis of the knee. *Clin Exp Rheumatol.* 22:699–706.

25. Woo, J, Lau, E, Lau, CS, Lee, P, Zhang, J, Kwok, T, et al. (2003) Socioeconomic impact of osteoarthritis in Hong Kong: utilization of health and social services, and direct and indirect costs. *Arthritis Rheum.* 49:526–534.

26. Maetzel, A, Li, LC, Pencharz, J, Tomlinson, G, Bombardier, C (2004) For the Community Hypertension and Arthritis Project Study Team. The economic burden associated with osteoarthritis, rheumatoid arthritis, and hypertension: a comparative study. *Ann Rheum Dis.*; 63:395–401.

27. Ryan, Bitton. (2009) The economic burden of osteoarthritis. *Am J Manag Care.* 15:S230–S235.

28. Engineering Institute for Nanoscale Science and Technology, University of Cincinnati, OH 45221 (2006). *Smart Materials Nanotechnology* Lab. www.min.uc.edu/~mschulz/smartlab/smartlab.html.

29. Stolz, M, et al. (2009) Early detection of aging cartilage and osteoarthritis in mice and patient samples using atomic force microscopy. *Nat Nanotechnol.* 4, 186–192.

30. Loparic, M, et al. (2010) Micro- and nanomechanical analysis of articular cartilage by indentation-type atomic force microscopy: 2005 validation with a gel-microfiber composite. *Biophys J.* 2;98(11):2731–2740.

31. Rozendaal, RM, Koes, BW, Van Osch, GJ, Uitterlinden, EJ, et al. (2008) Effect of glucosamine sulfate on hip osteoarthritis: a randomized trial. *Ann Intern Med.* 148:268–277.
32. Galea, MP, Levinger, P, Lythgo, N, et al. (2008) A targeted home- and center-based exercise program for people after total hip replacement: a randomized clinical trial. *Arch Phys Med Rehabil.* 89:1442–1447.
33. Jong, WHDE., Borm, PJA (2008) Drug delivery and nanoparticles: Applications and hazards. *Int J Nanomed.* 3(2): 133–149.
34. Tasker, LH, et al. (2007) Applications of nanotechnology in orthopaedics. *Clin Orthop Relat Res.* 456: 243–249.
35. Bozic, KJ, Kurtz, SM, Lau, E, Ong, K, et al. (2009) The epidemiology of revision total hip arthroplasty in the United States. *J Bone Joint Surg Am.* 91:128–133.
36. Price, RL, Ellison, K, Haberstroh, KM, Webster, TJ (2004) Nanometer surface roughness increases select osteoblast adhesion on carbon nanofiber compacts. *J. Biomed. Mater. Res.* 70:129–138.
37. Webster, TJ (2003) In *Advances in Chemical Engineering* (J Ying, ed.), Academic Press Inc., CA, pp. 125–166.
38. Mustafa, K, Oden, A, Wennerberg, A, et al. (2005) The influence of surface topography of ceramic abutments on the attachment and proliferation of human oral fibroblasts. *Biomaterials*, 26: 373–381.
39. Schakenraad, JM. (1996), in *Biomaterial Science* (BD Ratner, AS Hoffman, FJ Schoen, JE Lemons, ed.), Academic Press, Inc., San Diego, pp. 140–141
40. Gong, D, et al. (2001) Titanium oxide nanotube arrays prepared by anodic oxidation. *J. Mater. Res.* 16: 3331–3334.
41. Webb, K, et al. (2000) Relationship among cell attachment, spreading, cytoskeletal organization, and migration rate for anchorage-dependent cells on model surfaces. *Biomed Mater Res.* 49:362–368.
42. Takebe, J, et al. (2000) Anodic oxidation and hydrothermal treatment of titanium results in a surface that causes increased attachment and altered cytoskeletal morphology of rat bone marrow stromal cells in vitro. *J Biomed Mater Res.* 51: 398–407.
43. Webster, TJ, et al. (2000) Specific proteins mediate enhanced osteoblast adhesion on nanophase ceramics. *J Biomed Mater Res.* 51: 475–483.
44. Chun, AL, Moralez, JG, et al. (2005) Helical rosette nanotubes: a biomimetic coating for orthopaedics. *Biomaterials.* 26: 7304–7309.
45. Li, WJ, et al. (2002) Electrospun nanofibrous structure: a novel scaffold for tissue engineering. *J Biomed Mater Res.* 60: 613–621.

46. Nukavarapu, SP et al. (2008) In *Biomedical Nanostructures* (Gonsalves KE, Laurencin CT, Halberstadt C, Nair LS, ed.), New York: Wiley, pp. 377–407

47. Kumbar, SG. et al. (2008) Electrospun nanofiber scaffolds: engineering soft tissues. *Biomed Mater*. 3: 1–15.

48. Park, G, Webster, T (2005) A review of nanotechnology for the development of better orthopaedic implants. *J Biomed Nanotechnol*. 1: 18–29.

49. Minas, T, Nehrer, S (1997) Current concepts in the treatment of articular cartilage defects. *Orthopedics*. 20: 525–538.

50. Hartgerink, JD, Beniash, E, Stupp, SI (2001) Self-assembly and mineralization of peptide-amphiphile nanofibers. *Science*. 294: 1684–1688.

51. Hosseinkhani, H., Hosseinkhani, M, Tian, F, et al. (2006) Osteogenic differentiation of mesenchymal stem cells in self-assembled peptide-amphiphile nanofibers. *Biomaterials*. 27: 4079–4086.

52. Kisiday, J, Jin, M, Kurz, B, et al. (2002) Self-assembling peptide hydrogel fosters chondrocyte extracellular matrix production and cell division: Implications for cartilage tissue repair. *Proc Natl Acad Sci*. 99: 9996–10001.

53. Zhang, L, Ramsaywack, S, Fenniri, H, Webster, TJ (2008) Enhanced osteoblast adhesion on self-assembled nanostructured hydrogel scaffolds. *Tissue Eng*. 14(8): 1353–1364.

54. Zhang, L, Hemraz, UD, Fenniri, H, Webster, TJ (2010) Tuning cell adhesion on titanium using osteogenic rosette nanotubes. *J. Biomed. Mater. Res. Part A*. 95A: 550–563.

55. Fecek, D, Yao, A, Kaçorri, A, Vasquez, S, et al. (2008) Chondrogenic derivatives of embryonic stem cells seeded into 3D polycaprolactone scaffolds generated cartilage tissue in vivo. *Tissue Eng Part A*. 14(8): 1403–1413.

56. Li, WJ, Tuli, R, Okafor, C, Derfoul, A, Danielson, KG, Hall, DJ, et al. (2005) A three-dimensional nanofibrous scaffold for cartilage tissue engineering using human mesenchymal stem cells. *Biomaterials*. 26(6): 599–609.

57. Zhang, L, Hemraz, UD, Fenniri, H, Webster, TJ (2010) Tuning cell adhesion on titanium using osteogenic Rosette nanotubes. *J Biomed Mater Res Part A*. 95A:550–563.

58. Xin, X, et al. (2007) Continuing differentiation of human mesenchymal stem cells and induced chondrogenic and osteogenic lineages in electrospun PLGA nanofiber scaffold. *Biomaterials*. 28(2): 316–325.

Chapter 6

Nanopharmaceutics: Structural Design of Cationic Gemini Surfactant–Phospholipid–DNA Nanoparticles for Gene Delivery

Marianna Foldvari
School of Pharmacy, Waterloo Institute of Nanotechnology, University of Waterloo, Waterloo, Ontario, Canada N2L 3G1
foldvari@uwaterloo.ca

Development of nano-sized delivery systems requires the understanding of a comprehensive set of parameters and the construction of fine-tuned particles capable of significantly more "intelligent" functions compared to traditional dosage forms. Among the many important properties, morphology and internal structure of nanoparticles play a significant role in their functional performance. Lipid and surfactant composite nanoparticles can assume various polymeric configurations and have the capability to flexibly change from one form to another. This flexibility may be a necessary factor in their cellular and subcellular interactions.

Horizons in Clinical Nanomedicine
Edited by Varvara Karagkiozaki and Stergios Logothetidis

ISBN 978-981-4411-56-1 (Hardcover), 978-981-4411-57-8 (eBook)
www.panstanford.com

6.1 Introduction

Pharmaceutics is a long-standing discipline defined as dosage form design for the most beneficial administration of a drug compound in humans and animals. It includes the use of excipient, physical chemistry, pharmacology and physiology knowledge to design formulations and incorporate drugs into dosage forms for the most optimal absorption and effect. Advancements in pharmaceutics include the development of highly sophisticated functions within the framework of dosage forms as well as novel administration methods. The potential for applications of nanotechnology tools in pharmaceutics revolutionised the field and presented a new era for treatment of diseases and maintenance of wellness.

Generally, during development of nanoparticles for drug delivery, regardless of the route of administration, a multitude of parameters are of interest for fine-tuned customisation. Important considerations in individual nanoparticle design for enhanced cellular interaction and drug delivery are: size, surface charge, surface area, shape, surface coatings, stability and structure of particles (Fig. 6.1).

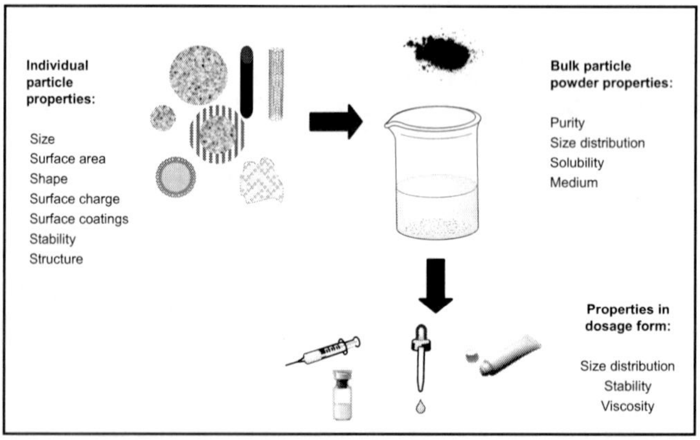

Figure 6.1 Nanoparticle properties, at three levels, important in achieving optimum effect: individual particles, bulk and dosage forms.

The understanding of nanoparticle systems through rational classification was pioneered by Tomalia [1]. Within this system, the nanoperiodic properties can be linked to their in vivo behaviour. An understanding of the periodic property patterns, and their dramatic influence on physicochemical properties of nanoparticles formed from a variety of molecules, is developing and may provide a systematic design template for drug delivery system design [1–3].

The importance of the thorough understanding of particle properties is widely recognised [4]; however, in practice, there is still much to be done to fully detail out all relevant characteristics of old and new systems. Nanoparticle biocompatibility, including pharmacokinetics of absorption, targetability, clearance and toxicity, is the direct result of the sum of the physicochemical features (Fig. 6.2) [5].

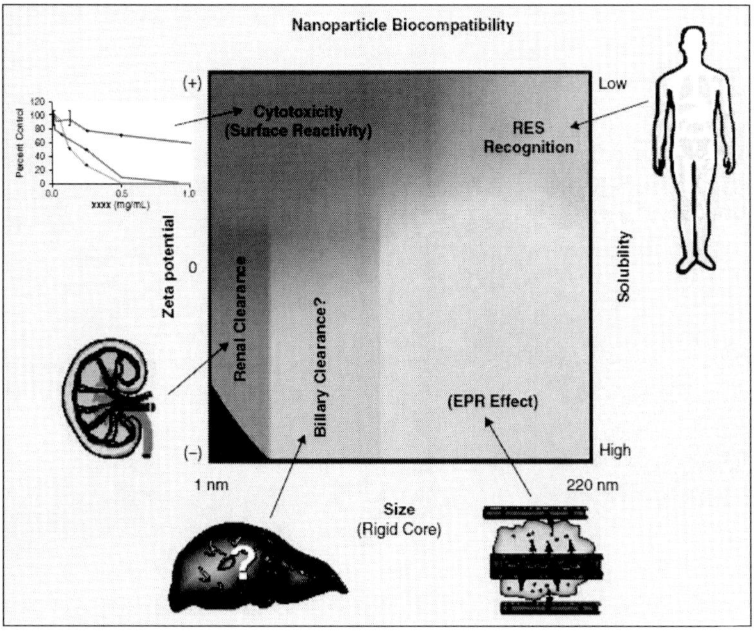

Figure 6.2 Relationship between some of the main physicochemical properties of nanoparticles and their in vivo fate. Reprinted with permission from McNeil [5].

Smaller particles <150 nm with positive zeta potential are typically taken up to a greater extent into cells, which was shown for nanoparticles made from various biomaterials [6]. Surface area influences exposure to recognition elements in the blood or tissues, and this may lead to adverse events.

Until recently, all delivery systems studied were mostly spherical. Non-spherical particle shape (filamentous, ellipsoid, cubes, rods, triangles, pentagons, discs) may provide significantly improved interactions with cells [7–12]. In this regard, CNTs provide a potential option for developing needle-shaped particles due to the high aspect ratio of carbon nanofibres. It was demonstrated recently that CNTs were able to penetrate into cells not only by clathrin- and caveolae-mediated pathways, but also by a non-energy dependent process, possibly passing through between the lipid molecules of membranes [13, 14].

The relationship between the structure of DNA complexes and gene delivery has been of considerable interest in the past few years from both the theoretical and experimental standpoint [15–24]. DNA complexes prepared with lipids and surfactants may contain a variety of lipid-packing arrangements such as lamellar, hexagonal and cubic phase structures (Fig. 6.3) [25, 26]. The presence and proportions of these polymorphic structural forms within particles have been investigated by synchrotron, magnetic resonance and electron microscopic methods. The role of polymorphic configurations in transfection efficiency of DNA nanoparticles has been investigated for nanoparticles made from a number of different lipids and surfactants. In addition to polymorphic molecular arrangements, however, other factors, such as membrane charge density and membrane elasticity may also be important [15, 18, 27]. Earlier studies suggested that for nanoparticles with lamellar structure (liposome type vesicles) to yield efficient transfection, their membrane charge density needed to be greater than a critical minimum of 1.04×10^{-2} e/Å^2 [15]. More recently, new data on the role of polymorphism in DNA complex formation and its effect on intracellular function are accumulating. For example, transfection efficiency could be increased by using non-lamellar complexes with lower cationic charge densities, such as a hexagonal complex [28] that provided fusogenic action within the endosomes. Hexagonal H_{II} and cubic phase configurations were implicated in enhanced intracellular delivery by gemini nanoparticles [26, 29], cationic nanoparticles made from tetraalkyl lipids [30], stable nucleic acid lipid nanoparticles (SNALPs) made from ionisable cationic lipid 1,2-dilinoleyloxy-3-dimethylaminopropane (DLinDMA) [31], and double-gyroid cationic complexes formed from multivalent cationic lipid and glycerol monooleate mixtures [17, 32].

These results suggest that the ability of a given DNA-lipid/surfactant complex to transfect DNA efficiently may depend on the flexibility of packing, that is, the ability to change from one polymorphic form to another.

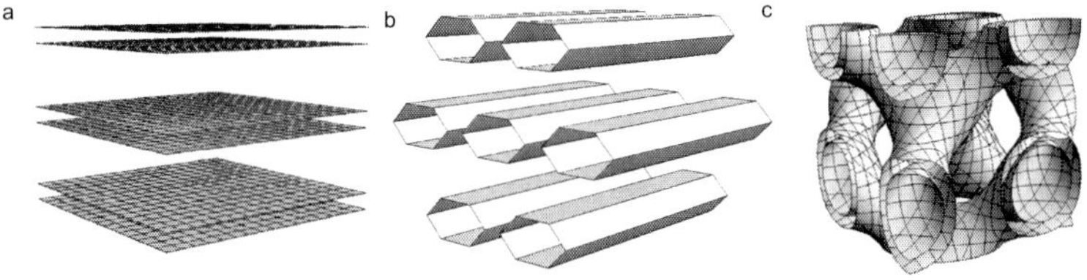

Figure 6.3 Computer-generated models of bilayer (a), inverted (H_{II}) hexagonal (b) and cubic (Pn3m) (c) phase packing of lipids and surfactants.

The focus in this chapter will be the exploration of relationship between morphology and internal structure of soft nanoparticles assembled from phospholipids and gemini surfactants and their interaction with cells in culture.

6.2 Gemini Nanoparticles

A new class of surfactants, dicationic *N*,*N*-bis(dimethylalkyl)-α,ω-alkene-diammonium surfactants, known as gemini surfactants, are among the promising building blocks for non-viral gene delivery systems [33–37]. Gemini surfactants (two interlinked monomeric surfactants) are amphiphilic molecules, composed of at least two hydrophobic tails and two hydrophilic head groups, linked by a spacer group. This basic design serves as a template for a number of possible structures [38, 39]. Through variations in the hydrophobic chain-length, hydrophilic head groups, and spacer groups, a wide range of tailored compounds can be designed [34, 40, 41].

Cationic gemini molecules form stable complexes with nucleic acids and improve gene transfection in vitro [36, 40, 42, 43]. As such, gemini surfactants show potential as multimodal, tunable carriers for nucleic acids. Their structural versatility offers great potential to overcome transfection barriers, including cell binding, intracellular trafficking, endosomal escape and nuclear entry [44]. The first generation of gemini compounds we have synthesised has the general structure of $[C_mH_{2m+1}(CH_3)_2N^+(CH_2)_sN^+(CH_3)_2\ C_mH_{2m+1}.2X^-]$, and abbreviated as the *m-s-m* series, where *m* and *s* refer to the number of carbon atoms in the alkyl tails and in the polymethylene spacer group, respectively, and X is the counter ion, without additional substituents (Fig. 6.4). One of the simplest members of the first generation gemini surfactants is the 12-3-12 surfactant, which has two dodecyl tails and a trimethylene spacer. The second generation of gemini surfactants have the same general structure but with additional substitutions in the spacer or tail. Modification of the tail region was carried out with pyrenyl moieties [45], and modification of the spacer was done with either a secondary amine (pH-sensitive function to facilitate endosomal escape) [45, 46] or amino acids and di-peptides on the spacer N atom such as glycine, lysine, histidine, glycyl-lysine, glycyl-glycine, and lysyl-lysine [47, 48].

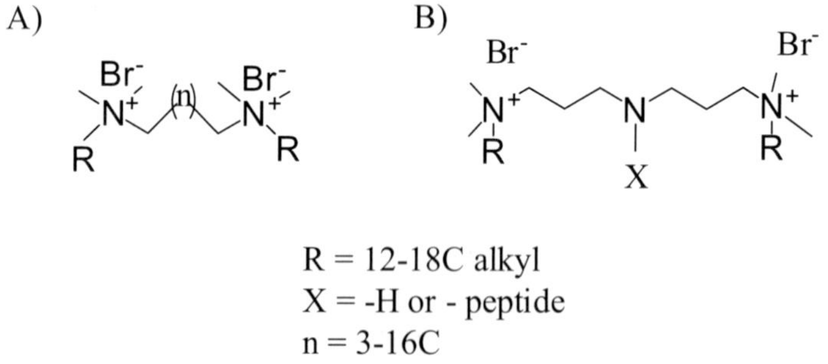

Figure 6.4 General structure of the *m-s-m* (first generation) gemini surfactants; R = 12–18 C, *n* = 2–16 C and *m-sX-m* (second generation) gemini surfactants, X= –H or peptide residue.

Nanoparticles can be formed from DNA and gemini surfactant with and without helper lipids. Plasmid and gemini surfactant complexes (PG) or plasmid–gemini surfactant–dioleoylphosphatidyl ethanolamine (DOPE) as helper lipid (PGL) have different structures and effects in cell culture [29, 37, 40].

Compared to conventional surfactants that are composed of a single hydrophobic tail connected to an ionic or polar head group, the unique dimeric structure of the gemini surfactant provides for improved surface activity and low critical micellar concentration (CMC), the minimum concentration above which surfactants form micelles. This is of particular interest in gene therapy research, where it is essential to maximise the safety profile of delivery agents by minimising their concentration in vivo [38].

6.3 Nanoparticle Structure Analysis

Gemini surfactants have advantageous properties allowing them to bind and condense DNA optimally; the complexes then attach to cell membrane surfaces and facilitate membrane translocation by endocytosis (or other mechanisms), followed by endosomal escape and transport of plasmid DNA into the nucleus. These essential properties strongly depend on the structural features of the gemini nanoparticles. A marked increase in transfection efficiency was found for gemini surfactants bearing amine, amino-acid, or carbohydrate substitutions on the spacer group [29, 44, 48]. DNA nanoparticles assembled from symmetric and dissymmetric dicationic surfactants of the gemini surfactant family possess a unique characteristic in that they can adapt a combination of phases simultaneously, or can transform from one form to another under specific conditions. A leading technique for structural characterisation of nanoparticles is small-angle X-ray scattering (SAXS), which provides information on the internal molecular arrangement within nanoparticles. For example, gemini surfactants *m*-4-*m* (m = 10–16) complexed with plasmid DNA mainly showed hexagonal packing [49], whereas complexes with bis(alkyl-1,3-diene)-based phosphonium gemini amphiphiles exhibited a cubic phase configuration [50]. We have shown that the 12-*s*-12 series PGL exhibit mostly hexagonal and lamellar morphologies; however, a Pn3m cubic phase component can also be detected [51]. This phase is one of several cubic phases that have been previously observed in lipid systems [52]. For 12-3-12 PGL nanoparticles having DNA-gemini surfactant cationic charge ratio $\rho_{+/-}$ = 0.5, the scattering profiles are characteristic of a hexagonal morphology (Table 6.1), while for the 16-3-16 and 18:1-3-18:1 PGL, it is suggestive of having both Pn3m hexagonal structures [51]. As the cationic charge ratio is increased, various structures can be observed, depending on the structure of the gemini surfactant used. In a mostly lamellar morphology, the Pn3m phase persists up to $\rho_{+/-}$ = 10 for the 12-3-12, 16-3-16 and 18:1-3-18:1 PGL systems (Fig. 6.5).

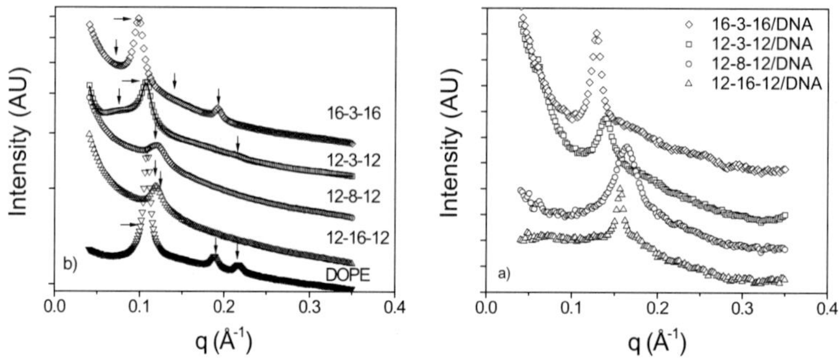

Figure 6.5 SAXS profiles for a) DNA-gemini, and b) DNA/gemini/DOPE (PGL) systems: 16-3-16 (◇); 12-3-12 (□); 12-8-12 (○); 12-16-12 (△); DOPE (▽). Reproduced from [26] by copyright permission.

6.4 Nanoparticle Structural Responsiveness and Transfection Efficiency

To better understand the reason for enhanced transfection efficiency, we compared both PG and PGL systems in terms of particle size and zeta potential. Since both these parameters are similar in these nanoparticles, the ability to transfect cells can be attributed to the structural differences between these two groups. PG complexes showed particle-like morphology, no specific polymorphic arrangement and no transfection. On the other hand, the PGL systems appeared vesicle-like, and the gemini surfactants formed mixed polymorphic systems in the presence of DNA and DOPE, and have the ability to induce polymorphic structures other than hexagonal (H_{II}). This may facilitate the eventual release of the DNA, resulting in increased transfection, which was specifically observed for the surfactants having short spacer groups. This is an interesting observation in the light of previous reports where a hexagonal structure is thought by many to be required for transferring DNA to cells [53–55]. Plasmid-DOPE complexes without any gemini surfactant show a typical hexagonal profile (q = 0.109, 0.188 and 0.218 $Å^{-1}$) with a lattice constant a = 66.9 Å; however, these complexes do not have the ability to transfect PAM212 cells [56].

We have also shown that more than one type of polymorphic phase (lamellar, hexagonal, cubic phases) can coexist in some of the gemini nanoparticles depending on the gemini surfactant-DNA charge ratio [26, 51], for example, hexagonal-cubic and lamellar-cubic mixed polymorphic structures (Fig. 6.6).

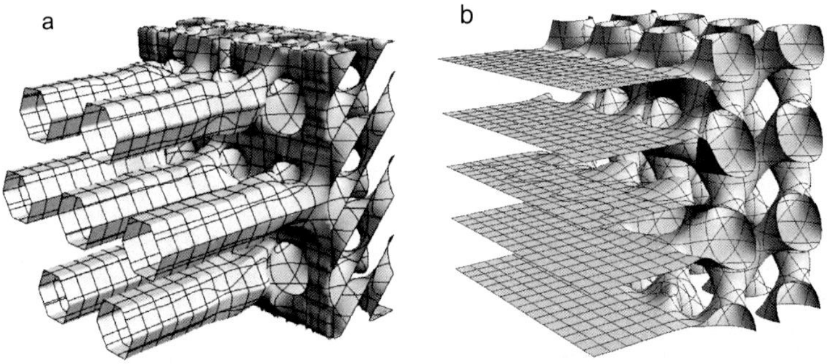

Figure 6.6 Computer-generated images of the mixed polymorphic domains within gemini nanoparticles. Hexagonal (H_{II})-cubic (a) and lamellar-cubic phase (b).

We have identified the Pn3m cubic phase in DNA-gemini-DOPE complexes with gemini surfactants having 12, 16 and 18:1 alkyl tail length. Increasing gemini/DNA charge ratios ($\rho_{+/-}$) from 0.5 to 10 resulted in increasingly mixed (Pn3m and H_{II} or Pn3m and L) polymorphic systems with lamellar (L) features becoming more predominant. DNA-gemini complexes exhibited very weak single scattering peaks representative of gemini-plasmid particles with no long range order and low transfection efficiency.

The importance of structural polymorphism has been demonstrated by using analogues of gemini surfactants with different lengths of spacer and hydrocarbon tail groups [26]. Transfection efficiencies were found to decrease with increasing spacer length up to C10, followed by an increase of up to C16. Interestingly, the highest transfection was shown for nanoparticles displaying mixed polymorphic structures [26]. The increasing transfection efficiency trend within the *m*-3-*m* series (m = 12, 16 or 18) corresponds to lamellar and Pn3m cubic structure combination (Fig. 6.7).

High transfection efficiencies in lamellar phase only nanoparticle systems are attributed to relatively high membrane charge density (σ_M) greater than 1.04×10^{-2} e/Å^2 [55, 57]. Interestingly, for our systems, only the 12-2-12 and 12-3-12 PGL systems have a calculated membrane charge density equivalent to or greater than this critical value (Table 6.1). Although the PGL system with 12-3-12 gemini surfactant has the highest transfection efficiency compared to the other PGL systems with longer spacers in this series, the PGL system with the 16-3-16 surfactant, exhibits the greatest transfection even though it has a lower σ_M value (9.22×10^{-3} e/Å^2).

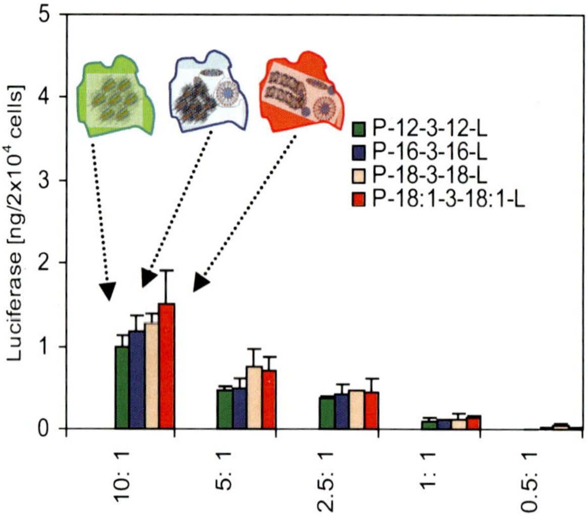

Figure 6.7 Transfection efficiency and corresponding polymorphic structure of *m*-3-*m* series DNA-gemini-DOPE (PGL) nanoparticles prepared from four different gemini surfactants with increasing alkyl chain length. Reproduced from [58] by permission of The Royal Society of Chemistry (RSC) on behalf of the Centre National de la Recherche Scientifique (CNRS) and the RSC.

These results provide additional support for the hypothesis that it is the polymorphic flexibility of DNA complexes (gemini surfactants and other helper lipids) in adapting cubic phase geometry that may facilitate DNA transfection.

6.5 Conclusions

The success of transfection using non-viral gene delivery systems is dependent on complex factors and is poorly understood. The fine tuning of physicochemical factors is necessary to develop an optimal system. One of the emerging factors for further consideration is the molecular organisation of particle internal structure and its responsiveness in different environments. In lipid systems, polymorphic composition and the morphological activity of the system may play a significant role in cell interactions.

Table 6.1 Structural parameters for the DNA/gemini surfactant/DOPE (PGL) systems determined from small angle X-ray scattering measurements

PGL systems	Q (Å^{-1})	d (Å)	σ_M# (e/ Å^2)
12-2-12/ DNA/DOPE	0.080 0.106	79.0 59.0	1.14×10^{-2}
12-3-12/ DNA/DOPE	**0.080** **0.107** **0.215**	**78.7** **58.7** **29.3**	1.03×10^{-2}
12-4-12/ DNA/DOPE	0.110 0.122	57.2 51.4	9.65×10^{-3}
12-6-12/ DNA/DOPE	0.053 0.116	118.5 51.4	8.38×10^{-3}
12-8-12/ DNA/DOPE	0.120	52.3	7.19×10^{-3}
12-10-12/ DNA/DOPE	0.120	52.4	6.30×10^{-3}
12-12-12/DNA/DOPE	0.124	50.7	6.32×10^{-3}
12-16-12/DNA/DOPE	0.113 0.120	55.4 52.5	8.06×10^{-3}
16-3-16/ DNA/DOPE	**0.070** **0.098** **0.138** **0.193**	**89.5** **64.1** **45.5** **32.6**	9.22×10^{-3}
DOPE*	0.109 0.189 0.218	57.9[a] 33.3 28.8	—

Source: Reproduced from [26] with permission.

[a]a, the lattice spacing [$=4\pi/(\sqrt{3}q)$], = 66.9 Å; complexes **in bold** exhibit a lamellar phase with a possible second phase; * exhibit hexagonal phase; # average membrane charge density (σ_M) was calculated based on the equation: $\sigma_M = eZN_{cl}/(N_{nl}A_{nl} + N_{cl}A_{cl}) = [1-\Phi_{nl}/\Phi_{nl}+r\Phi_{cl})]\ \sigma_{cl}$ where the ratio of the headgroup area of the cationic to neutral lipid is $r = A_{cl}/N_{cl}$; $\sigma_{cl} = eZ/A_{cl}$ is the charge density of the cationic lipid with valence Z; $\Phi_{nl} = N_{nl}/(N_{nl}+N_{cl})$ and $\Phi_{cl} = N_{cl}/(N_{nl}+N_{cl})$ are the mole fractions of the neutral and cationic lipids, respectively, from [15].

Acknowledgements

The author thanks Damien Baboolal for generating the computer-generated images of polymorphic phases and Marina Ivanova for editing the manuscript. The research was supported by the Canadian Institutes of Health Research and the Natural Sciences and Engineering Research Council of Canada.

References

1. Tomalia, D. A. (2009) In quest of a systematic framework for unifying and defining nanoscience, *J Nanopart Res* **11**, 1251–1310.
2. Percec, V., Won, B. C., Peterca, M., and Heiney, P. A. (2007) Expanding the structural diversity of self-assembling dendrons and supramolecular dendrimers via complex building blocks, *J Am Chem Soc* **129**, 11265–11278.
3. Kobayashi, H., and Brechbiel, M. W. (2003) Dendrimer-based macromolecular MRI contrast agents: characteristics and application, *Mol Imaging* **2**, 1–10.
4. Gentleman, D. J., and Chan, W. C. W. (2009) A systematic nomenclature for codifying engineered nanostructures, *Small* **5**, 426–431.
5. McNeil, S. E. (2009) Nanoparticle therapeutics: a personal perspective, *Wiley Interdiscip Rev Nanomed Nanobiotechnol* **1**, 264–271.
6. Rejman, J., Oberle, V., Zuhorn, I. S., and Hoekstra, D. (2004) Size-dependent internalisation of particles via the pathways of clathrin- and caveolae-mediated endocytosis, *Biochem J* **377**, 159–169.
7. Champion, J. A., Katare, Y. K., and Mitragotri, S. (2007) Particle shape: a new design parameter for micro- and nanoscale drug delivery carriers, *J Control Release* **121**, 3–9.
8. Gratton, S. E., Napier, M. E., Ropp, P. A., Tian, S., and DeSimone, J. M. (2008) Microfabricated particles for engineered drug therapies: elucidation into the mechanisms of cellular internalisation of PRINT particles, *Pharm Res* **25**, 2845–2852.
9. Gratton, S. E., Ropp, P. A., Pohlhaus, P. D., Luft, J. C., Madden, V. J., Napier, M. E., and DeSimone, J. M. (2008) The effect of particle design on cellular internalization pathways, *Proc Natl Acad Sci U S A* **105**, 11613–11618.
10. Chithrani, B. D., Ghazani, A. A., and Chan, W. C. (2006) Determining the size and shape dependence of gold nanoparticle uptake into mammalian cells, *Nano Lett* **6**, 662–668.

11. Perry, J. L., Herlihy, K. P., Napier, M. E., and Desimone, J. M. (2011) PRINT: a novel platform toward shape and size specific nanoparticle theranostics, *Acc Chem Res* **44**, 990–998.

12. Galloway, A. L., Murphy, A., Rolland, J. P., Herlihy, K. P., Petros, R. A., Napier, M. E., and DeSimone, J. M. (2011) Micromolding for the fabrication of biological microarrays, *Methods Mol Biol* **671**, 249–260.

13. Bhirde, A. A., Patel, V., Gavard, J., Zhang, G., Sousa, A. A., Masedunskas, A., Leapman, R. D., Weigert, R., Gutkind, J. S., and Rusling, J. F. (2009) Targeted killing of cancer cells in vivo and in vitro with EGF-directed carbon nanotube-based drug delivery, *ACS Nano* **3**, 307–316.

14. Kostarelos, K., Lacerda, L., Pastorin, G., Wu, W., Wieckowski, S., Luangsivilay, J., Godefroy, S., Pantarotto, D., Briand, J. P., Muller, S., Prato, M., and Bianco, A. (2007) Cellular uptake of functionalised carbon nanotubes is independent of functional group and cell type, *Nat Nanotechnol* **2**, 108–113.

15. Ewert, K., Slack, N. L., Ahmad, A., Evans, H. M., Lin, A. J., Samuel, C. E., and Safinya, C. R. (2004) Cationic lipid-DNA complexes for gene therapy: understanding the relationship between complex structure and gene delivery pathways at the molecular level, *Curr Med Chem* **11**, 133–149.

16. Chan, C. L., Majzoub, R. N., Shirazi, R. S., Ewert, K. K., Chen, Y. J., Liang, K. S., and Safinya, C. R. (2012) Endosomal escape and transfection efficiency of PEGylated cationic liposome-DNA complexes prepared with an acid-labile PEG-lipid, *Biomaterials* **33**, 4928–4935.

17. Leal, C., Ewert, K. K., Shirazi, R. S., Bouxsein, N. F., and Safinya, C. R. (2011) Nanogyroids incorporating multivalent lipids: enhanced membrane charge density and pore forming ability for gene silencing, *Langmuir* **27**, 7691–7697.

18. Dittrich, M., Heinze, M., Wolk, C., Funari, S. S., Dobner, B., Mohwald, H., and Brezesinski, G. (2011) Structure-function relationships of new lipids designed for DNA transfection, *Chemphyschem* **12**, 2328–2337.

19. Almeida, J. A., Pinto, S. P., Wang, Y., Marques, E. F., and Pais, A. A. (2011) Structure and order of DODAB bilayers modulated by dicationic gemini surfactants, *Phys Chem Chem Phys* **13**, 13772–13782.

20. Piest, M., and Engbersen, J. F. (2010) Effects of charge density and hydrophobicity of poly(amido amine)s for non-viral gene delivery, *J Control Release* **148**, 83–90.

21. Creusat, G., Rinaldi, A. S., Weiss, E., Elbaghdadi, R., Remy, J. S., Mulherkar, R., and Zuber, G. (2010) Proton sponge trick for pH-sensitive

disassembly of polyethylenimine-based siRNA delivery systems, *Bioconjug Chem* **21**, 994–1002.

22. Caracciolo, G., Marchini, C., Pozzi, D., Caminiti, R., Amenitsch, H., Montani, M., and Amici, A. (2007) Structural stability against disintegration by anionic lipids rationalises the efficiency of cationic liposome/DNA complexes, *Langmuir* **23**, 4498–4508.
23. Safinya, C. R., Ewert, K., Ahmad, A., Evans, H. M., Raviv, U., Needleman, D. J., Lin, A. J., Slack, N. L., George, C., and Samuel, C. E. (2006) Cationic liposome-DNA complexes: from liquid crystal science to gene delivery applications, *Philos Transact A Math Phys Eng Sci* **364**, 2573–2596.
24. Gurtovenko, A. A., Miettinen, M., Karttunen, M., and Vattulainen, I. (2005) Effect of monovalent salt on cationic lipid membranes as revealed by molecular dynamics simulations, *J Phys Chem B* **109**, 21126–21134.
25. Hoekstra, D., Rejman, J., Wasungu, L., Shi, F., and Zuhorn, I. (2007) Gene delivery by cationic lipids: in and out of an endosome, *Biochem Soc Trans* **35**, 68–71.
26. Foldvari, M., Badea, I., Wettig, S., Verrall, R., and Bagonluri, M. (2006) Structural characterization of novel gemini non-viral DNA delivery systems for cutaneous gene therapy, *J Exp Nanosci* **1**, 165–176.
27. Leal, C., Bouxsein, N. F., Ewert, K. K., and Safinya, C. R. (2010) Highly efficient gene silencing activity of siRNA embedded in a nanostructured gyroid cubic lipid matrix, *J Am Chem Soc* **132**, 16841–16847.
28. Amenitsch, H., Caracciolo, G., Foglia, P., Fuscoletti, V., Giansanti, P., Marianecci, C., Pozzi, D., and Lagana, A. (2011) Existence of hybrid structures in cationic liposome/DNA complexes revealed by their interaction with plasma proteins, *Colloids Surf B Biointerfaces* **82**, 141–146.
29. Cardoso, A. M., Faneca, H., Almeida, J. A., Pais, A. A., Marques, E. F., de Lima, M. C., and Jurado, A. S. (2011) Gemini surfactant dimethylene-1,2-bis(tetradecyldimethylammonium bromide)-based gene vectors: a biophysical approach to transfection efficiency, *Biochim Biophys Acta* **1808**, 341–351.
30. Gaucheron, J., Wong, T., Wong, K. F., Maurer, N., and Cullis, P. R. (2002) Synthesis and properties of novel tetraalkyl cationic lipids, *Bioconjug Chem* **13**, 671–675.
31. Semple, S. C., Akinc, A., Chen, J., Sandhu, A. P., Mui, B. L., Cho, C. K., Sah, D. W., Stebbing, D., Crosley, E. J., Yaworski, E., Hafez, I. M., Dorkin, J. R., Qin, J., Lam, K., Rajeev, K. G., Wong, K. F., Jeffs, L. B., Nechev, L., Eisenhardt,

M. L., Jayaraman, M., Kazem, M., Maier, M. A., Srinivasulu, M., Weinstein, M. J., Chen, Q., Alvarez, R., Barros, S. A., De, S., Klimuk, S. K., Borland, T., Kosovrasti, V., Cantley, W. L., Tam, Y. K., Manoharan, M., Ciufolini, M. A., Tracy, M. A., de Fougerolles, A., MacLachlan, I., Cullis, P. R., Madden, T. D., and Hope, M. J. (2010) Rational design of cationic lipids for siRNA delivery, *Nat Biotechnol* **28**, 172–176.

32. Ahmad, A., Evans, H. M., Ewert, K., George, C. X., Samuel, C. E., and Safinya, C. R. (2005) New multivalent cationic lipids reveal bell curve for transfection efficiency versus membrane charge density: lipid-DNA complexes for gene delivery, *J Gene Med* **7**, 739–748.
33. Karaborni, S., Esselink, K., Hilbers, P. A., Smit, B., Karthauser, J., van Os, N. M., and Zana, R. (1994) Simulating the self-assembly of gemini (dimeric) surfactants, *Science* **266**, 254–256.
34. Menger, F. M., and Keiper, J. S. (2000) Gemini Surfactants, *Angew Chem Int Ed Engl* **39**, 1906–1920.
35. Karlsson, L., van Eijk, M. C., and Soderman, O. (2002) Compaction of DNA by gemini surfactants: effects of surfactant architecture, *J Colloid Interface Sci* **252**, 290–296.
36. Kirby, A. J., Camilleri, P., Engberts, J. B., Feiters, M. C., Nolte, R. J., Soderman, O., Bergsma, M., Bell, P. C., Fielden, M. L., Garcia Rodriguez, C. L., Guedat, P., Kremer, A., McGregor, C., Perrin, C., Ronsin, G., and van Eijk, M. C. (2003) Gemini surfactants: new synthetic vectors for gene transfection, *Angew Chem Int Ed Engl* **42**, 1448–1457.
37. Bombelli, C., Borocci, S., Diociaiuti, M., Faggioli, F., Galantini, L., Luciani, P., Mancini, G., and Sacco, M. G. (2005) Role of the spacer of cationic gemini amphiphiles in the condensation of DNA, *Langmuir* **21**, 10271–10274.
38. Kirby, A. J., Camilleri, P., Engberts, J. B. F. N., Feiters, M. C., Nolte, R. J. M., Soderman, O., Bergsma, M., Bell, P. C., Fielden, M. L., Garcia Rodriguez, C. L., Guedat, P., Kremer, A., McGregor, C., Perrin, C., Ronsin, G., and van Eijk, M. C. P. (2003) Gemini surfactants: New synthetic vectors for gene transfection, *Angew Chem Int Ed* **42**, 1448–1457.
39. Wettig, S. D., Verrall, R. E., and Foldvari, M. (2008) Gemini surfactants: A new family of building blocks for non-viral gene delivery systems, *Curr Gene Ther* **8**, 9–23.
40. Badea, I., Verrall, R., Baca-Estrada, M., Tikoo, S., Rosenberg, A., Kumar, P., and Foldvari, M. (2005) In vivo cutaneous interferon-gamma gene delivery using novel dicationic (gemini) surfactant-plasmid complexes, *J Gene Med* **7**, 1200–1214.

41. Zhang, Q. S., Guo, B. N., and Zhang, H. M. (2004) Development and application of gemini surfactants, *Prog Chem* **16**, 343–348.

42. Singh, J., Michel, D., Chitanda, J. M., Verrall, R. E., and Badea, I. (2012) Evaluation of cellular uptake and intracellular trafficking as determining factors of gene expression for amino acid-substituted gemini surfactant-based DNA nanoparticles, *J Nanobiotechnol* **10**, 7.

43. Sugiyasu, K., Tamaru, S., Takeuchi, M., Berthier, D., Huc, I., Oda, R., and Shinkai, S. (2002) Double helical silica fibrils by sol-gel transcription of chiral aggregates of gemini surfactants, *Chem Commun (Camb)* **11**, 1212–1213.

44. Wettig, S. D., Badea, I., Donkuru, M., Verrall, R. E., and Foldvari, M. (2007) Structural and transfection properties of amine-substituted gemini surfactant-based nanoparticles, *J Gene Med* **9**, 649–658.

45. Wang, C. Z., Wettig, S. D., Foldvari, M., and Verrall, R. E. (2007) Synthesis, characterisation, and use of asymmetric pyrenyl-gemini Surfactants as emissive components in DNA - Lipoplex systems, *Langmuir* **23**, 8995–9001.

46. Wettig, S. D., Badea, I., Donkuru, M., Verrall, R. E., and Foldvari, M. (2007) Structural and transfection properties of amine-substituted gemini surfactant-based nanoparticles, *J Gene Med* **9**, 649–658.

47. Singh, J., Yang, P., Michel, D., Verrall, R. E., Foldvari, M., and Badea, I. (2011) Amino acid-substituted gemini surfactant-based nanoparticles as safe and versatile gene delivery agents, *Curr Drug Deliv* **8**, 299–306.

48. Yang, P., Singh, J., Wettig, S., Foldvari, M., Verrall, R. E., and Badea, I. (2010) Enhanced gene expression in epithelial cells transfected with amino acid-substituted gemini nanoparticles, *Eur J Pharm Biopharm* **75**, 311–320.

49. Uhrikova, D., Zajac, I., Dubnickova, M., Pisarcik, M., Funari, S. S., Rapp, G., and Balgavy, P. (2005) Interaction of gemini surfactants butane-1,4-diyl-bis(alkyldimethylammonium bromide) with DNA, *Colloids Surf B Biointerfaces* **42**, 59–68.

50. Pindzola, B. A., Jin, J., and Gin, D. L. (2003) Cross-linked normal hexagonal and bicontinuous cubic assemblies via polymerisable gemini amphiphiles, *J Am Chem Soc* **125**, 2940–2949.

51. Foldvari, M., Wettig, S., Badea, I., Verrall, R., and Bagonluri, M. (2006) Dicationic gemini surfactant gene delivery complexes contain cubic-lamellar mixed polymorphic phase., *NSTI-Nanotech* **2**, 400–403.

52. Winter, R., and Kohling, R. (2004) Static and time-resolved synchrotron small-angle x-ray scattering studies of lyotropic lipid mesophases, model biomembranes and proteins in solution, *J Phys Condensed Matter* **16**, S327-S352.

53. Koltover, I., Salditt, T., Radler, J. O., and Safinya, C. R. (1998) An inverted hexagonal phase of cationic liposome-DNA complexes related to DNA release and delivery, *Science* **281**, 78–81.

54. Kennedy, M. T., Pozharski, E. V., Rakhmanova, V. A., and MacDonald, R. C. (2000) Factors governing the assembly of cationic phospholipid-DNA complexes, *Biophys J* **78**, 1620–1633.

55. Ewert, K., Slack, N. L., Ahmad, A., Evans, H. M., Lin, A. J., Samuel, C. E., and Safinya, C. R. (2004) Cationic lipid-DNA complexes for gene therapy: understanding the relationship between complex structure and gene delivery pathways at the molecular level, *Curr Med Chem* **11**, 133–149.

56. Foldvari, M., Badea, I., Wettig, S., Verrall, R., and Bagonluri, M. (2005) Structural characterisation of novel micro- and nano-scale non-viral DNA delivery systems for cutaneous gene therapy, *NSTI-Nanotech* **1**, 128–131.

57. Lin, A. J., Slack, N. L., Ahmad, A., George, C. X., Samuel, C. E., and Safinya, C. R. (2003) Three-dimensional imaging of lipid gene-carriers: membrane charge density controls universal transfection behavior in lamellar cationic liposome-DNA complexes, *Biophys J* **84**, 3307–3316.

58. Donkuru, M., Wettig, S. D., Verrall, R. E., Badea, I., and Foldvari, M. (2012) Designing pH-sensitive gemini nanoparticles for non-viral gene delivery into keratinocytes, *J Mater Chem* **22**, 6232–6244.

Chapter 7

Nanomedicine and Embryology: Causative Embryotoxic Agents Which Can Pass the Placenta Barrier and Induce Birth Defects

Elpida-Niki Emmanouil-Nikoloussi
Faculty of Medicine, Aristotle University of Thessaloniki, 54124 Thessaloniki, Greece
emmanik@med.auth.gr

7.1 Introduction

Nanomaterials as a science tool are widely discussed for their impact and have been estimated as having conflicting, i.e. positive and negative effects, on living tissues and organs in diverse report papers [1–9].

Nanotechnology consists of a new scientific sector in biomedical sciences that is developing very rapidly. Nanoparticles and their behaviour, and their biocompatibility and toxicity with diverse tissues, especially at the developing embryo stage, is one of the most recent challenging research targets. Dimensions, chemical composition and consistency of nanoparticles are widely discussed

Horizons in Clinical Nanomedicine
Edited by Varvara Karagkiozaki and Stergios Logothetidis

ISBN 978-981-4411-56-1 (Hardcover), 978-981-4411-57-8 (eBook)
www.panstanford.com

in the international literature, as those parameters render their biocompatibility in diverse tissues and organs and determine the applications of nanoparticles for therapeutic and diagnostic purposes.

One of the major challenges discussed for nanoparticles is their placenta barrier permeability and biocompatibility through this action to the developing embryonic/foetal tissues and organs.

The features of nanoparticles, which are constructed at the size of 1–100 nm dimension range, are various and discussion has been arisen about the type of nanoparticles in which new drugs could be incorporated. These issues are determinant for new drug development and for their future treatment applications. Karakoti et al. (2006), consider that the physical parameters of nanomaterials are very important for their biocompatibility and their toxicity when applied to living structures [6]. As for the general pathways of possible toxicity of nanomaterials versus human health, it has been described that inhaled ultrafine particles and nanoparticles can be toxic to respiratory system and especially to lungs and lung vesicles, as well as to several organs and systems as the gastrointestinal, urinary and renal tubules and cardiovascular systems [9–14]. Data on yet unpublished research activities concerning heavy metal admission to newborn mice, after 'in vivo' treatment during pregnancy, indicate that even via drinking water solutions, heavy metals can accumulate to alveolar and liver neonatal tissues [15]. Nanoparticles can accumulate in cellular organelles and for this reason coatings of their surface are determinant for their invasion to cellular membrane and organelles. Transportation of nanomaterials can be easily directed, although it is not yet affirmed that they can be concentrated to their target cellular populations in organs and systems [9, 11, 16, 17, 18]. Hoet et al. (2004) are discussing the biocompatibility and available data of beneficial or/and hazardous effects of nanomaterials and support that nanomaterial toxicity and properties to living tissues mainly depend on their size [11]. Recently, increased attention has been raised on the potential toxicity of nanoparticles and nanomaterials in general; and much discussion has been developed on their composition, size and surface properties [19, 20]. Parameters that influence their toxicity are nanoparticles' charges adhering to the cellular membrane and also their chemical construction which can cause damage to the cellular cytoplasm intruding it [21].

According to Dawson et al. (2009), charged nanoparticles can alter the local physical properties of lipid membranes [22]. The authors suggest that nanoparticles with a diameter of less than 100 nm are able to enter cells via the cellular membrane, nanoparticles with a diameter less than 40 nm can enter the cellular nuclei, and nanoparticles with a diameter less than 35 nm can pass through the blood–brain barrier and enter the brain interfering with the neural brain cells, thus constituting the brain area target for those nanoparticles. The latest report is compatible with the placenta barrier penetration of nanomaterials as this barrier is a very strong one, consisting of a coherent endothelium and basement membrane and surrounded by syncytiotrophoblastic cells with tight epithelial junctions. Those criteria are similar to those of the coherent blood brain barrier. Therefore, particles from the environment or from drugs encountered below a certain size are considered to be able to pass through the placental barrier, but this issue has to be further studied [9,23].

Understanding the way nanoparticles interact with living matter will open up fundamentally new opportunities in medicine and diagnostics. Charged nanoparticles can alter the local physical properties of cellular membranes, which could shed new light on the interactions between living cells and nanomaterials.

Modern literature is focusing on the way that nanoparticles can interact with living tissues and will give the opportunity to biomedical sciences for new therapeutic methods and new diagnostic fields [3,24]. New drug-delivery systems raise enormous scientific interest, giving hope for disease treatment and disease control, using technologies with novel nanoparticles and nanodevices [25].

Placenta and nanotoxicity is in their very preliminary status of international research discussion and very recently this type of research has been developed [9,23], as until now little information has been provided on whether maternal organism exposed to nanoparticles during pregnancy, either via the environmental pollution or via drug treatment, can pass those nanoparticles to their embryos/foetuses via the placenta barrier. Researchers, though, would like to determine how nanoparticles might be used for therapeutic purposes in the future and a solution for this issue could be considered if they could be charged with new drug molecules as vehicles assigned to targeted embryonic/foetal organs and systems via the placenta barrier, without any harmful effects to the maternal organism. A very useful query of today's research on nanomaterials is to determine according to their structure why nanomaterials might exert toxicity on the placenta and on embryonic/foetal tissues and organs. This query is related to the international literature reports about the possible benefits versus toxicity of diverse nanomaterials and how this balance can have an inclinator to the benefits decreasing the toxicity of them [9, 23].

7.2 Drugs, Environmental Pollution and Embryotoxicity

For the treatment of some diseases like cystic acne, psoriasis, keratinising dermatoses, oral lichen planus, leucoplakia and epilepsy, several drugs as retinoids , antiepileptic drugs such as difenyntoin and valproic acid are used [26]; they are to be avoided in women of child bearing age because of their powerful teratogenic effects as notably demonstrated and described histologically and by SEM examination in extended studies [26–28], in which we observed diverse cranio-facial malformations, particularly exophthalmos and excencephaly [26]. Additionally to them, many drugs used for several diseases, as multiple sclerosis, chemotherapy, arthritis, autoimmune system diseases, etc. are involved when it is necessary to be used even in pregnancy for mothers, although the embryos/foetuses are in the control of many developmental processes. Furthermore, excess and deficiency of several drugs can be teratogenic.

Although several potential risk assessments in administrating drugs during gestation is known, reasons for treatment of pregnant women are balanced when beneficial effects of drug administration to them are validated to surpass drug risk assessment to the embryonic/foetal organism. Embryological developmental studies support the theory that the role of teratogenic drugs is mediated by different cellular receptors using several molecular ligands. For example, neural tube malformations are developed due to genetic code disturbances, and/or diverse teratogenetic drugs and environmental toxicants' influence producing apoptosis [15, 28–37].

Additional to drug toxicity, environmental pollution of several toxic substances as heavy metals and pesticides are discussed as environmental pollutants. And among these pollutants, nanoparticles contained in those elements and those existing in natural sources at the environment have to be considered. Those nanoparticles can be present in water, sea water, volcanos, or by chemical composed materials as in chemical industries, chemical exhausts, petrol and diesel oil fuels and substances contained in the atmosphere and enter human organism via inhalation , food and water consumption.

All these nanoparticles can enter the human organism and become toxic to the host, especially during pregnancy, when they pass the placenta barrier and cause diverse problems to the developing embryo/foetus.

Accumulation, localisation and penetration of nanoparticles or pollutants in maternal and embryonic/foetal organs via the placenta barrier establish potential ranges of tissue damage in those organs, as besides nutritional substances which have the capability to cross the placenta barrier, those toxic substances can also cross it and cause toxic and harmful effects to the developing organism (Figs. 7.1–7.5 and 7.19–7.27) [15, 38–40].

Drug Molecules Interfering with Limb Development

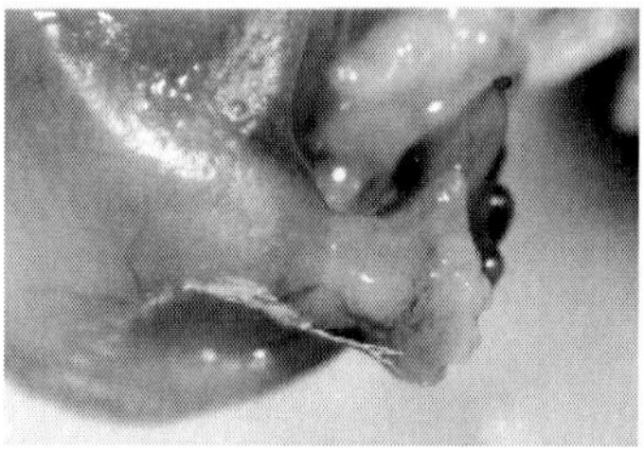

Figure 7.1 Stereoscopical observations of syndactyly and limb shortening after retinoid analogues treatment in rats.

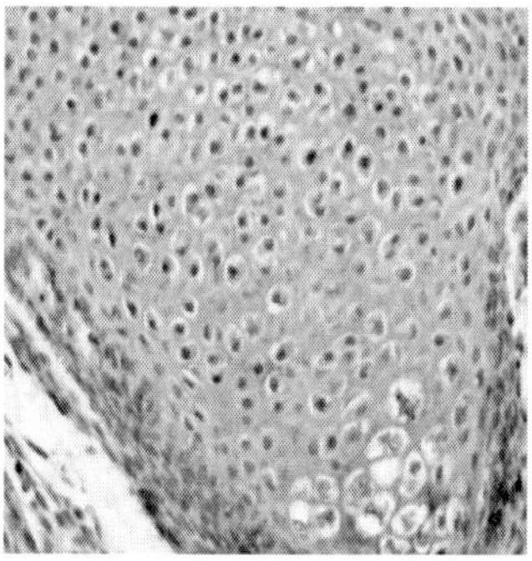

Figure 7.2 Light microscopical observations in foetal mice femur treated with retinoic acid. Positive expression of bone morphogenetic Protein 6-BMP-6 protein in hindlimb.

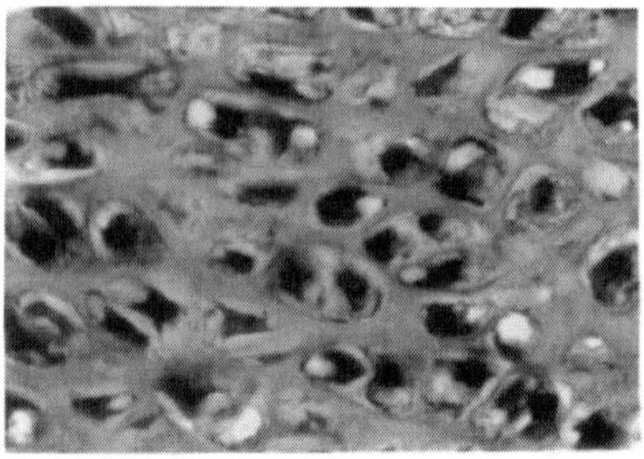

Figure 7.3 Light microscopical observations in foetal mice femur treated with retinoic acid. Positive expression of heat shock protein antibody.

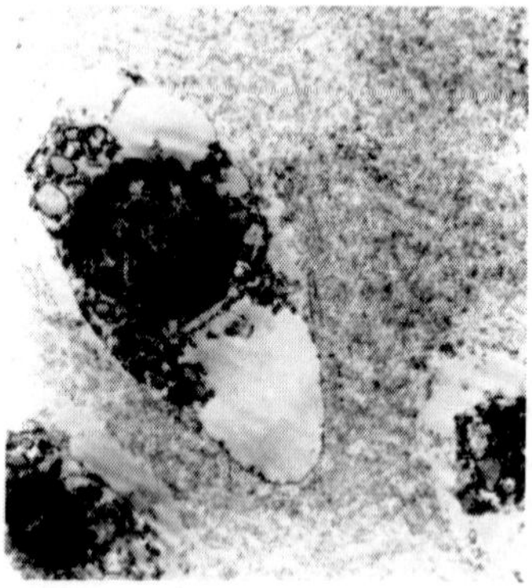

Figure 7.4 Transmission electron microscopical observation of apoptotic chondroblasts in foetal mice femur treated with retinoic acid.

Drugs and Cellular Apoptosis

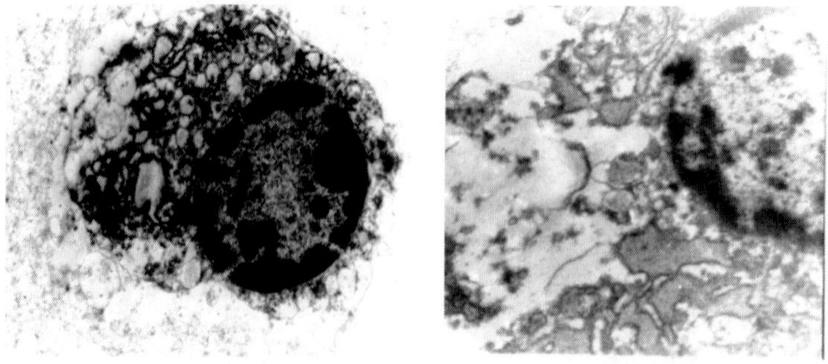

Figure 7.5 Transmission electron microscopical observation of apoptotic chondroblasts in foetal mice femur treated with retinoic acid.

Drugs Interfering with Central Neural System Development: Neural Tube Malformations

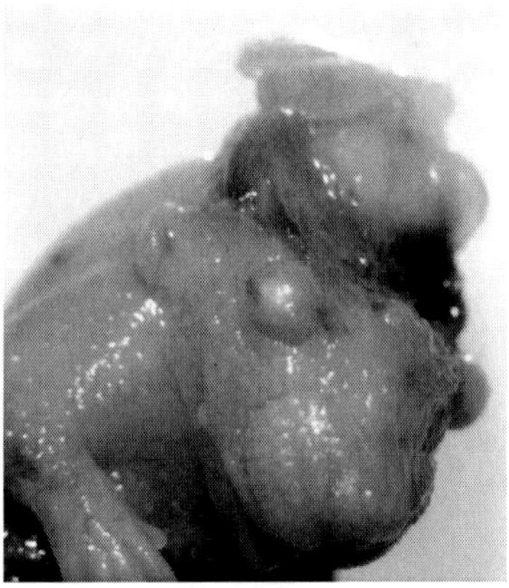

Figure 7.6 Stereoscopical observations of excencephaly after retinoid analogues treatment in rats.

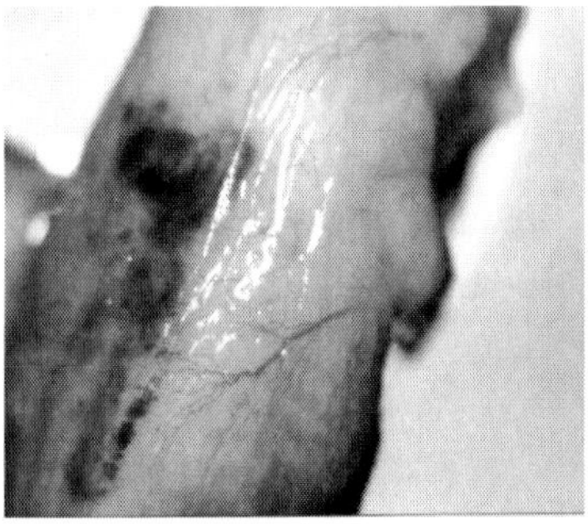

Figure 7.7 Stereoscopical observations of spina bifida after antiepileptic drugs treatment in rats.

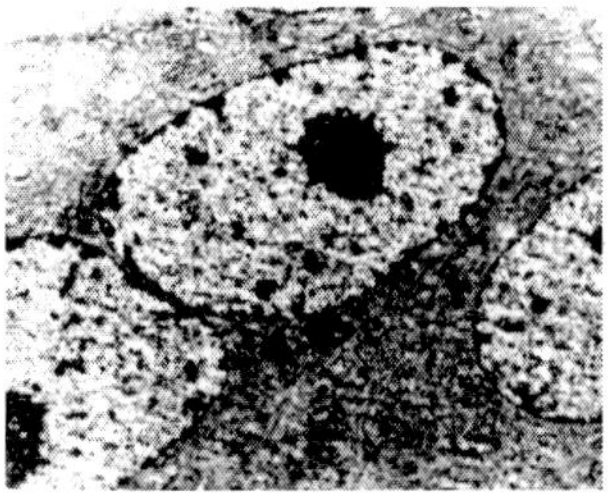

Figure 7.8 Transmission electron microscopical observation of normal neural crest cells migrating for differentiation.

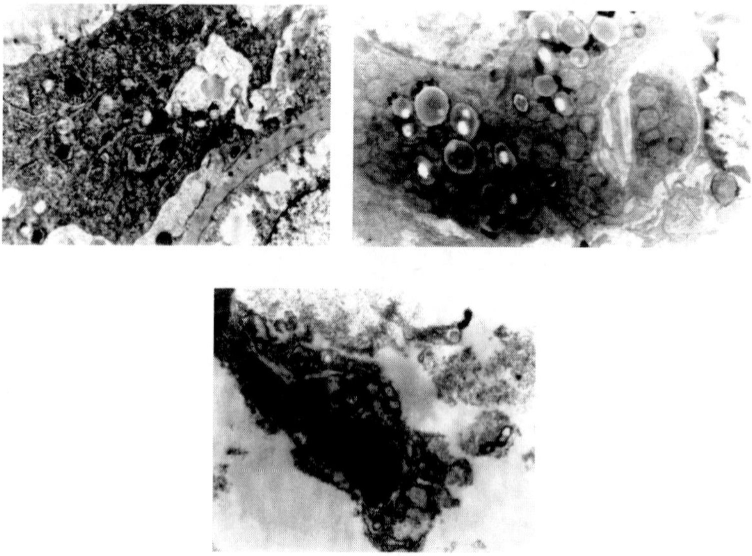

Figure 7.9 Transmission electron microscopical observation of neural crest cells migrating for differentiation and presenting degenerative lesions due to retinoid (top) and antiepileptic analogues (middle and bottom).

Teratomas and Placenta during Toxicity Status

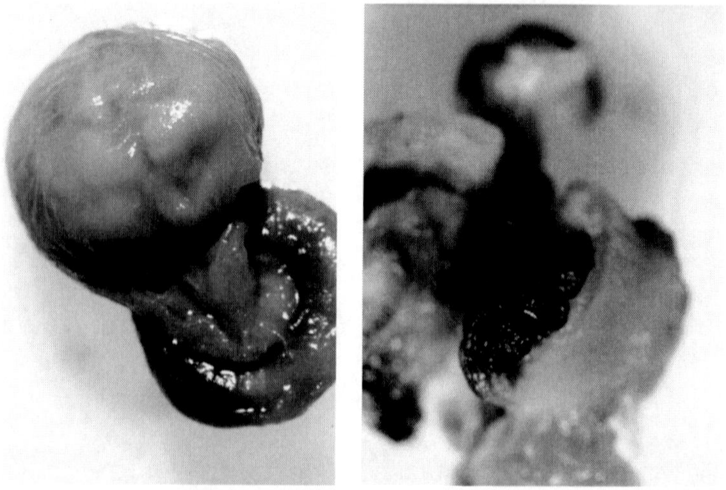

Figure 7.10 Stereoscopical observations of placenta and teratoma after retinoid analogue treatment.

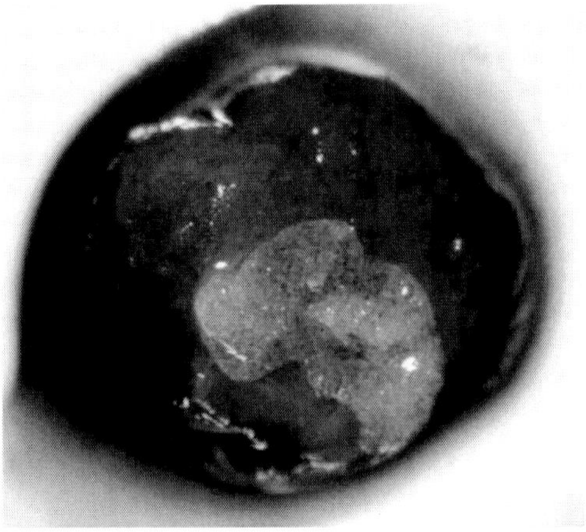

Figure 7.11 Stereoscopical observations of teratoma and placenta after retinoid analogue treatment.

Disruption of Placenta Blood Barrier

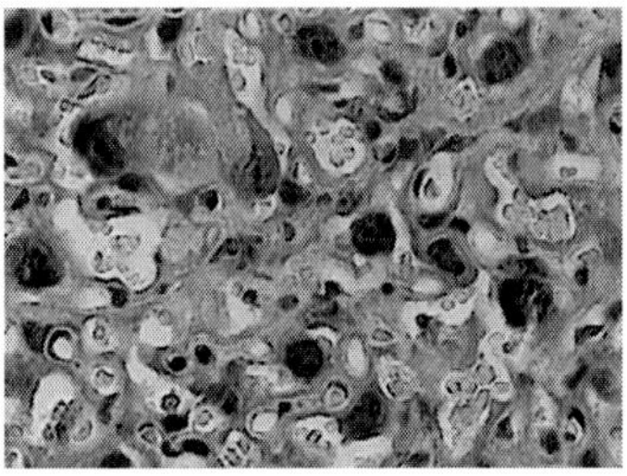

Figure 7.12 Light microscopical observations in normal foetal mice placenta at the final stage of gestation.

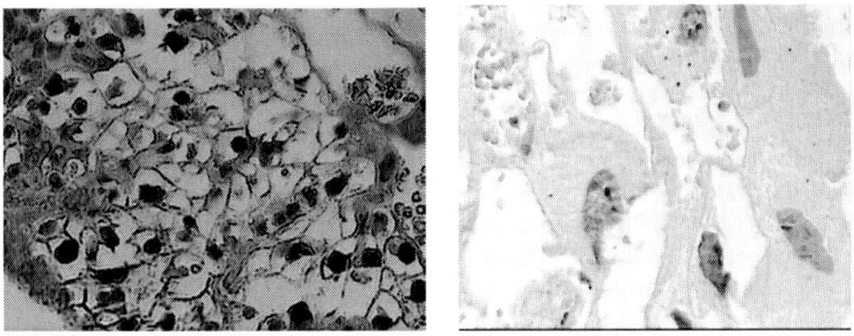

Figure 7.13 Light microscopical observations in foetal mice placenta blood barrier at final stage of gestation. Disruption of chorionic villi and placenta blood barrier after retinoic analogue treatment.

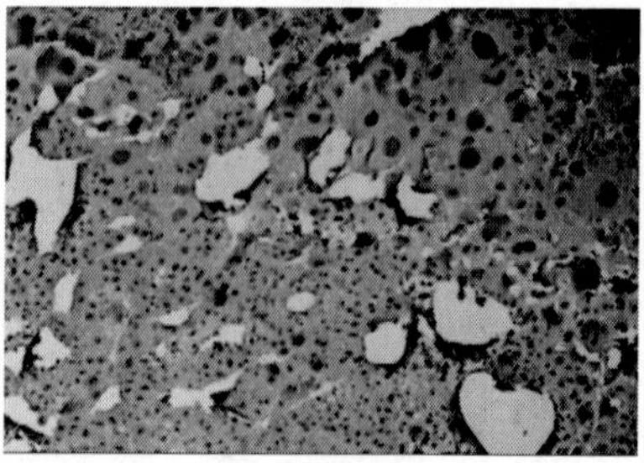

Figure 7.14 Light microscopical observations in foetal mice placenta blood barrier at final stage of gestation . Normal foetal mice chorionic villi and placenta blood barrier. Heat shock protein antibody identification.

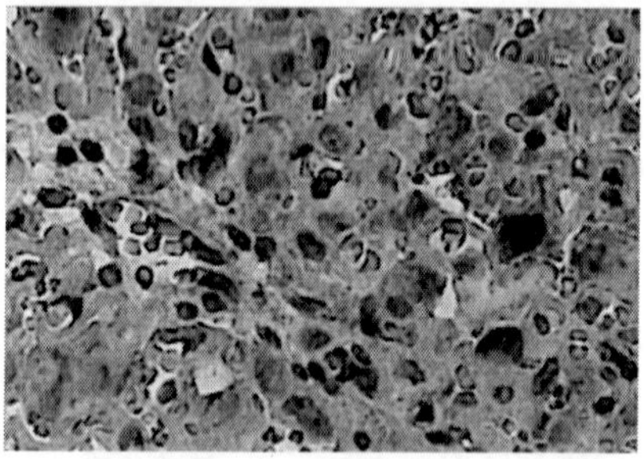

Figure 7.15 Light microscopical observations in foetal mice placenta blood barrier at final stage of gestation. Disruption of chorionic villi and placenta blood barrier after retinoic analogue treatment. Heat Shock Proteins (HSPs) antibody identification.

Placenta Blood Barrier Disruption: Transmission and Scanning Electron Microscopical Observations

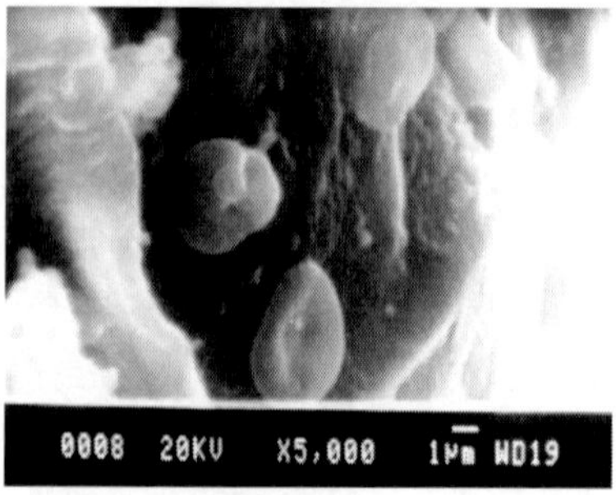

Figure 7.16 Normal placenta blood barrier: scanning electron microscopical observations.

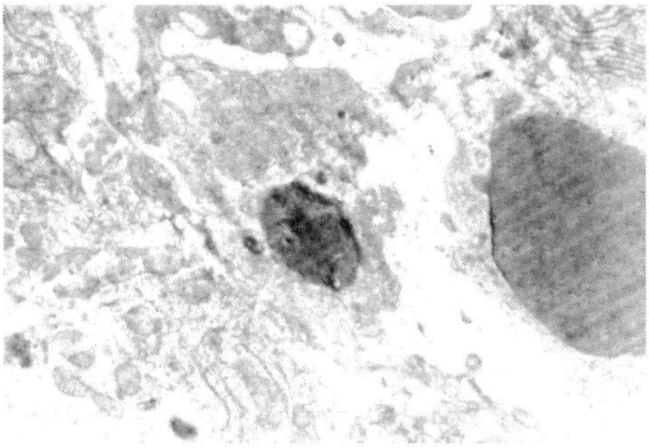

Figure 7.17 Placenta chorionic villi and blood barrier disruption after thalidomide treatment. Transmission electron microscopical observations.

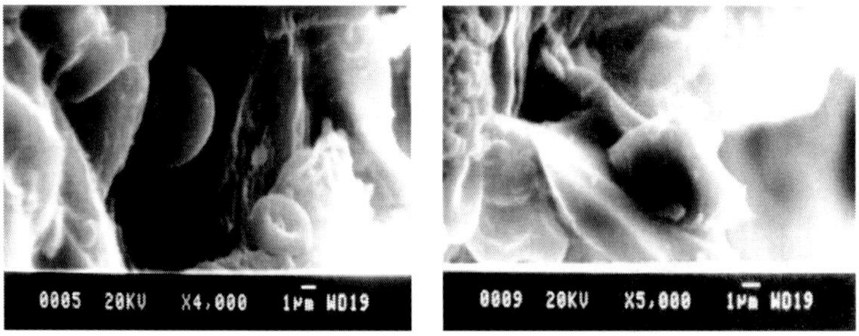

Figure 7.18 Placenta chorionic villi and blood barrier disruption after thalidomide treatment. scanning electron microscopical observations.

Liver Toxicity

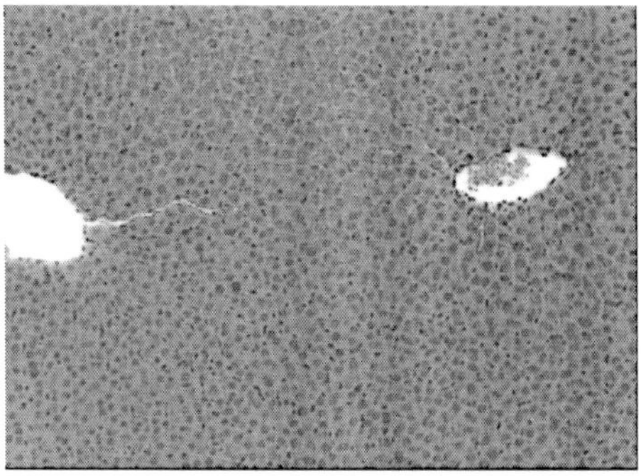

Figure 7.19 Light microscopical observations in normal foetal mice liver.

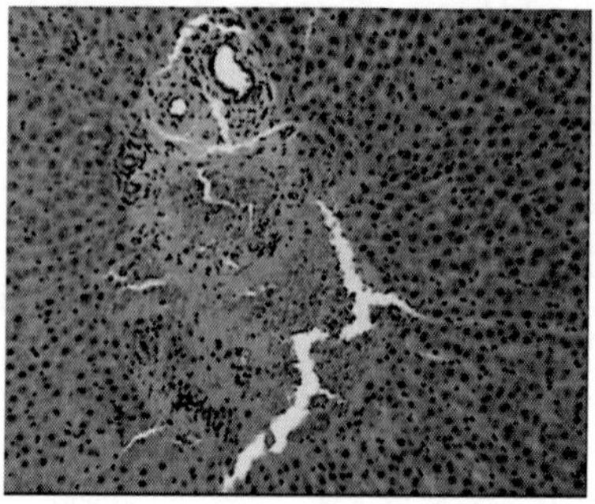

Figure 7.20 Light microscopical observations in foetal mice liver after retinoic analogue treatment. Liver fibrosis.

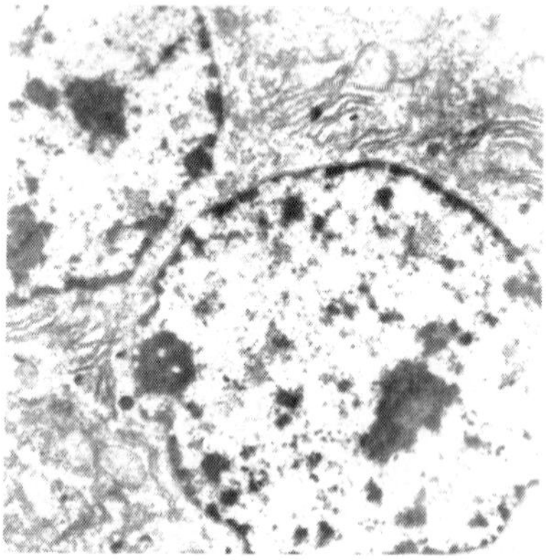

Figure 7.21 Transmission electron microscopical observations in normal foetal mice liver.

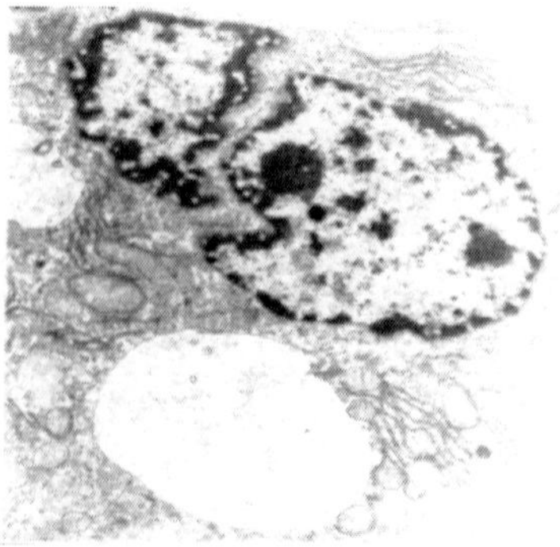

Figure 7.22 Transmission electron microscopical observations in foetal mice liver after retinoic analogue treatment.

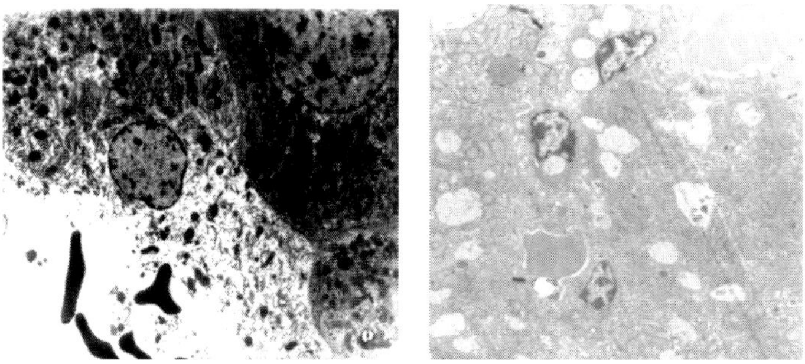

Figure 7.23 Transmission electron microscopical observations in foetal mice liver after hydroxyurea treatment.

Heavy Metal Toxicity

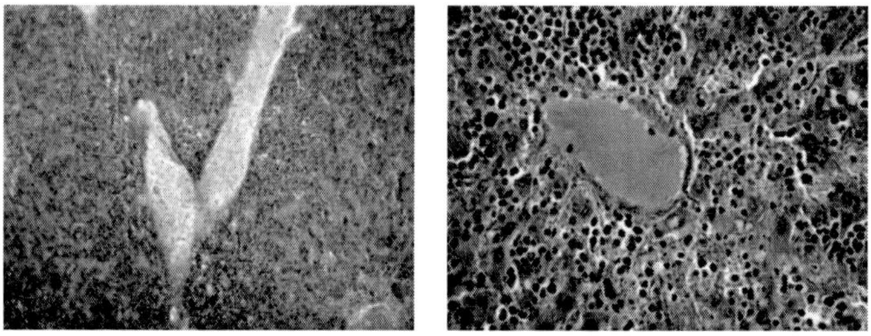

Figure 7.24 Light microscopical observations in foetal mice liver after cadmium treatment. Liver inflammation and leucocyte infiltration.

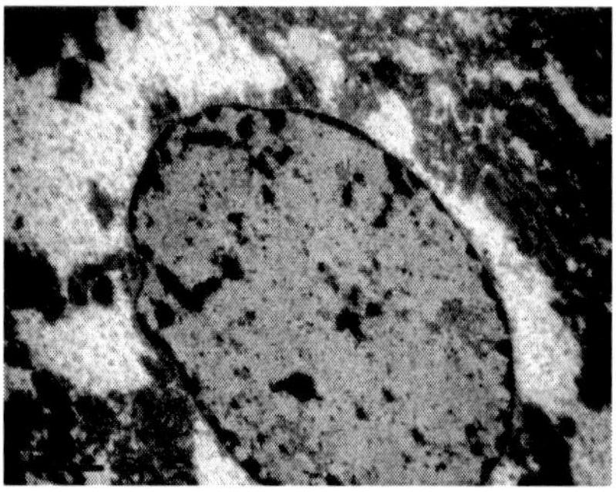

Figure 7.25 Transmission electron microscopical observations in foetal mice liver after lead treatment. Apoptotic nuclei and chromatin condensation. Lead inclusions into the nuclei.

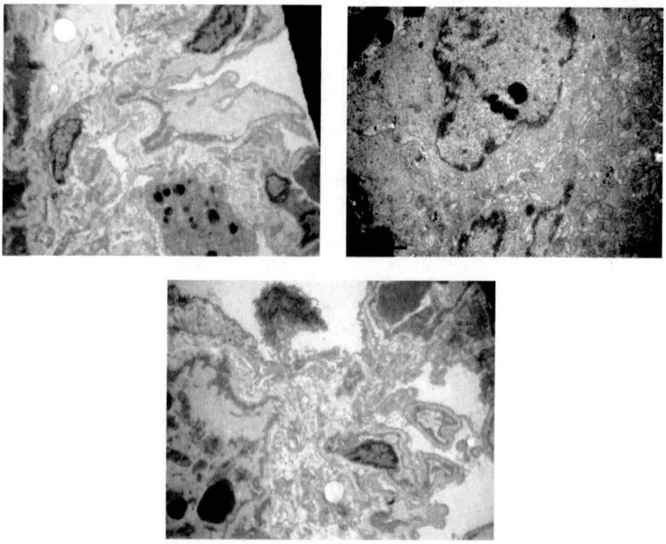

Figure 7.26 Transmission electron microscopical observations in foetal mice lung alveoli after cadmium treatment. Apoptotic alveolar cell type II in the figure on the left and cadmium inclusions into the cytoplasm and nuclei of alveolar cells.

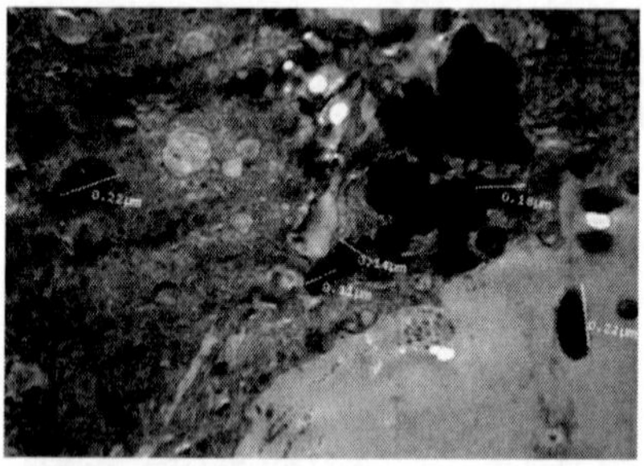

Figure 7.27 Transmission electron microscopical observations in foetal mice lung alveoli after cadmium treatment. Cadmium inclusions into alveolar cell type I.

7.3 Drugs, Nanoparticles and Embryology

The biologic activity and biokinetics of nanoparticles are dependent on many parameters: size, shape, chemistry, crystallinity, surface properties such as area, porosity, charge, surface modifications, weathering of coating, agglomeration state, biopersistence, and dose. For this reason, nanomechanics, which can be introduced to pharmaceutical companies, can handle nanoproducts such as drugs incorporated into nanoparticles which can interfere with the target organ and protect the embryo-maternal organisms from damage. It is of good understanding that nanomedicine products must be well tested before introduction into the marketplace, especially as this concerns the susceptible developing organism. Consequently, for the manufacturers of most current nanotechnology products, regulations requiring nanomaterial-specific data on toxicity before introduction into the marketplace are an evolving area and presently under discussion [9, 11, 23, 41]. Development of standardised approaches stands to reason for the evaluation of whether exposures to specific types of nanomaterials can be related and operate genotoxic effects [42].

The diversity of engineered nanomaterials and their potential effects represent major challenges and research needs of nanotoxicology, including the need for assessing human exposure during manufacture and use and especially during embryonic/foetal development.

The diversity of engineered nanomaterials and the potential effects that this research topic can produce in the coming years in manufacturing new drugs, represent major challenges and research needs for nanotoxicology, including the need for assessing human exposure during manufacture and use [41, 43–46]. The goal to exploit positive aspects of engineered nanomaterials and avoid potential toxic effects can best be achieved through a multidisciplinary team effort involving researchers in toxicology, materials science, medicine, molecular biology, bioinformatics, and their subspecialties. Nanoparticles' coating for new drug target molecules scheduled to cure embryonic malformations must show bioavailability as they have to pass the placenta barrier and concentrate on the target damaged embryonic/foetal organ. The diversity of engineered nanomaterials and the potential effects that this research topic can produce in the coming years in manufacturing new drugs represent major challenges and research needs for nanotoxicology, including the need for assessing human exposure during manufacture and use.

7.3.1 Causative Embryotoxic Agents Inducing Cellular Apoptosis and Creating Birth Defects

According to the stage of embryonic/foetal development, a drug or a harmful environmental agent can have different effects at various stages of gestation.

Thus, during early gestation, it can have either a complete cells' destruction which leads to abortion, or have no effects and the embryo can be rescued and remain alive unaffected. At the organogenesis period in which we have organ differentiation, the noxious and teratogenic agents can produce mostly anatomical congenital defects. Lastly, at the foetogenic period in which the foetus is growing rapidly, the noxious and teratogenic agents can produce mostly functional disorders that are interpreted as congenital defects during life. Most often during that period, central neural system disorders and cognitive reactions of the developing child are expressed (Figs. 7.6–7.9). All those issues deal with the drug or environmental agent dosage level and the maternal health condition as well. Therefore, as referred above, several embryotoxic agents passing the placenta barrier can cause malformations, birth defects, automatic abortions or, additionally, embryonic absorption. Major teratogenic defects can be major or minor anatomical malformations, while more minor teratogenic defects are functional defects occurring mostly at the late foetal period. Researchers occupied with nanotechnology and nanomedicine issues have to answer several queries such as the following:

- What can causative embryotoxic agents induce to the developing mammalian organism?
- Could those side effects be inhibited by Nanomechanics charged with new drug molecules?
- Could those side effects be restored by Nanomechanics charged with new drug molecules?

7.3.2 Embryonic Development and Nanomolecules

A new research area created recently is the issue of embryonic development and treatment of any sudden influence of embryo-maternal toxication. There are parameters studying those effects and examining if they can be related to Nanomedicine, in order for this new science to be able, via the placenta barrier, to treat and cure with biocompatible targeted new drug nanoparticle molecules, embryonic/foetal organs and systems influenced by noxious agents. Developmental toxicity and nanoparticles is one of the most challenging research areas discussed nowadays.

7.3.3 Mechanics and Nanomolecules during Development

Mechanics is shown to be an important and perhaps a central component to the differentiation and development of embryos. Especially the mechanics of the nucleus may also be involved in determining which genes are expressed in a given cell. At present, there are two major approaches to the mechanics of differentiation in embryos: the morphomechanics, and the waves of differentiation. Both of them are compared in detail, to provide a starting point for future experimental work to bring them into one fundamental structure of research. The morphomechanics and differentiation waves provide a network of molecular interactions between genes and proteins which drives animal embryonic development. Nevertheless, much less is known about the role of endogenous low molecular weight metabolites and their mechanics, including mostly the differention waves in the early developing embryo and the ability to introduce, remove, or modify molecules in the intracellular environment. These issues are considered to be fundamental and important to the understanding of cellular structure and function, especially for embryological/foetal mechanisms of development. Oxidative stress is one of the most common causative factors of cellular apoptosis and cellular abolition. Antioxidant substances can protect against free radical attacks of metal toxicity [47]. Therefore, case antioxidant treatment with nanoparticles charged with antioxidant material can be a useful tool for therapeutical treatment. Cerium oxide nanoparticles (nanoceria) may act as antioxidants and auto-regenerative free radical scavengers [48] and can behave as antioxidants with catalytic activities mimicking the neuroprotective enzymes superoxide dismutase and catalase [49]. Modified fullerenes nanoparticles and nanoceria have been proved to protect the mammalian cells from oxidative stress damage [50,51]. Neuroprotective effects of nanoceria can be a useful tool for Central Nervous Tissue (CNS) diseases [52]. Nanoceria can be internalised into mouse tissues and inside the cytoplasm of several organs without any signs of pathology and toxicity, reducing Oxidative Stress in inflammation and, therefore, can be used as a tool for therapy in inflammatory status [48]. Researchers propose that nanoparticles charged with cerium oxide can possibly be a useful therapeutic tool for antioxidant treatment when oxidative stress is induced [53, 54]. They report that cerium oxide nanoparticles exert catalytic properties as antioxidants and, therefore, those nanoparticles can be a useful tool for protecting and treatment in several disorders which are associated with inflammation and oxidative stress. Nevertheless, although nanoceria are exerting their non-toxic activities on normal myofibroblasts, they exert cytotoxic effects on squamous epithelium tumour cells giving, therefore, a promising message that if they are used as charged with chemotherapeutic drugs nanoparticles for cancer therapy, they can give positive treatment expectations to patients [55].

7.3.4 Nanoparticles as a Study Tool in Developmental Processes and Reproductive Toxicology

The introduction of nanostructured materials for biomedical applications opens tremendous opportunities as therapeutic and diagnostic tools in the fields of engineering for Embryology and Congenital Malformations. Handling of nanoproducts as drugs which can protect the embryo-maternal organism from damage is the target of the immediate future in biomedicine. Nanoparticles charged with new drugs can be employed as a vehicle in transporting those drugs in targeted mode of actions into the circulatory system of the embryonic/foetal organism, without that action inducing health problems in pregnant women.

Nanoparticles have recently been studied for their prospective applications in biomedicine. Although their commercialisation is progressing, their health impact has not been yet well understood. Zebra fish embryos are a suitable model for nanoparticle toxicity applications during the sensible embryonic/foetal development [56]. Toxicity studies of several types of nanoparticles as silver, gold, and platinum were applied to zebra fish embryos by Asharani et al. (2011). From them, according to the writers, Ag-NPs were considered as more toxic, while Au-NPs were considered as non-toxic [24].

In the international literature, the zebra fish embryo is considered as one of the best animal models for testing nanomaterial's toxicity, especially in embryology, as zebra fish possess the most homologous genome degree with higher resemblance to the human genome, thus corresponding to human embryo functional biological behaviour [57].

Truong et al. (2011,) in their experimental paper, support that in zebra fish embryos, ionic levels of nanoparticles induce the response of biological samples according to their media ionic strength [58].

Samples of lead engineered nanoparticles demonstrated that their surface appearance and material type discernment is a basic element for the biological responses to their constructive material [59]. An important issue to be considered is that nanoparticles decompose and precipitate upon exposure to environment and air, subsequently inhalation of them must also be considered for their risk assessment to human health and developmental toxicity [59]. In the international literature, 'in vitro' studies discuss that Ag nanoparticles are toxic to several types of mammalian cells, especially parenchyma cells from liver, brain, respiratory tract system, cardiovascular system and especially to the reproductive system. Ag nanoparticles can cause cellular apoptosis and DNA damage to those systems and organs. Developmental toxicity is discussed for those kinds of nanoparticles and among the most vulnerable target areas described for this type of toxicity are the cellular membranes, the mitochondria and the genetic cells of spermatids and oocytes and their genetic code [60]. Lee et al. (2007) and Browning et al. (2009) examined the 'in vivo' effects of Ag and Au nanoparticles in zebra fish embryos and found that Au nanoparticles can pass the chorionic vessels barrier and can be accumulated into the chorionic space and through chorionic vessels again intrude into the embryonic body and developmental embryonic systems. Evidence for biocompatibility and toxicity of Ag nanoparticles in comparison with Au nanoparticles is discussed in the international literature that refers to the zebra fish embryo [61, 62]. The authors support that the embryos can serve as effective in vivo assays for screening nanoparticles' biocompatibility and prove that the toxicity of those nanoparticles is highly dependent on their chemical properties. Additionally, they declare that Au nanoparticles are much more biocompatible to zebra fish embryonic organism than Ag nanoparticles [61, 62]. Further studies mention that DNA stores in its molecule genetic information which render this molecule a useful engineered material for nanoparticle construction giving biocompatibility opportunities to gold nanoparticles [14]. Menezes et al. (2011), in their review, scrutinise the potential risk assessment of nanoparticles when administrated in pregnancy and the possibilities of the placenta to uptake and transfer nanoparticles to the embryonic/foetal organism in possible nanoparticle-based therapeutical treatment [63]. The study of Juch et al. (2011) support that there are nanotoxic effects in 'in vitro' studies of cultivated placenta explants and also an early evidence of apoptotic effects in placenta treated 'in vitro' with nanoparticles of Min-U-Sil5 in comparison to the control placenta. Their results give evidence that due to potential placental transfer of nanoparticles, there are possibilities of birth defects, augmented rates of abortions and IUGR [23].

7.4 Placenta Permeability and Nanoparticles

The placenta consists of a transportation channel to the embryo/foetus and thus can be considered as a target organ for embryonic/foetal toxicity of drugs and harmful substances in general. It is also considered as the barrier of any drug intruding the maternal organism. The degree to which a drug is bound to plasma proteins may also affect the rate of transfer and amount transferred from the maternal organism to the embryo/foetus via the placenta barrier (Figs. 7.10–7.18).

Nanoparticles charged with drugs form the most modern topic of the new international literature for their kinetics into transplacental passage [63]. The placenta supplies the embryonic/foetal organism with oxygen and all nutrients needed and acts as a filter of sorting sequencing between the maternal and the embryonic/foetal organism having the commitment to keep the maternal and embryonic/foetal organism circulatory systems not to be mixed among them. Studies on 'ex vivo' human placental studies support that fluorescent polystyrene nanoparticles of up to 240 nm could cross the placenta blood barrier without causing any damage to the 'ex vivo' placental tissue, while all reports support that further studies are needed to determine placental functions and nanoparticles activity [64, 65]. Dendrimers are considered as important nanocarriers for directed drug delivery. Designation of 'ex vivo' placental studies have been recently performed to determine if dendrimers can be used as drug nanocarriers for maternal and foetal treatment and gave evidence that dendrimers were found in the cytoplasm and nuclei of syncytiotrophoblastic cells and inside the villous core; they were, however, sparsely found in the intravillous capillaries. Modern therapies indicate that dendrimers, due to their selective permeability, can be used for novel methods of nanomedicine applications in pregnancy [66]. Chu et al. (2011) in their study enquire the transfer of Quantum Dots (QDs) from pregnant mice to foetuses and support that those nanoparticles can be transferred through the placenta blood barrier even if they are capped with organic polyethylene glycol or inorganic silica shell, suggesting that the therapeutical treatment availability of QDs have to be limited in pregnancy [67]. A future perspective will be to introduce new drug molecules to nanoparticles, which will behave as vehicles to embryonic/foetal affected organs without any harmful effects to the maternal organism.

7.5 Conclusions

Nanoparticles' coating for new drugs' target molecules, scheduled to cure embryonic malformations, must show bioavailability as they have to pass the placenta barrier and concentrate on the target: damaged embryonic cells and organs. The goal to exploit the positive aspects of engineered nanomaterials and avoid their potential toxic effects can best be achieved through a multidisciplinary team effort involving researchers in toxicology, materials science, medicine, molecular biology, bioinformatics, and their subspecialties.

Among this multidisciplinary group of researchers, embryologists and teratologists are one of the most challenging new groups to be established and are immediately included in the other teams in order to be able to produce compatible nanocarrier systems. "NANO-REPROTOX" research is a very promising and challenging target research area for reproductive toxicology, birth defects and neonatology diseases in the future.

Researchers across the world would like to determine how, in the future, nanoparticles might be used for therapeutic purposes. Their target research topic is to confirm how those nanoparticles could be essentially margined to be used as a coating vehicle for transporting drugs to the circulatory system of an embryo/foetus via the circulatory system of the placenta barrier, without inducing any noxious effects to the maternal organism. By these manipulations, application of new target drugs, which will be incorporated in nanoparticles to treat maternal and embryonic/foetal diseases, can be directed to cross the placenta and treat therapeutically mothers and embryos/foetuses . This research object creates a weapon to introduce new trends in obstetrics-gynaecology and to 'in utero' foetal microsurgery and rehabilitation.

Research, therefore, must focus on the development of biocompatible nanoparticles, which will be charged or incorporate new drugs and which can cross the placenta barrier without damaging it and without causing any danger to the embryonic/foetal tissues targeting for treatment of diseases. Such type of diseases could be Central Neural System (CNS) birth defects and those can be rehabilitated. This rehabilitation can be achieved by pharmaceutical treatment and 'in utero' embryonic surgery rehabilitation and stabilisation of results. Additionally to those charged with pharmaceutical substances nanoparticles application, or targeted with charged nanoparticles gene therapy to extirpate future generation epigenetic diseases especially for 'in vitro' fertilisation applications, seems to be the optimal future application formula for nanoparticles in embryology.

References

1. Maynard, A. D., Baron, P. A., Foley, M., Shvedova, A. A., Kisin, E. R., and Castranova, V. (2004) Exposure to carbon nanotube material: aerosol release during the handling of unrefined single-walled carbon nanotube material, *J. Toxicol. Environ. Health A*, **67**(1), 87–107.
2. Oberdorster, G., Maynard, A., Donaldson, K., Castranova, V., Fitzpatrick, J., Ausman, K., Carter, J., Karn, B., Kreyling, W., Lai, D., Olin, S., Monteiro-Riviere, N., Warheit, D., and Yang, H. (2005) Principles for characterizing the potential human health effects from exposure to nanomaterials: elements of a screening strategy, *Part. Fibre Toxicol.* **2**(8).
3. Oberdörster, G., Oberdörster, E., and Oberdörster, J. (2005) Nanotoxicology: an emerging discipline evolving from studies of ultrafine particles, *Environ. Health Perspect.*, **113**(7), 823–839.
4. Karakoti, S. A., Hench, L. L., and Seal, S. (2006) The potential toxicity of nanomaterials—the role of surfaces, *J. Mater.*, **58**(7), 77.

5. Nel, A., Xia T., Mädler, L., and Li, N. (2006) Toxic potential of materials at the nanolevel, *Science*, **311**(5761), 622–627.

6. Isakovic, A., Markovic, Z., Todorovic-Markovic, B., Nikolic, N., Vranjes-Djuric, S., Mirkovic, M., Dramicanin, M., Harhaji, L., Raicevic, N., Nikolic, Z., and Trajkovic, V. (2006) Distinct cytotoxic mechanisms of pristine versus hydroxylated fullerene, *Toxicol. Sci.*, **91**(1), 173–183.

7. Seetharam, R. J., and Sridhar, K. R. (2007) Nanotoxicity: threat posed by nanoparticles, *Curr. Sci. India*, **93**(6), 769–770.

8. Meng, H., Xia, T., George, S., and Nel, A. E. (2009) A predictive toxicological paradigm for the safety assessment of nanomaterials, *ACS Nano*, **3**(7), 1620–1627.

9. Hougaard K. S., Fadeel B., Gulumian M., Kagan V. E., and Savolainen, K. M. (2011) Developmental toxicity of engineered nanoparticles, in *Reproductive and Developmental Toxicology* (Gupta R. C., ed.), Elsevier, Academic Press, London, pp. 269–290.

10. Rao K. M., Porter D. W., Meighan T., and Castranova, V. (2004) The sources of inflammatory mediators in the lung after silica exposure, *Environ. Health Perspect.*, **112**(17), 1679–1686.

11. Hoet, P. H., Brüske-Hohlfeld, I., and Salata, O. V. (2004) Nanoparticles - known and unknown health risks, *J. Nanobiotechnology*, **2**(1), 12.

12. Meng, H., Chen, Z., Xing, G., Yuan, H., Chen, C., Zhao, F., Zhang, C., and Zhao, Y. (2007) Ultrahigh reactivity provokes nanotoxicity: explanation of oral toxicity of nano-copper particles, *Toxicol. Lett.*, **175**(1–3), 102–110.

13. Fischer, H. C., and Chan, W. C. (2007) Nanotoxicity: the growing need for in vivo study, *Curr. Opin. Biotechnol.*, **18**(6), 565–571.

14. Capek, I. (2011) Dispersions based on noble metal nanoparticles-DNA conjugates, *Adv. Colloid. Interface Sci.*, **163**(2), 123–143.

15. Makaronidis, I., Nikoloussis, E., Papamitsou, T., Sioga, A., Kaidoglou K, . and Emmanouil-Nikoloussi, E. N.(2012) Ultrastructural observations in neonatal micelungs after "in vivo" exposure to cadmium chloride-1-hydrate during gestation. *Preliminary study. RTX Journal*, **34**(2), 164–165.

16. Maynard, A. D. (2006) Safe handling of Nanotechnology, *Nature*, **444**(7117), 267–269.

17. Jan, E., Byrne, S. J., Cuddihy, M., Davies, A. M., Volkov, Y., Gun'ko Y. K., and Kotov, N. A. (2008) High-content screening as a universal tool for fingerprinting of cytotoxicity of nanoparticles, *ACS Nano*, **2**(5), 928–938.

18. Kroll, A., Pillukat, M. H., Hahn, D., and Schnekenburger, J. (2009) Current in vitro methods in nanoparticle risk assessment: limitations and challenges, *Eur. J. Pharm. Biopharm.*, **72**(2), 370–377.

19. Walker, N. J., and Bucher, J. R. (2009) A 21st century paradigm for evaluating the health hazards of nanoscale materials? *Toxicol. Sci.* **110**(2), 251–254.

20. Park, M. V., Lankveld, D. P., van Loveren, H., and de Jong, W. H. (2009) The status of in vitro toxicity studies in the risk assessment of nanomaterials, *Nanomedicine (Lond),* **4**(6), 669–685.

21. Lee, K. P., Kelly, D. P., Schneider, P. W., and Trochimowicz, H. J. (1986) Inhalation toxicity study on rats exposed to titanium tetrachloride atmospheric hydrolysis products for two years, *Toxicol. Appl. Pharmacol.* **83**(1), 30–45.

22. Dawson, K. A., Salvati, A., and Lynch, I. (2009) Nanotoxicology: nanoparticles reconstruct lipids, *Nat.Nanotechnol.* **4**(2), 84–85.

23. Juch, H., Krassing, S., Gauster, M., Holzapfel-Bauer, M., and Dohr, G. (2011) Nanotoxicity to the placenta, *Reprod. Toxicol.*, **31**(2), 260–261.

24. Asharani, P. V., Lianwu, Y., Gong, Z., and Valiyaveettil, S. (2011) Comparison of the toxicity of silver, gold and platinum nanoparticles in developing zebrafish embryos, *Nanotoxicology*, **5**(1), 43–54.

25. Logothetidis, S. (2006) Nanotechnology in medicine: the medicine of tomorrow and nanomedicine, *Hippokratia* 10(10), 7–21.

26. Emmanouil-Nikoloussi, E. N., Goret-Nicaise, M., Foroglou, C., Katsarma, E., Dhem, A., Dourov, N., Persaud, T.,V., and Thliveris, J. A. (2000) Craniofacial abnormalities induced by retinoic acid: A preliminary histological and scanning electron microscopic (SEM) study, *Exp. Toxicol. Pathol.*, **52**(5), 445–453.

27. Emmanouil-Nikoloussi, E. N., Goret-Nicaise, M., Foroglou, P., Kerameos-Foroglou, C., Persaud, T. V., Thliveris, J. A.,and Dhem, A. (2000) Histological observations of palatal malformations in rat embryos induced by retinoic acid treatment, *Exp. Toxicol. Pathol.*, **52**(5), 437–444.

28. Emmanouil-Nikoloussi, E. N., Goret-Nicaise, M., Kerameos-Foroglou, C., and Dhem, A. (2000) Anterior neural tube malformations induced after all-trans retinoic acid administration in white rats embryos. Macroscopical observations, *Morphologie*, **84**(264), 5–11.

29. Emmanouil-Nikoloussi, E. N., Goret-Nicaise, M., Manthos, A., and Foroglou, C. (2003) Histological study of anopthalmia observed in exencephalic rat embryos after all-trans-retinoic acid administration. *J. Toxicol.-Cutan. Ocul.*, **22**(1and 2), 33–46.

30. Emmanouil-Nikoloussi, E. N., Foroglou, N. G., Kerameos-Foroglou, C., and Thliveris, J. A. (2004) Effect of valproic acid on fetal and maternal organs in the mouse. A morphological study, *Morphologie*, **88**(180), 41–45.

31. Emmanouil-Nikoloussi, E. N., Frangou-Massourides, H., Nikolussis, M., Massouridou, S., Szabova, E., Zelzenkova, D., and Navarova, J. (2005) Apoptotic effect of 13-cis retinoic acid and hydroxyurea on caspase-3 activity in mice placenta: an early marker of congenital malformations, *Slov. Anthropol.*, **8**(1), 129–132.

32. Emmanouil-NIkoloussi, E. N. Goret-Nicaise, M., Foroglou, P., Thliveris, J. A., and Kerameos-Foroglou, C. (2008) All-trans-retinoic acid-induced disturbance of forelimb digital apoptosis in mice embryos: a preliminary scanning electron microscope (SEM) study, *Eur. J. Anat.*, **12**(1), 25–32.

33. Emmanouil-Nikoloussi, E. N., Goula, O. C., Manthou, M. E., Nikoloussis, E., Likartsis, C., Massouridou, S., and Frangou, H. (2009) Ultrastructural study of hepatic stellate cells (Ito cells) after teratogenic drug treatment in pregnancy: experimental study in balb/c mice. *Aristotle Univ. Med. J.*, **36**(3), 19–25.

34. Emmanouil-NIkoloussi, E. N., Nikoloussis, E., Manthou, M. E., Goula, O. C., Likartsis, C., Papamitsou, T., Frangou, H., Massouridou, S., Lazaridis, C., and Manthos, A. (2010) Breast tumor developed in a pregnant rat after treatment with the teratogen cycloheximide, *Hippokratia*, **14**(2), 136–138.

35. Gunston, E., Emmanouil-Nikoloussi, E. N., and Moxham, B. J. (2005) Palatal abnormalities in the developing rat induced by retinoic acid, *Eur. J. Anat.*, **9**(1), 1–16.

36. Wise, M., Emmanouil-Nikoloussi, E. N., and Moxham, B. J. (2007) Histological examination of major craniofacial abnormalities produced in rat fetuses with a variety of retinoids, *Eur. J. Anat.*, **11**(1), 17–26.

37. Ritsardson, L., Emmanouil-Nikoloussi, E. N., and Moxham, B. J. (2009) Calcium folinate diminishes the teratogenic effects of all-trans retinoic acid in the developing craniofacial region and neural tube of the rat, *Eur. J. Anat.*, **13**(2), 49–66.

38. Zelzenkova, D., Szabova, E., Emmanouil-Nikoloussi, E. N., Navarova, J., Ujhazy, E. and Nikoloussis, E. (2005) Speciment banking for biomedical research, *Slov. Anthropol.*, **8**(1), 195–197.

39. Mirilas, P., Mentessidou, A., Kontis, E., Asimakidou, M., Moxham, B. J., Petropoulos, A. S., and Emmanouil-Nikoloussi, E. N. (2011). Parental

exposures and risk of nonsyndromic orofacial clefts in offspring: a case-control study in Greece. *Int. J. Pediatr. Otorhinolaryngol.*, **75**(5), 695–699.

40. Nday, C. M., Georgiou, A., Emmanouil-NIkoloussi, E. N., Frangou-Massourides, E., Papamitsou, T., and Salifoglou, T (2012) In vivo acute pleiotropic toxicity of well-defined binary Cd(II)- and V(V)-citrate compounds on mice (Submitted for publication).

41. Buzea, C., Pacheco, I. I., and Robbie, K. (2007) Nanomaterials and nanoparticles: Sources and toxicity, *Biointerphases* **2**(4), MR17–MR71.

42. Warheit, D. B., and Donner, E. M. (2010) Rationale of genotoxicity testing of nanomaterials: regulatory requirements and appropriateness of available OECD test guidelines. *Nanotoxicology*. **4**, 409–413.

43. Britton, R. S. (1996) Metal-induced hepatotoxicity, *Semin. Liver Dis.*, **16**(1), 3–12.

44. Deshpande, A., Narayanan, P. K., and Lehnert, B. E. (2002) Silica-induced generation of extracellular factor(s) increases reactive oxygen species in human bronchial epithelial cells, *Toxicol. Sci.*, **67**(2), 275–283.

45. Colvin, V. L. (2003) The potential environmental effect of nanomaterials, *Nat. Biotechnol.*, **21**(10), 1166–1170.

46. Belyanskaya, L., Manser, P., Spohn, P., Bruinink, A., Wick, P. (2007) The reliability and limits of the MTT reduction assay for carbon nanotubes–cell interaction, *Carbon* **45**(13), 2643–2648.

47. Valko, M., Morris, H., and Cronin, M. T. (2005) Metals, toxicity and oxidative stress, *Curr. Med. Chem.*, **12**(10), 1161–1208.

48. Hirst, S. M., Karakoti, A. S., Tyler, R. D., Sriranganathan, N., Seal, S., and Reilly, C. M. (2009) Anti-inflammatory properties of cerium oxide nanoparticles, *Small*, **5**(24), 2848–2856.

49. Kong, L., Cai, X., Zhou, X., Wong, L. L., Karakoti, A. S., Seal, S., and McGinnis, J. F. (2011) Nanoceria extend photoreceptor cell lifespan in tubby mice by modulation of apoptosis/survival signaling pathways, *Neurobiol. Dis.*, **42**(3), 514–523.

50. Spohn, P., Hirsch, C., Hasler, F., Bruinink, A., Krug, H. F., and Wick, P. (2009) C60 fullerene: A powerful anti oxidant or a damaging agent? The importance of an in-depth material characterization prior to toxicity assays, *Environ. Pollut.*, **157**(4), 1134–1139.

51. Karakoti, A., Singh, S., Dowding, J. M., Seal, S., and Self, W. T. (2010) Redox-active radical scavenging nanomaterials, *Chem. Soc. Rev.*, **39**(11), 4422–4432.

52. Estevez, A. Y., Pritchard, S., Harper, K., Aston, J. W., Lynch, A., Lucky, J. J., Ludington, J. S., Chatani, P., Mosenthal, W. P., Leiter, J. C., Andreescu, S., and Erlichman, J. S. (2011) Neuroprotective mechanisms of cerium oxide nanoparticles in a mouse hippocampal brain slice model of ischemia, *Free Radic. Biol. Med.*, **51**(6), 1155–1163.

53. Hirst, S. M., Karakoti, A., Singh, S., Self, W., Tyler, R., Seal, S., and Reilly, C. M. (2011) Bio-distribution and in vivo antioxidant effects of cerium oxide nanoparticles in mice, *Environ. Toxicol.*, 2011 May 26 (Epub ahead of print, doi: 10.1002/tox.20704).

54. Celardo, I., Traversa, E., and Ghibelli, L. (2011) Cerium oxide nanoparticles: a promise for applications in therapy, *J. Exp. Ther. Oncol.*, **9**(1), 47–51.

55. Alili, L., Sack, M., Karakoti, A. S., Teuber, S., Puschmann, K., Hirst, S. M., Reilly, C. M., Zanger, K., Stahl, W., Das, S., Seal, S., and Brenneisen, P. (2011) Combined cytotoxic and anti-invasive properties of redox-active nanoparticles in tumor-stroma interactions, *Biomaterials,* **32**(11), 2918–2929.

56. Fako, V. E., and Furgeson, D. Y. (2009) Zebrafish as a correlative and predictive model for assessing biomaterial nanotoxicity, *Adv. Drug Deliver. Rev.*, **61**(6), 478–486.

57. Bar-Ilan, O., Albrecht, R. M., Fako, V. E., and Furgeson, D. Y. (2009) Toxicity assessments of multisized gold and silver nanoparticles in zebrafish embryos, *Small*, **5**(16), 1897–1910.

58. Truong, L., Zaikova, T., Richman, E. K., Hutchison, J. E., and Tanguay, R. L. (2011) Media ionic strength impacts embryonic responses to engineered nanoparticle exposure, *Nanotoxicology.* 2011 August 2 (Epub ahead of print, doi: 10.3109/17435390.2011.604440).

59. Truong, L., Moody, I. S., Stankus, D. P., Nason, J. A., Lonergan, M. C., and Tanguay, R. L. (2011) Differential stability of lead sulfide nanoparticles influences biological responses in embryonic zebrafish, *Arch. Toxicol.*, **85**(7), 787–798.

60. Ahamed, M., Alsalhi, M. S., and Siddiqui, M. K. (2010) Silver nanoparticle applications and human health, *Clin. Chim. Acta,* **411**(23–24), 1841–1848.

61. Lee, K. J., Nallathamby, P. D., Browning, L. M., Osgood C. J., and Xu, X. H. (2007) In vivo imaging of transport and biocompatibility of single silver nanoparticles in early development of zebrafish embryos, *ACS Nano.,* **1**(2), 133–143.

62. Browning, L. M., Lee, K. J., Huang, T., Nallathamby, P. D., Lowman, J. E., and Xu, X. H. (2009) Random walk of single gold nanoparticles in

zebrafish embryos leading to stochastic toxic effects on embryonic developments, *Nanoscale,* **1**(1), 138–152.

63. Menezes, V., Malek, A., and Keelan, J. A. (2011) Nanoparticulate drug delivery in pregnancy: placental passage and fetal exposure, *Curr. Pharm. Biotechnol.*, **12**(5), 731–742.

64. Myllynen, P. K., Loughran, M. J., Howard, C. V., Sormunen, R., Walsh, A. A., and Vähäkangas, K. H. (2008) Kinetics of gold nanoparticles in the human placenta, *Reprod Toxicol.*, **26**(2), 130–137.

65. Wick, P., Malek, A., Manser, P., Meili, D., Maeder-Althaus, X., Diener, L., Diener, P. A., Zisch, A., Krug, H. F., and von Mandach, U. (2010) Barrier capacity of human placenta for nanosized materials, *Environ. Health Perspect.*, **118**(3), 432–436.

66. Menjoge, A. R., Rinderknecht, A. L., Navath, R. S., Faridnia, M., Kim, C. J., Romero, R., Miller, R. K., and Kannan, R. M. (2011) Transfer of PAMAM dendrimers across human placenta: prospects of its use as drug carrier during pregnancy, *J. Control Release*, **150**(3), 326–338.

67. Chu, M., Wu, Q., Yang, H., Yuan, R., Hou, S., Yang, Y., Zou, Y., Xu, S., Xu, K., Ji, A., and Sheng, L. 2010 Transfer of quantum dots from pregnant mice to pups across the placental barrier, *Small*, **6**(5), 670–678.

Chapter 8

Nanomedicine and HIV/AIDS

Enikő R. Tőke and Julianna Lisziewicz
Emmunity Inc., 4400 East West Hwy, Bethesda, MD 20814, USA
julianna.lisziewicz@emmunityinc.com

8.1 Introduction

In 1981, the emergence of AIDS was first reported, followed by the identification of HIV as the cause of the disease in 1983.[1–4] Since then, HIV/AIDS has become a global pandemic, the leading infectious killer, affecting more than 33 million people worldwide.[5] HIV infects immune cells, i.e., T-cells, macrophages and dendritic cells[6] and spreads by the transfer of body fluids, for example by injections with contaminated needles, sexual contact and from mother to child.

In 1987, an early movement was already appealing for a cure for HIV/AIDS that spurred expedited approval by the US Food and Drug Administration (FDA) of the first antiretroviral drug—Zidovudine (AZT)—and made new investigational drugs available for desperately ill patients early in the drug development process.[7,8] To date, HIV/AIDS is treated with the combination of 25 antiretroviral drugs (ARV) that are divided into six classes according to their interference

Horizons in Clinical Nanomedicine
Edited by Varvara Karagkiozaki and Stergios Logothetidis

ISBN 978-981-4411-56-1 (Hardcover), 978-981-4411-57-8 (eBook)
www.panstanford.com

with the HIV life-cycle: fusion/entry inhibitors, integrase inhibitors, protease inhibitors, non-nucleoside reverse trancriptase inhibitors, nucleoside analog reverse transcriptase inhibitors and multidrug combination products. HIV/AIDS treatment using any single class of ARV has not been efficient in controlling infection due to the development of resistant strains of the virus. Hence, three or more ARVs are used in combination (combination antiretroviral therapy, cART) to treat the disease.

cART has been effective in decreasing morbidity and mortality associated with HIV infection.[9] Currently available cART are efficacious in providing protection against AIDS with only minimum side effects. Key products used for the treatment of the majority of patients such as Atripla (efavirenz/tenofovir/emtricitabine), Truvada (tenofovir/emtricitabine), Sustiva (efavirenz), Kaletra (lopinavir/ritonavir), Reyataz (atazanavir) and Isentress (raltegravir) satisfy the ARV drug demand in the market.

Successful management of HIV-infected patients is challenging, requiring highly experienced physicians due to resistance and overlapping toxicities of the complex daily ARV regimen that needs to be taken life-long. However, even optimal cART, characterised by suppression of viral load to undetectable levels for years, has not provided a cure to the disease. Patients on optimal cART have 12 years shorter life expectancy than HIV negative people.[10,11] In addition, increased AIDS-related and non-AIDS-related morbidity and mortality has been described in a significant proportion of individuals on optimal cART because the lack of normalisation of their $CD4^+$ T cell counts.[12] Optimal cART failed to decrease the viral reservoirs, especially in the gut mucosa, where the residual low-level viral replication may be the cause of persistent immune activation that facilitates the progression to AIDS and death.[13] HIV producing cells in the reservoirs that are not eliminated by ARV drugs would be susceptible to immune clearance, but long-term optimal cART diminishes HIV-specific T cell responses.[14] Therefore, the immune system of successfully treated HIV-infected people is not prepared to decrease viral reservoirs and control the virus replication, if ARVs are irregularly taken.

Presently, there are two major unmet needs in HIV treatment: (i) The development of single-tablet regimens that are effective, safe and well tolerated, but do not aim to cure the disease. This approach, pursued by large pharmaceutical companies, utilises conventional drug formulation; (ii) Nanomaterials are exploited in the development of highly innovative immunotherapies aiming to stop disease progression and cure HIV/AIDS.

Nanomedicine stands for the application of nanomaterials in medical technology. The European Commission has adopted cross-cutting definition of nanomaterials to be used for regulatory purposes as being 50% or more of the particles with 1–100 nm size in at least one dimension.[15] Structures measuring several hundred nanometers or a few micrometers, however, are also considered under nanotechnology applications.[16]

Here we review novel disease-modifying treatment approaches that exploit nanomedicine for either drug delivery or induction of HIV-specific immunity with therapeutic vaccines.

8.2 Nanomedicine in Antiretroviral Drug Development

Application of nanotechnology to antiretroviral drug delivery holds promise in the treatment of HIV/AIDS, because it could modify tissue distribution by targeting drugs to HIV reservoirs and by increasing the half-lives of drugs. The use of nano-delivery systems has been extensively reviewed previously[17–19]; therefore, we only highlight some of the recent advances in the field that could play a role in achieving a cure.

To target antiretroviral drugs to the lymphoid organs, a pH-dependent nanomedicine formulation of indinavir was investigated in macaques,[20] This nanomedicine formulation increased indinavir concentration in the lymph nodes.[21] These 50–80 nm nanoparticles were trapped in lymph nodes as they circulated through the lymphatic system. The authors concluded that the targeting effect of nanoparticles to the lymphoid tissues was mainly particle size dependent.

Improvement of bioavailability is a major challenge for oral drug development. For example, protease inhibitors indinavir, ritonavir and navirapine have different, but still acceptable oral bioavailability of 39%, 60–70% and 92%, respectively. In contrast, saquinavir only possesses

4% of bioavailability that influenced the efficacy and toxicity of this drug and the company decided to withdraw the drug from the market after obtaining marketing authorisation.[7] To improve bioavailability, nanomedicine formulation of efavirenz (EFV) was investigated. The drug was incorporated into the core of linear and branched poly(ethylene oxide)–poly(propylene oxide) block copolymer micelles. In rats, EFV-nanomedicine had up to 88% higher plasma concentrations compared to EFV suspensions.[22]

ARV resistance occurs often in case of non-adherence or when the viral load is insufficiently suppressed. The presence of resistant HIV then limits the treatment options of patients. To improve adherence with longer dosing intervals, nanomedicine formulation of rilpivirine was investigated in rats and dogs. This nanomedicine demonstrated a sustained and dose-proportional release over two months and a significant half-life enhancement compared to the 38 h of the free drug.[23]

One promising strategy for decreasing the use of ARVs would be the prevention of vaginal and rectal HIV transmission by topical microbicides. Proof of concept has been recently achieved in a Phase II clinical trial using a vaginal gel containing an ARV drug, tenofovir. This hallmark study demonstrated partial protection, suggesting that achieving sustainable concentrations of an ARV at the genital mucosal tissue is a crucial step toward efficacy.[24] The perspective of using nanotechnology for microbicides could be significant because nanomedicine formulation could enhance epithelial penetration and improve targeting without increasing the toxicity.

To increase the epithelial penetration of antiretroviral-based microbicides specific mucoadhesive and non-mucoadhesive nanomedicine formulations were investigated.[25] The interaction of the nanomedicine with the mucus fluids covering the vaginal mucosal epithelium can work either as docking point or as barrier for diffusion. The combination of size and surface properties substantially influences this interaction. The surface chemistry of the nanomedicine determines attraction/repulsion with mucin fibres, whereas the diameter controls their ability to 'fit' within the mucin mesh pores. In particular, positively charged polymers, such as chitosan and derivatives, seem to substantially increase mucoadhesion by means of ionic interaction with negatively charged mucin chains. Furthermore, chitosan modification with thiol side groups was shown to enhance mucoadhesion of nanoparticles when compared with systems that were surface-functionalised with nonmodified chitosan.[26] The use of chitosan, however, is a concern in microbicides, because its permeation enhancement properties could stimulate the mucosal translocation of HIV and other viruses.[25]

Some of the nanomedicines might have an intrinsic activity to inhibit viral entry. For example, either empty or drug-loaded (2-RANTES, MC1220) liposomes showed partial protection against infection in a macaque model after vaginal instillation.[27,28] Liposomes are lipid vesicles with an aqueous core used to encapsulate hydrophilic drugs, while hydrophobic and amphiphilic drugs can be solubilised within the phospholipid bilayers. Liposomes are phagocytosed by macrophages and quickly cleared from the circulation after uptake by the reticuloendothelial system. Various liposomal formulations are already commercially available. Both polyvinyl pyrrolidone-coated silver nanoparticles and mannose-coated gold nanoparticles inhibited the interaction of HIV envelop (gp120) with different mucosal or epithelial cells, therefore inhibiting the fusion or entry of HIV into the host cell.[29,30] Nanomedicine consisting of L-Lysine dendrimers formulated as a carbomer gel (VivaGel®, Starpharma) and applied as topical microbicide in female macaques showed dose-dependent resistance to viral challenge.[31] The mechanism of action is based on the surface chemistry of the nanomedicine: the polylysine branches of the dendrimer are terminally derivatised with naphthalene disulphonate groups that are responsible for the direct interaction with HIV envelope glycoproteins.[32] Safety of Vivagel (SPL7013) has been demonstrated in human subjects (Phase I).[32]

8.3 Nanomedicine in Vaccine Development

Immunotherapeutic nanomedicines are new, complex, multi-modular vaccines that provide superior therapeutic effects compared to all previous approaches. Their physical size is usually over 50 nm, the approximate threshold of immune recognition.[33] In fact, the size range of immunotherapeutic nanomedicine corresponds to the size of pathogenic viruses. Against viruses, nature developed a sensitive and specific immune surveyance. Consequently, triggering the immune system by nanomedicines provides unprecedented immunogenicity, since our body considers nanomedicine as a harmful virus that needs to be eliminated.[34] The best examples for the superior immune recognition of nanomedicine are the human papillomavirus (HPV) vaccines, Gardasil and Cervarix. These vaccines composed from one surface protein of the HPV, L1, which self-assembles to a virus-like particle (VLP, a nanomedicine). These VLPs, morphologically similar to the wild type HPV, induce potent immune responses in the absence of adjuvants. In contrast, the L1 protein purified from bacteria remains soluble, does not assemble to a VLP, and does not induce immune responses.[35] These VLP vaccines are safe and effective in young uninfected people; none of them developed thus far have proven to be useful as therapeutic vaccines since they unsuccessfully induce therapeutically beneficial T cell-mediated immune responses.[36]

Creating "particulate vaccines" has recently been recognised in the HIV field to improve the immunogenicity of small soluble antigens. This approach involves increasing the physical size of the antigen to the size of pathogens. There are so-called "natural" particulate vaccines, based on VLPs in the 40 nm range that induce both humoral and cellular immune responses against HIV.[37,38] HIV VLPs are essentially non-infective viruses consisting of self-assembled viral envelope proteins without the accompanying genetic material. A different approach is to use an adjuvant that increases the size of the antigen. One of the several proposed mechanisms of aluminium salts, the adjuvant approved in the US and EU, is attributed to their particulate nature, although concerns have been raised recently regarding their safety.[38]

An alternative approach is the use of a plasmid DNA (pDNA) that can express one or more antigens in the body. These intracellularly expressed protein antigens are processed and presented on the host MHC molecules. These features, combined with recently improved large-scale manufacturing capabilities, make pDNA attractive for immunotherapeutic nanomedicine development. pDNA has an excellent safety profile because it does not replicate or cause disease and the cells expressing the pDNA-encoded antigens are eliminated by the immune responses. Unfortunately, the very promising animal studies demonstrating the induction of immune responses with naked pDNA injected intramuscularly or intradermally were not reproduced in human subjects. One of the reasons of the weak immunogenicity is that the expressed soluble antigens were not recognised by the immune system, similarly to the previously described soluble L1 protein of the HPV. Soluble antigens much less than 50 nm in size are generally not recognised by the immune system as particles and are not immunogenic.

Various biodegradable and non-biodegradable polymeric and liposomal systems have been explored for transforming HIV-antigens to synthetic nanoparticles in order to increase their immunogenicity and to protect them against extra- and intracellular degradation. Targeting dendritic cells (DC) that are essential for initiating immune responses, can be achieved by different nanomedicine sizes; >100 nm nanomedicine targets the peripheral immature DCs, and the smaller size ~20–50 nm nanomedicine drain to the lymph node resident DCs.[39] Modification on the surface of the nanomedicine with DC-specific receptor ligands has been shown to increase the targeting specificity.[40] Several challenges, including crossing physical barriers like the cell and nuclear membranes or adhesion to non-target tissues, however, still need to be overcome during the development of a synthetic delivery system. Table 8.1 compares HIV-specific immunotherapeutic nanomedicines that are most advanced in development.

Table 8.1 Features of the advanced HIV/AIDS immunotherapeutic nanomedicines

Features	**DermaVir pathogen-like nanoparticle**[41]	**Gold/DNA microparticle**[42]	**PLA/p24 nanoparticle**[43]
Development stage	Phase II	Preclinical	Preclinical
API	pDNA	pDNA	p24 protein
Delivery system	pDNA/PEIm nanoparticle	pDNA adsorbed on the surface of gold particle	p24 adsorbed on the surface of PLA nanoparticle
Particle	70–300 nm (virus size)	1–3 μm (bacteria size)	~500 nm (virus size)
Administration	Topical, DermaPrep (targeting LC)	Gene gun to skin (targeting LC and keratinocytes)	Injection of autologous DC
DC targeting element	Sugar residues	No targeting element	Co-cultured with DCs
Cellular entry	Endocytosis (virus-like)	"Forced" entry	Endocytosis by cultured DCs
Post-entry	Nuclear targeting of pDNA	Not regulated	N/A

API, active pharmaceutical ingredient; PEIm, mannosylated polyethyleneimine; PLA, poly(D,L-lactic acid); DC, dendritic cells; LC, Langerhans cells, precursors of DCs.

DermaVir is the first nanomedicine for HIV/AIDS that has reached encouraging Phase II clinical safety, immunogenicity and efficacy results.[44] We provide here a detailed overview on the development of the technology of DermaVir, because it represents a new immunotherapy platform with application against a wide range of pathogens and chronic conditions. In the following sections, we use the DermaVir example to introduce the elements of this nanomedicine platform, describing its major features and emphasising on the essential role of nanotechnology.

8.4 DermaVir Clinical Nanomedicine Product Candidate for HIV Immunotherapy

DermaVir is a novel nanomedicine. It features three soft-particle nano-elements, according to the classification proposed by Tomalia[45] (Fig. 8.1): The active pharmaceutical ingredient (API) is a pDNA (S-6) that expresses 15 HIV proteins, these assemble to a complex virus-like particle (S-5).[48] To deliver the pDNA to dendritic cells and achieve effective protein expression the pDNA is condensed and packaged into a mannosylated polyethylenimine "envelop" (S-3).[46,47] Here, we present a rational, target product profile-oriented design of DermaVir nanomedicine, including the importance of detailed physico-chemical analysis of the components and their effect on the product quality.

The objective of the antigen design was to preserve the structure and the epitope content of the wild-type virus and to create a safe immunogen. We constructed a single pDNA to drive the expression of fifteen HIV proteins in a cell. These proteins self-assemble to replication-, reverse transcription- and integration—defective complex viruses like particles (VLP^+).[48] This pDNA is inherently safe because the molecular modifications prevent the replication and integration of the VLP^+. The expression of 15 HIV proteins from the single pDNA supports the presentation of the highest number of HIV epitopes and the induction of HIV-specific T cell responses with the broadest specificity.

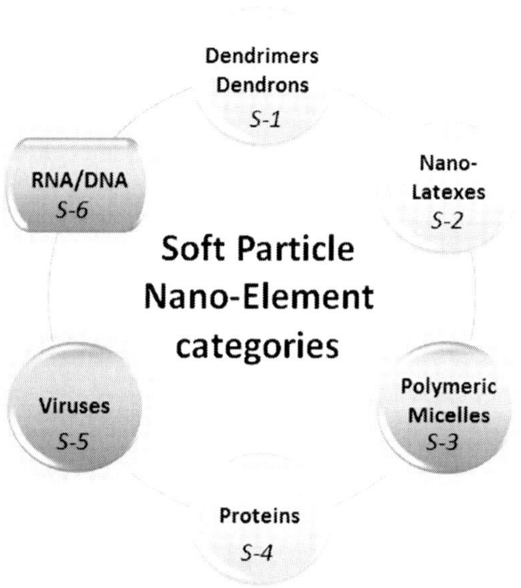

Figure 8.1 Proposed soft particle nanoelement categories[45]: According to Critical Nanoscale Design Parameters (CNDPs) such as (a) size, (b) shape, (c) surface chemistry, (d) flexibility, and (e) elemental composition, nanomaterials can be classified as either hard (H), i.e. inorganic-based, or soft (S), i.e., organic-based, particles. Colored shapes: soft-particle nanoelements included in the DermaVir nanomedicine.

The next objective was to express pDNA-encoded antigens in DCs. pDNA delivery to DCs is a complex challenge involving DC binding, antigen uptake, expression, processing and presentation to naïve T-cells.[49] We designed a "pathogen-like" nanomedicine in order to encapsulate the pDNA within the positively charged linear mannosylated polyethylenimine (PEIm). The DermaVir nanomedicine is similar in size, appearance and DNA-delivery features to viruses that naturally evolved to deliver genetic materials to cells. DermaVir's PEIm "envelop" protects the condensed pDNA "core" from extra- and intracellular degradations. Its particle size (70–300 nm) is optimal for receptor-mediated endocytosis into cells and it has sufficient stability to support the release of the pDNA from the endosomal compartment and the delivery of the pDNA to the nucleus. These features, unique to DermaVir nanomedicine, are essential for potent expression of antigens.[46,50] Consequently, the biological activity of DermaVir is dependent on its inherent structure and binding.[47,51,52] The endosomal escape is only part of the efficacy-limiting processes: for efficient gene expression, the nanomedicine must release the pDNA in the proximity of the nucleus. These two competing processes require an optimal degree of association between the pDNA and PEIm. (Fig. 8.2). Hyperchromicity (Hc), obtained from absorption spectroscopy measurements, controls this feature.[47] The ionic strength of the DermaVir solution also influences the binding affinity of the pDNA to the PEIm.

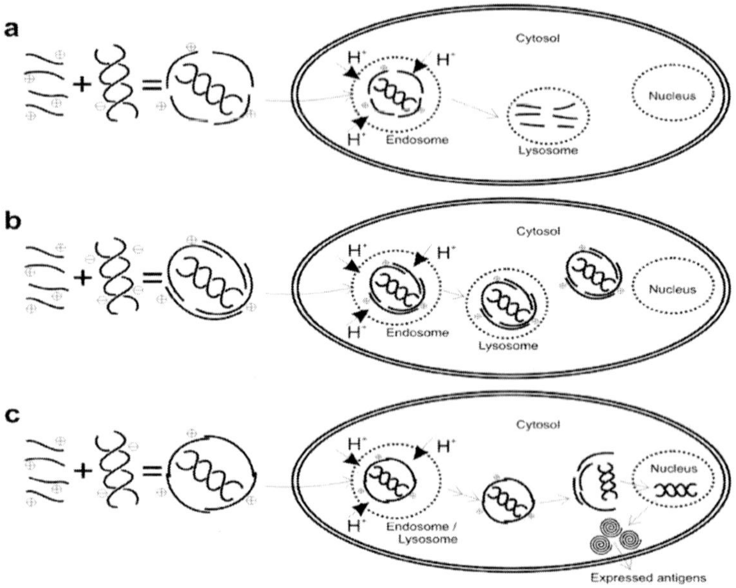

Figure 8.2 Effect of the degree of association between the pDNA and PEIm on gene expression: (a) When degree of association is too low, DermaVir disintegrates at low pH in the endosome. (b) When degree of association is too high, DermaVir survives the low pH in the endosome and in the lysosome, and remains in the cytosol; therefore the pDNA cannot reach the nucleus. (c) When degree of association is optimal, DermaVir escapes from endosomal degradation, releases the pDNA in the cytosol near the nucleus, where the pDNA-encoded antigens are expressed.

The third objective was to deliver DermaVir in vivo to the lymph node DCs. Therefore, we developed DermaPrep, the first DC-targeting administration device. DermaPrep employs a skin preparation method that interrupts the stratum corneum, facilitating nanomedicine penetration and providing the essential "danger" signal to the LCs residing just below this protective layer.[53,54] Once activated, LCs are naturally looking for pathogens and capturing the "pathogen-like" DermaVir nanomedicine applied to the prepared skin surface under a semi-occlusive patch. After DermaVir has been captured, LCs mature to DCs and migrate to the local lymph nodes. Here DCs express pDNA-encoded antigens and present most HIV epitopes to the passing naïve T-cells. HIV-specific precursor/memory T-cells primed by DCs further differentiate into HIV-specific effector T-cells, circulating out of the lymph node to seek virus-infected targets. Each killer effector cell can destroy several HIV-infected cells (Fig. 8.3).

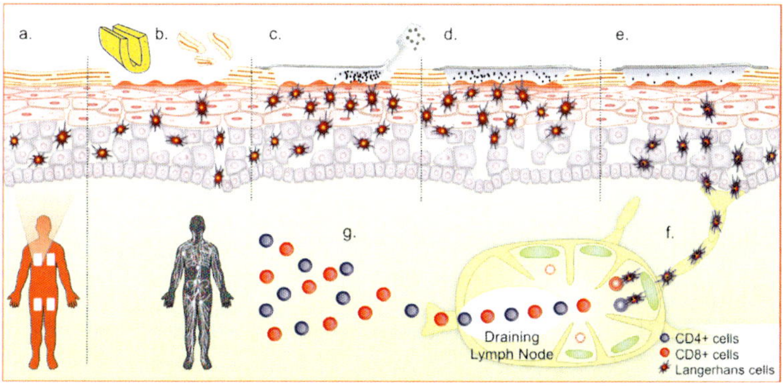

Figure 8.3 Mechanism of action of DermaVir immunotherapy: (a) Skin sites are selected; (b) skin preparation using DermaPrep; (c) patch applied and DermaVir administered for 3 h; (d) activated Langerhans cells (LC) capture DermaVir nanoparticles; (e) LCs migrate to lymph nodes, mature to DCs; (f) DCs present DermaVir-encoded epitopes to naïve $CD4^+$ and $CD8^+$ T-cells; (g) HIV-specific $CD4^+$ and $CD8^+$ T precursor/memory T-cells proliferate and search for HIV-infected cells throughout the body.

The main advantage of DermaPrep is the natural targeting of a large number of LCs (eight million) that form a horizontal 900 to 1,800 cells/mm^2 network under the skin surface.[55] These LCs then migrate to the draining lymph node and mature to antigen-presenting DCs, achieving the optimal in vivo expression of the antigens to DCs.

Prior to initiation of human trials with a new medicinal product, regulatory agencies require the demonstration of their quality, safety and efficacy in appropriate animal models. DermaVir was designed to express the pDNA-derived antigens specifically in the DCs of the lymph nodes to prime and boost HIV-specific T-cells in infected hosts. To demonstrate the mechanism of action first we established in human primary cells in vitro and in macaques ex vivo that DermaVir-expressing DCs prime naïve T-cells and induce both HIV-specific helper and cytotoxic T-cells.[56] Preclinical animal studies conducted in mice, rabbits and macaques consistently demonstrated that *topical* DermaVir immunisation targets and expresses the pDNA-encoded antigens in the DCs of the lymph nodes.

Proof of concept efficacy studies in chronically SIV_{251}-infected macaques, some of them with AIDS, suggested that repeated DermaVir immunisations, alone or in combination with antiretroviral drugs, result in viral load reduction and survival benefit.[57] DermaVir administered in combination with ART boosted SIV-specific T-cells that possessed significant antiretroviral activity in both chronically infected and late stage macaques. These primate experiments provided the rationale to investigate repeated DermaVir immunisations in combination with ART in HIV-infected human subjects.

The Phase I clinical study, conducted in Hungary, was designed to evaluate the safety and immunogenicity of a single DermaVir immunisation in HIV-infected subjects on fully suppressive combination antiretroviral therapy (cART). This first-in-human dose escalation study was conducted with three topical DermaVir doses: 0.1 mg DNA targeted to two, 0.4 mg and 0.8 mg DNA targeted to four lymph nodes. The 28-day study with a 48-week safety follow-up evaluated HIV-specific T cell responses. DermaVir-associated side effects were mild, transient and not dose-dependent. The striking result was the dose-dependent expansion of HIV-specific precursor/memory T-cells with high proliferation capacity.[58] In low, medium and high dose cohorts, this T cell population increased up to 1866-fold after four weeks, and up to 667-fold after one year compared to baseline.[59] These findings suggested that DermaVir could practically boost T cell responses to all the 15 HIV antigens expressed from the single DNA. We concluded that for durable immune reactivity, repeated DermaVir immunisation might be required since the frequency of DermaVir-boosted HIV-specific T-cells decreased during the 48-week follow up.

A Phase I/II clinical trial conducted by the AIDS Clinical Trials Group (ACTG) in several U.S. clinical centres used multiple administrations of escalating DermaVir doses (0.2 or 0.4 mg pDNA three times or 0.4 mg pDNA six times) or placebo on 26 HIV-infected adults receiving fully suppressive ART. The primary endpoint was any possibly or definitely vaccine-related grade ≥3 adverse event (AE) appearing up to 28 days after the final study vaccination. No primary safety endpoints or AE-related study treatment changes/discontinuations occurred. AE incidence was similar across groups.[60] Immunogenicity data demonstrated the boosting of HIV-specific memory/precursor T-cells. The optimal immunogenicity was induced in the 0.4 mg pDNA group.[61] This trial further confirmed the safety and immunogenicity of repeated DermaVir immunisations in combination with ART.

Repeated DermaVir immunisations in chronically SIV_{251}-infected macaques, in the absence of ART, transiently suppressed virus replication—leading to improvement of median survival time from 18 to 38 weeks, compared to no treatment.[57] These primate experiments provided the rationale to investigate repeated DermaVir immunisations prior to initiation of ART in HIV-infected individuals. As DermaVir immunisations in combination with ART did not show any product- or administration-related AE higher than grade 2, we developed a Phase II protocol to evaluate the safety and to test the immunogenicity and antiviral efficacy of repeated DermaVir immunisations. Thirty-six HIV-infected adults were randomised to receive one of three DermaVir doses (0.2, 0.4 or 0.8 mg pDNA) or placebo. The primary endpoint of the trial was safety at Week 24 and secondary endpoints were HIV RNA and immunogenicity.[61] No subject stopped vaccinations due to an AE and only one subject initiated ART. Only one Grade 2 AE occurred in the 0.2 mg DermaVir group judged to be possibly related to treatment. Based on secondary analyses, the 0.4 mg DermaVir dose was superior to the others. In this group, the HIV-specific memory/precursor T-cells measured by PHPC increased from 5,055 to 9,978 cells/million PBMC ($P = 0.07$) and the median log10 HIV-RNA decreased from 4.5 to 4.0, significantly different from the placebo ($P = 0.045$). Viral load suppression by DermaVir vaccinations occurs slowly, as predicted by its mechanism of action, similar to cancer vaccines.[62] Consistent with the primate results, DermaVir immunisations alone did not suppress viral load to an undetectable level and did not increase $CD4^+$ T cell counts. These results suggested that repeated DermaVir immunisation boosted HIV-specific precursor/memory T-cells and the improved immunity contributed to the preservation of the health of HIV-infected individuals. Larger and longer studies are required to demonstrate that boosting of HIV-specific memory/precursor T-cells slows disease progression in HIV-infected individuals.

8.5 Conclusion

As described above, nanotechnology offers opportunities to further improve drugs and microbicides and develop safe, immunogenic and effective immunotherapeutics towards the cure of HIV. The clinical feasibility and significance of such approaches, however, need to be put in perspective. The question is whether the potential benefit achieved by the increased efficacy would prevail over the risk of developing nanomedicine formulations for the treatment of HIV. We should consider that systemic administration of nanomedicine formulations for the treatment of cancer induced a new immunotoxicity that has not been observed with drugs.[34] Would the improvement in half-life or the increase of bioavailability of the ARV presently used justify a new potential toxicity? We should also consider the need of sophistication in manufacturing and quality control that is usually accompanied by more difficult scale-up and higher production costs resulting in an increase in price. Price of an HIV drug that needs to be taken life-long is an important challenge during marketing to both the patients and the insurance companies. Therefore, researchers who are working hard to solve scientific problems in antiviral nanomedicine development should consider the clinical and market feasibility—our biggest challenge when we want to meet our commitment to cure HIV. Therefore, our available resources should advance new technologies and treatment approaches that could satisfy the need to modify HIV disease and decrease antiretroviral drug exposure, while maintaining undetectable HIV-RNA levels.

References

1. Barré-Sinoussi, F., Chermann, J. C., Rey, F., Nugeyre, M. T., Chamaret, S., Gruest, J., Dauguet, C., Axler-Blin, C., Vézinet-Brun, F., Rouzioux, C., Rozenbaum, W., Montagnier, L., (1983) Isolation of a T-lymphotropic retrovirus from a patient at risk for acquired immune deficiency syndrome (AIDS), *Science*, 220, 868–871.
2. Blattner, W., Gallo, R. C., Temin, H. M. (1988) HIV causes AIDS, *Science*, 241(4865), 515–516.
3. Gallo, R. C. (2002) Historical essay. The early years of HIV/AIDS, *Science*, 298(5599), 1728–1730.
4. Gallo, R. C., Montagnier, L. (2003) The discovery of HIV as the cause of AIDS, *New Engl. J. Med.*, 394(24), 2283–2285.
5. UNAIDS *Report on the Global AIDS Epidemic 2010*, http://www.unaids.org/globalreport/Global_report.htm. Published November 2010 (accessed December 12, 2011).
6. Shattock, R. J., Moore, J. P. (2003) Inhibiting sexual transmission of HIV-1 infection, *Nat. Rev. Microbiol.*, 1(1), 25–34.
7. Ma, X., Wang, D., Wu, Y., Ho, R. J. Y., Jia, L., Guo, P., Hu, L., Xing, G., Zeng, Y., Liang, X.-J. (2010) AIDS treatment and novel anti-HIV compounds improved by nanotechnology, *The AAPS Journal*, 12(3), 272–278.
8. Gould, S. J. (1989) A potential paradigm shift in health care policy, *JPP&M*, **8**, 40–52.
9. Palella, F. J., Delaney, K. M., Moorman, A. C., Loveless, M. O., Fuhrer, J., Satten, G. A., Aschman, D. J., Holmberg, S. D., the HIV Outpatient Study investigators, (1998) Declining morbidity and mortality among patients with advanced human immunodeficiency virus infection, *N. Engl. J. Med.*, 338(13), 853–860.
10. Holtgrave, DR. (2005) Causes of the decline in AIDS deaths, United States, 1995–2002: Prevention, treatment or both? *Int. J. STD AIDS*, 16(12), 777–781.
11. Losina, E., Schackman, B. R., Sadownik, S. N., Gebo, K. A., Walensky, R. P., Chiosi, J. J., Weinstein, M. C., Hicks, P. C., Aaronson, W. H., Moore, R. D., Paltiel, A. D., Freedberg. K. A. (2009) Racial and sex disparities in life expectancy losses among HIV-infected persons in the United States: impact of risk behavior, late initiation, and early discontinuation of antiretroviral therapy, *Clin. Infect. Dis.*, 49, 1570–1578.
12. Baker, J. V., Peng, G., Rapkin, J., Abrams, D. I., Silverberg, M. J., MacArthur, R. D., Cavert, W. P., Henry, W. K., Neaton, J. D., Terry Beirn community

programs for clinical research on AIDS (CPCRA) (2008) CD4+ count and risk of non-AIDS diseases following initial treatment for HIV infection, *AIDS*, 22(7), 841–848.

13. Mavigner, M., Delobel, P., Cazabat, M., Dubois, M., L'faqihi-Olive, F. E., Raymond, S., Pasquier, C., Marchou, B., Massip, P., Izopet, J. (2009) HIV-1 residual viremia correlates with persisitent T-cell activation in poor immunological responders to combination antiretroviral therapy, *PLoS One*, 4(10), e7658.
14. Casazza, J. P., Betts, M. R., Picker, L. J., Koup, R. A. (2001) Decay kinetics of Human Immunodeficiency Virus-specific CD8+ T-cells in peripheral blood after initiation of highly active antiretroviral therapy, *J. Virol.*, 75, 6508–6516.
15. *Commission Recommendation of XXX on the Definition of Nanomaterial*, published on 18 October 2011, http://ec.europa.eu/environment/chemicals/nanotech/pdf/commission_recommendation.pdf(accessed 17 December 2011).
16. Farokhzad, O. C. (2008) Nanotechnology for drug delivery: the perfect partnership, *Expert. Opin. Drug. Deliv.*, 5(9), 927–929.
17. Vyas, T. K., Shah, L., Amiji, M. M. (2006) Nanoparticulate drug carriers for delivery of HIV/AIDS therapy to viral reservoir sites, *Expert. Opin. Drug Deliv.*, **3**(5), 613–628.
18. Amiji, M. M., Vyas, T. K., Shah, L. K. (2006) Role of nanotechnology in HIV/AIDS treatment: potential to overcome the viral reservoir challenge, *Discov. Med.*, 6(34), 157–162.
19. Nowacek, A., Gendelman, H. E. (2009) NanoART, neuroAIDS and CNS drug delivery, *Nanomedicine*, 4(5), 557–574.
20. Choi, S. U., Bui, T., Ho, R. J. (2008) pH-Dependent interactions of indinavir and lipids in nanoparticles and their ability to entrap a solute, *J. Pharm. Sci.*, 97(2), 931–943.
21. Kinman, L., Brodie, S. J., Tsai, C. C., Bui, T., Larsen, K., Schmidt, A., Anderson, D., Morton, W. R., Hu, S. L., Ho, R. J. (2003) Lipid-drug association enhanced HIV-1 protease inhibitor indinavir localization in lymphoid tissues and viral load reduction: a proof of concept study in HIV-2287-infected macaques, *J. Acquir. Immune Defic. Syndr.*, 34(4), 387–397.
22. Chiappetta, D. A., Hocht, C., Taira, C., Sosnik, A. (2010) Efavirenz-loaded polymeric micelles for pediatric anti-HIV pharmacotherapy with significantly higher oral bioavailabilty, *Nanomedicine*, 5(1), 11–23.
23. Baert, L., van't Klooster, G., Dries, W., François, M., Wouters, A., Basstanie, E., Iterbeke, K., Stappers, F., Stevens, P., Schueller, L., van Remoortere, P.,

Kraus, G., Wigerinck, P., Rosier, J. (2009) Development of a long-acting injectable formulation with nanoparticles of rilpivirine (TMC278) for HIV treatment, *Eur. J. Pharm. Biopharm.* 72(3), 502–508.

24. Abdool Karim, Q., Abdool Karim, S. S., Frohlich, J. A., Grobler, A. C., Baxter, C., Mansoor, L. E., Kharsany, A. B., Sibeko, S., Mlisana, K., Omar, Z., Gengiah, T. N., Maarschalk, S., Arulappan, N., Mlotshwa, M., Morris, L., Taylor, D., CAPRISA 004 Trial Group (2010) Effectiveness and safety of tenofovir gel, an antiretroviral microbicide, for the prevention of HIV infection in women, *Science*, 329(5996), 1168–1174.
25. Das Neves, J., Amiji, M., Sarmento, B. (2011) Mucoadhesive nanosystems for vaginal microbicide development: friend or foe? *Wiley Interdiscip. Rev. Nanomed. Nanobiotechnol.*, 3, 389–399.
26. Grabovac, V., Bernkop-Schnurch, A. (2007) Development and in vitro evaluation of surface modified poly (lactide-co-glycolide) nanoparticles with chitosan-4-thiobutylamidine, *Drug. Dev. Ind. Pharm.*, 33, 767–774.
27. Kish-Catalone, T., Pal, R., Parrish, J., Rose, N., Hocker, L., Hudacik, L., Reitz, M., Gallo, R., Devico, A. (2007) Evaluation of -2 RANTES vaginal microbicide formulations in a nonhuman primate simian/human immunodeficiency virus (SHIV) challenge model, *AIDS Res. Hum. Retroviruses*, 23, 33–42.
28. Caron, M., Besson, G., Etenna, S. L., Mintsa-Ndong, A., Mourtas, S., Radaelli, A., Morghen, C. de G., Loddo, R., La Colla, P., Antimisiaris, S. G., Kazanji, M. (2010) Protective properties of non-nucleoside reverse transcriptase inhibitor (MC1220) incorporated into liposome against intravaginal challenge of Rhesus macaques with RT-SHIV, *Virology*, 405(1), 225–233.
29. Lara, H. H., Ixtepan-Turrent, L., Garza-Trevino, E. N., Rodriguez-Padilla, C. (2010) PVP-coated silver nanoparticles block the transmission of cell-free and cell-associated HIV-1 in human cervical culture, *J. Nanobiotechnol*, 8(15), 1–11.
30. Martinez-Avila, O., Bedoya, L. M., Marradi, M., Clavel, C., Alcami, J., Penades, S. (2009) Multivalent manno-glyconanoparticles inhibit DC-SIGN-mediated HIV-1 transinfection of human T-cells, *ChemBioChem*, 10(11), 1806–1809.
31. Mamo, T., Moseman, E. A., Kolishetti, N., Salvador-Morales, C., Shi, J., Kuritzkes, D. R., Langer, R., von Andrian, U., Farokhzad, O. C. (2010) Emerging nanotechnology approaches for HIV/AIDS treatment and prevention, *Nanomedicine*, **5**(2), 269–285.

32. Rupp, R., Rosenthal, S. L., Stanberry, L. R. (2007) VivaGel (SPL 7013 Gel): a candidate dendrimer-microbicide for the prevention of HIV and HSV infection, *Int. J. Nanomed.*, 2(4), 561–566.

33. Xiang, S. D., Scholzen, A., Minigo, A., David, C., Apostolopoulos, V., Mottram, P. L., Plebanski, M. (2006) Pathogen recognition and development of particulate vaccines: Does size matter? *Methods*, 40, 1–9.

34. Lisziewicz, J., Szebeni, J. (2010) The Janus face of immune stimulation by nanomedicines: examples for the good and the bad, *Eur. J. Nanom.*, DOI 10.3884/0003.1.5.

35. Christensen, N. D., Höpfl, R., DiAngelo, S. L., Cladel, N. M., Patrick, S. D., Welsh, P. A., Budgeon, L. R., Reed, C. A., Kreider, J. W. (1994) Assembled Baculovirus-expressed human papillomavirus type 11 L1 capsid protein virus-like particles are recognized by neutralizing monoclonal antibodies and induce high titres of neutralizing antibodies, *J. Gen. Virol.*, 75(9), 2271–2276.

36. Hildesheim, A., Herrero, R., Wacholder, S., Rodriguez, A. C., Solomon, D., Bratti, M. C., Schiller, J. T., Gonzalez, P., Dubin, G., Porras, C., Jimenez, S. E., Lowy, D. R., Costa Rican HPV Vaccine Trial Group (2007) Effect of human papillomavirus 16/18 L1 viruslike particle vaccine among young women with preexisting infection: a randomised trial, *JAMA*, 298(7), 743–753.

37. Martin, S. J., Vyakarnam, A., Cheingsong-Popov, R., Callow, D., Jones, K. L., Senior, J. M., Adams, S. E., Kingsman, A. J., Matear, P., Gotch, F. M., et al. (1993) Immunization of human HIV-seronegative volunteers with recombinant p17/p24:Ty virus-like particles elicits HIV-1 p24-specific cellular and humoral immune responses, *AIDS*, 7, 1315–1323.

38. Peek, L. J., Middaugh, C. R., Berkland, C. (2008) Nanotechnology in vaccine delivery, *Adv. Drug Deliv. Rev.*, 60, 915–928.

39. Reddy, S. T., Rehor, A., Schmoekel, H. G., Hubbell, J. A., Swartz, M. A. (2006) In vivo targeting of dendritic cells in lymph nodes with poly(propylene sulfide) nanoparticles, *J. Con. Rel.*, 112, 26–34.

40. Diebold, S. S., Kursa, M., Wagner, E., Cotton, M., Zenke, M. (1999) Mannose polyethylenimine conjugates for targeted DNA delivery into dendritic cell, *J. Biol. Chem.*, 274(27), 19087–19094.

41. http://www.clinicaltrials.gov/ct2/results?term=dermavir (accessed 12 January, 2012).

42. Fuller, D. H., Loudon, P., Schmaljohn, P. (2006) Preclinical and clinical progress of particle-mediated DNA vaccines for infectious diseases, Methods, 40, 86–97.

43. Aline, F., Brand, D., Pierre, J., Roingeard, P., Séverine, M., Verrier, B., Dimier-Poisson, I. (2009) Dendritic cells loaded with HIV-1 p24 proteins adsorbed on surfactant-free anionic PLA nanoparticles induce enhanced cellular immune responses against HIV-1 after vaccination, Vaccine, 27(38), 5284–5291.

44. Natz, E., Lisziewicz, J. (2011) Rational design of formulated DNA vaccines: the DermaVir approach, in *Gene Vaccines* (ed. Thalhamer, J., Weiss, R., Scheiblhofer, S.), SpringerWienNewYork, pp. 109–125.

45. Tomalia, D. A. (2009) In quest of a systematic framework for unifying and defining nanoscience, *J. Nanopart. Res.*, 11(6), 1251–1310.

46. Toke, E. R., Lorincz, O., Somogyi, E., Lisziewicz, J. (2010) Rational development of a stable liquid formulation for nanomedicine products, *Int. J. Pharm.*, 392(1–2), 261–267.

47. Lorincz, O., Toke, E. R., Somogyi, E., Horkay, F., Chandran, P. L., Douglas, J. F., Szebeni, J., Lisziewicz, J. (2011) Structure and biological activity of pathogen-like synthetic nanomedicines, *Nanomedicine*, in press.

48. Somogyi, E., Xu, J., Gudics, A., Tóth, J., Kovács, A., Lori, F., Lisziewicz, J. (2011) A plasmid DNA immunogen expressing fifteen protein antigens and complex virus-like particles (VLP+) mimicking naturally occurring HIV, *Vaccine*, 29(4), 744–753.

49. Reddy, S. T., Swartz, M. A., Hubbell, J. A. (2006) Targeting dendritic cells with biomaterials: developing the next generation of vaccines, *Trends Immunol.*, 27(12), 573–579.

50. Schaffer, D. V., Fidelman, N. A., Dan, N., Lauffenburger, D. A. (2000) Vector unpacking as a potential barrier for receptor-mediated polyplex gene delivery, *Biotech. Bioeng.*, 67, 598–606.

51. Han X., Fang, Q., Yao, F., Wang, X., Wang, J., Yang, S., Shen, B. Q. (2009) The heterogeneous nature of polyethylenimine-DNA complex formation affects transient gene expression, *Cytotechnology*, 60(1–3), 63–75.

52. Suh, J., Wirtz, D., Hanes, J., (2003) Efficient active transport of gene nanocarriers to the nucleus,. *Proc. Natl. Acad. Sci.*, 100, 3878–3882.

53. Matzinger, P. (2002) The danger model: a renewed sense of self, *Science*, 296(5566), 301–305.

54. Nicolas, J. F., Guy, B. (2008) Intradermal, epidermal and transcutaneous vaccination: from immunology to clinical practice, *Expert. Rev. Vaccines*, **7**(8), 1201–1214.

55. Bauer, J., Bahmer, F. A., Worl, J., Neuhuber, W., Schuler, G., Fartasch, M. (2001) A strikingly constant ratio exists between Langerhans cells and other epidermal cells in human skin. A stereologic study using the

optical disector method and the confocal laser scanning microscope, *J. Invest. Dermatol.*, 116(2), 313–318.

56. Lisziewicz, J., Gabrilovich, D. I., Varga, G., Xu, J., Greenberg, P. D., Arya, S. K., Bosch, M., Behr, J. P., Lori, F. (2001) Induction of potent human immunodeficiency virus type 1-specific T-cell-restricted immunity by genetically modified dendritic cells, *J. Virol.*, **75**(16), 7621–7628.
57. Lisziewicz, J., Trocio, J., Xu, J., Whitman, L., Ryder, A., Bakare, N., Lewis, M. G., Wagner, W., Pistorio, A., Arya, S., Lori, F. (2005) Control of viral rebound through therapeutic immunization with DermaVir, *AIDS*, 19(1), 35–43.
58. Calarota, S. A., Foli, A., Maserati, R., Baldanti, F., Paolucci, S., Young, M. A., Tsoukas, C. M., Lisziewicz, J., Lori, F. (2008) HIV-1-specific T cell precursors with high proliferative capacity correlate with low viremia and high CD4 counts in untreated individuals, *J. Immunol.*, 180(9), 5907–5915.
59. Lisziewicz, J., Bakare, N., Calarota, S.A., Banhegyi, D., Szlavik, J., Ujhelyi, E., Toke, E.R., Molnar, L., Lisziewicz, Zs., Autran, B., Lori, F. (2012) Single DermaVir immunization: dose-dependent expansion of precursor/memory T cells against all HIV antigens in HIV-1 infected individuals, *PloS One*, **7**: e35416.
60. Rodriguez, B., Asmuth, D.M., Matining, R.M., Spritzler, J., Jacobson, J.M., Mailliard, R.B., Li, X.D., Martinez, A.I., Tenorio, A.R., Lori, F., Lisziewicz, J., Yesmin, S., Rinaldo, C.R., Pollard, R.B. (2013) Safety, tolerability, and immunogenicity of repeated doses of DermaVir, a candidate therapeutic HIV vaccine, in HIV-infected patients receiving combination antiretroviral therapy: results of the ACTG 5176 trial, *Journal of Acquired Immune Deficiency Syndromes*, **64**(4), 351–359.
61. van lunzen, J., Pollard, R., Stellbrink, H. J., Plettenberg, A., Natz, E., Lisziewicz, Z., Freese, R., Molnar, L., Calarota, S. A., Lori, F., Lisziewicz, J. (2010) *DermaVir for Initial Treatment of HIV-Infected Subjects Demonstrates Preliminary Safety, Immunogenicity and HIV-RNA Reduction versus Placebo Immunization*, paper presented at the XVIIIth International AIDS Conference, Vienna, Austria, Abstract #A-240-0111-12561.
62. Kantoff, P. W., Higano, C. S., Shore, N. D., Berger, E. R., Small, E. J., Penson, D. F., Redfern, C. H., Ferrari, A. C., Dreicer, R., Sims, R. B., Xu, Y., Frohlich, M. W., Schellhammer, P. F. (2010) Sipuleucel-T immunotherapy for castration-resistant prostate cancer, *N. Engl. J. Med.*, 363(5), 411–422.

Chapter 9

Nanoscaffolds and Other Nano-Architectures for Tissue Engineering–Related Applications

Paraskevi Kavatzikidou and Stergios Logothetidis
Lab for Thin Films—Nanosystems and Nanometrology, Department of Physics, Aristotle University of Thessaloniki, 54124 Thessaloniki, Greece
pkavatzi@physics.auth.gr, logot@auth.gr

9.1 Introduction

In case of tissue damage, the tissue recovery is often incomplete, leading to scarred tissue and impaired functionality. For tissues such as heart valves, arteries, myocardium, bone, cartilage and nerve, full regeneration is not possible and replacement strategies are frequently applied. Such strategies include the use of autografts, allografts, xenografts and artificial prostheses. Each of these methods, however, has its limitations, including limited donor tissue, immune rejection, pathogen transfer, anticoagulation therapy, and limited durability.[1] More importantly, they do not allow full regeneration and functional

Horizons in Clinical Nanomedicine
Edited by Varvara Karagkiozaki and Stergios Logothetidis

ISBN 978-981-4411-56-1 (Hardcover), 978-981-4411-57-8 (eBook)
www.panstanford.com

recovery and hence reduce the patient's life expectancy, as compared to that of age-matched healthy subjects.[2] The emerging fields of regenerative medicine and tissue engineering try to find solutions for tissue regeneration in the human body. As a result, they employ living cells, biomaterials, soluble mediators of tissue regeneration, or a combination of these to recapitulate normal tissue structure and function.[3] The success of these strategies generally depends on the comparison of the tissue regenerated structure with 'the original' or native tissue, and on the improvement of tissue function following implantation in preclinical studies.

This chapter will attempt to provide an overview on specific tissue engineering applications and, in particular, bone, cardiac and nerve tissues, focusing on the synthetic and biological substrates used as scaffolds for these tissues. The main scaffolds' properties to meet the specific tissue environment and methods to evaluate and model their contribution both in vitro and in vivo are discussed. Finally, we co-relate the cellular-material interactions at the nanoscale level and focus on the main challenges therein.

9.2 Nano-Architectures and Nanoscaffolds

9.2.1 Materials

In designing a scaffold, the selection of the right biomaterial is significant. The main requirements for the development of scaffolds will focus on creating synthetic nano- and micro-environments, providing 3D support to control and direct cellular behaviour, and to promote specific cell interactions. The main properties affecting scaffolds' performance include biocompatibility; porosity and pore size (to facilitate oxygen, nutrients, and waste transfer, as well as tissue integration and rapid vascularisation); appropriate surface chemistry (including pH and surface charge) to favour cellular adhesion, proliferation, and differentiation; mechanical properties; controlled biodegradability; and the ability to integrate in the implantation site and promote cell-substrate interaction.[4] These materials are categorised in different classes such as natural, synthetic and semi-synthetic materials.

9.2.1.1 Natural materials

Natural materials possess several properties that make them attractive for tissue engineering applications. They are obtained from natural sources, exhibiting similar properties to the tissues they are replacing, and many of them contain specific signals for cell adhesion, allowing for cell infiltration.[4] Natural materials such as collagen,[5] gelatine,[6] laminin,[7] and chitosan[8] have been used as electrospun scaffolds for tissue reconstruction (e.g. nerve).[9]

Collagen (triple-helix protein) is known as one of the major components of the extracellular matrix present in all connective tissues. Therefore, it is one of the most widely studied natural biomaterials employed in the field of tissue engineering.[4] Collagen possesses high mechanical strength, good biocompatibility, and low antigenicity; it can be processed into porous sponges, gels, and sheets, and it has the potential of being cross-linked with chemicals to make it stronger, or to alter its degradation rate.[10] Despite their advantages, natural materials may induce immunological and inflammatory responses due to undefined factors and pathogens, which are not possible to completely eliminate by purification before implantation. Also, homogeneity of products between lots can be an issue with natural materials as a result of residual growth factors and undefined and/or nonquantified constituents that remain.[14] Moreover, natural materials are less amenable to modification as compared with synthetic materials, which are cheaper and more easily characterised.[4]

9.2.1.2 Synthetic materials

Synthetic materials have several advantages for their use as tissue engineering scaffolds. These polymers have known compositions and can be designed to minimise the immune response. They can be tailored to produce a wide range of nerve prostheses by reacting different polymers to combine properties that are unique to each, so that it is possible to obtain tubular scaffolds with different mechanical properties by varying the relative amounts of each copolymer.[4] Clearly, biodegradable polymeric scaffolds are preferable to non-biodegradable polymers because of the advantage of avoiding a second surgery to remove the conduit.

An important class of biomaterials used to develop tissue engineering scaffolds is poly(α-hydroxy esters). These synthetic polymers are readily made into 3D scaffolds that biodegrade via hydrolysis in CO_2 and H_2O, resulting, therefore, in their bioresorption. Researchers have tested several polyester nanofibrous fibres and scaffolds that have shown negligible inflammatory response, made of poly(glycolic acid) (PGA),[11] poly(lactide acid) (PLA),[12] poly(L-lactic acid) (PLLA),[13] or of a blend of poly(L-lactic acid)-caprolactone (PLLA-PCL),[14] and poly(D,L-lactide-co-glycolide) and poly(εcaprolactone) (PLGA/PCL).[15]

Other synthetic materials used for tissue engineering are hydrogels, insoluble hydrophilic polymers with high water content and similar to tissue mechanical properties, which allow them to be attractive scaffolds for implantation or for direct application to the specific area to enhance cell attachment and growth. Self-assembling peptides also belong to the hydrogel class of biomaterials. They consist of a well-defined amino acid sequence that self-assembles under physiological conditions forming a fibrous scaffold within the nanoscale (~10 nm in diameter) [4]. In the last few decades, they have been studied as biomaterials not only useful for specific 3D tissue cell cultures, but also for tissue repair and regenerative therapies. These peptide scaffolds can be custom synthesised commercially and are readily modifiable inexpensively and quickly at the single–amino acid level. Furthermore, these designer self-assembling peptide scaffolds have recently become powerful tools for regenerative medicine to repair infarcted myocardial[16] to stop bleeding in seconds,[17] and to repair nervous tissue,[18] as well as being useful medical devices for slow drug release.[19–21]

The greatest disadvantage of synthetic materials, however, is the lack of cell recognition signals, resulting, therefore, in few cellular interactions. Towards this end, efforts are being made to incorporate cell adhesion peptide motifs into synthetic biomaterials.

9.2.1.3 Semi-synthetic materials

The concept of combining synthetic materials with cell recognition sites of naturally derived materials is very attractive. These new and versatile hybrid biomaterials could possess the favourable properties of synthetic materials, including widely varying mechanical characteristics, biodegradability, reproducible large-scale production, and good processability, as well as a biological activity similar to that of natural materials.[4] Many researchers have been focusing on the development of these semi-synthetic biomaterials which, because of their characteristics, hold clinical promise in serving as implants to promote wound healing and tissue regeneration. Placed at the site of a defect, such materials should actively and temporarily participate in the regeneration process by providing a scaffold on which cell triggered remodelling could occur.[22–23]

9.2.2 Fabrication Methods

The future of tissue engineering depends on optimisation of the scaffold geometry, mechanical properties, and cross-sectional area. The development of nanofibres has greatly enhanced the scope of fabricating scaffolds that can potentially meet this challenge. Moreover, loading the nanostructured prosthesis with various fillers like components of the ECM, specific growth factors, self-assembling peptide gels, and/or different cell types like should have a synergistic effect on nervous regeneration.

Polymeric nanofibre matrix is among the most promising ECM-mimetic biomaterials. In fact, its physical structure is similar to that of fibrous proteins in native ECM.[4] Currently, there are three major methods to fabricate nanofibres: electrospinning, self-assembly, and phase separation.

9.2.2.1 Electrospinning

Electrospinning offers the opportunity for control over thickness and composition of the nanofibres, along with porosity of the nanofibre meshes, using a relatively simple experimental setup. Although polymer melt can also be electrospun, the resultant fibre is generally above 1 μm in diameter,[24] whereas for electrospun polymer solution, an average fibre diameter of 19 nm has been obtained.[25] The material can be natural, synthetic, or semi-synthetic, and it has to be dissolved in a suitable solvent to obtain a viscous solution. The solution is first passed through a spinneret, and a high voltage supply is used to charge the solution. At a critical voltage (typically N200 V/m), the repulsive forces of the charged solution particles result in a jet of solution erupting from the tip of the spinneret.[25] In comparison to the conventional electrospinning, where nanofibres are randomly oriented, the use of an electric field on the charged polymer solution makes it possible, instead, to control its trajectory, enabling the production of oriented nanofibres that can be useful in the designing of scaffolds for tissue engineering. Both natural polymers (e.g. collagen,[26–27] chitosan,[28–29] hyaluronic acid,[30] and silk fibroin[31]) and synthetic polymers (e.g. PGA,[4] PLA,[32] PLGA,[33–34] PLLA,[35] and PCL-gelatin[4]) have been used to produce nanofibres that can form potential scaffolds for tissue engineering applications.

The electrospinning technique can be applied to a wide variety of natural and synthetic polymers, making it a very versatile technique, as well as producing highly porous scaffolds with high surface areas. The main limitations, however, focus on the wide range of fibre thickness, random orientation of nanofibres, and low mechanical properties of the fibre scaffolds. Overall, electrospinning is a relatively robust and simple technique to produce nanofibres from a wide variety of polymers.

9.2.2.2 Self-assembly

Molecular self-assembly has recently emerged as a new approach in engineering artificial scaffolding materials that emulate natural extra-cellular matrix (ECM) both structurally and functionally. The main advantage in comparison to other techniques is that it incorporates specific biological components of the ECM and mimics the process of ECM assembly from the bottom up.[36] Specific polypeptide sequences have the capacity to self-assemble into various structures, ranging from assembly of β-sheets via hydrogen bonding to cylindrical micelles via hydrophobic interactions.[4] Zhang and colleagues developed a scaffolding material that self-assembled from amphiphilic oligopeptides, which consist of alternating positively (lysine or arginine) and negatively (aspartate or glutamate) charged residues separated by hydrophobic residues (alanine or leucine).[37,38]

Furthermore, the combination of synthetic materials with cell recognition sites naturally found in living systems is very attractive. These hybrid materials could possess the properties of both synthetic favourable properties and specific biological activities. As to self-assembling peptides, the simplest method to incorporate functional motifs found in ECM proteins is to synthesise them sequentially with the self-assembling sequences themselves.[39] In solutions with neutral pH, the functionalised sequences self-assemble to form nanofibres flanked with the added motifs, creating a microenvironment with specific biological stimuli.[40]

The self-assembly technique enables a variety of self-assemblies, including layered and lamellar structures, allowing flexibility to the system. Thus, the self-assembly technique shows good potential for further exploration, with the goal of designing novel scaffolds for tissue engineering applications.

9.2.2.3 Phase separation

The motivation to mimic the 3D structure of collagen, present in the natural ECM, led Ma and Zhang to develop a new technique, called thermally induced liquid–liquid phase separation, for the formation of nanofibrous foam materials.[41,42] The nanofibrous foams produced using the phase separation technique are very similar in diameter to the natural collagen present in the ECM (50–500 nm).

Phase separation of a polymer solution can produce a polymer-rich domain and a solvent-rich domain, of which the morphology can be fixed by quenching under low temperature.[4] Removal of the solvent through freeze-drying or extraction can produce porous polymer scaffolds. Phase separation can be induced by changing the temperature or by adding non-solvents to the polymer solution—called thermal induced or non-solvent-induced phase separation, respectively. Polymer scaffolds obtained by the phase separation method usually have a sponge-like porous morphology with microscale spherical pores.[43] If conditions such as solvent, polymer concentration, gelation temperature, and gelation time are precisely controlled, however, micro- or nanoscale polymer fibres can be obtained. Yang and colleagues fabricated a nanostructured highly porous scaffold made of biodegradable poly(L-lactide acid), or PLLA, by the phase separation method. This nanofibrous scaffold had fibres' diameter ranging from 50 to 500 nm.[44] The porosity and shrinkage rate were decreased by increasing the PLLA concentration.

A highly porous scaffold is desirable to allow cell migration and nutrient supply throughout the scaffold. In cases wherein the regenerative response induced by the scaffold alone is insufficient, not only pore orientation but also pore size, pore volume fraction, and resulting surface area available for cell attachment must be controlled by the scaffold fabrication method.

9.2.3 Different Architectures and Their Properties

Nanomaterials are commonly defined as those materials with very small components and/or structural features (such as particles, fibres, and/or grains) with at least one dimension in the range of 1–100 nm.[45] They can be polymers, ceramics, metals, or a mixture of these known as composite materials, possessing novel properties compared to conventional materials because of their nanoscale features. Nanomaterial science approaches atomic level control of material assembly as their bulk and surface properties are influenced by quantum phenomena that do not govern traditional bulk material behaviour.[45] The growing exploration of nanotechnology has resulted in the identification of many unique properties of nanomaterials, such as enhanced magnetic, catalytic, optical, electrical, and mechanical properties when compared to conventional formulations of the same material.[46–50] Moreover, in recent years, researchers have exhibited an increased interest in exploring numerous biomedical applications of nanomaterials.[47,48] Nanomaterials can be divided into three major categories according to their geometry, such as equiaxed, one dimensional (or fibrous), and two dimensional (or lamellar) forms. Besides these forms, there are also nanocoatings, nanofilms, and nanostructured surfaces created by numerous modification techniques which are also being widely exploited for biomedical applications.[45] Nanomaterials are being used in a wide spectrum of applications such as bone, cartilage, vascular, and neural systems always depending on their interactions with the proteins and the specific type of cells. The novelty of nanomaterials focuses on their initial interactions with the proteins that control the specific cell functions of the required regenerated tissue area.

9.3 Tissue Engineering

9.3.1 Introduction

Tissue Engineering is an interdisciplinary field that applies the principles of engineering and life sciences to the development of biological substitutes that restore, maintain or improve tissue formation.[51] The main objective of tissue engineering is to produce patient-specific biological substitutes in order to avoid the limitations of existing clinical treatments for damaged tissues or organs. Such limitations are shortage of donor organs, chronic rejection and cell morbidity.[52] In the case of donor organs, it was reported that in early 2002, in the U.S. alone, the number of patients registered on the waiting list for donor organs totalled 79,446, a threefold increase in numbers compared to 1990.[53] To prolong and improve the quality of life of patients awaiting organ transplants, other therapies exist such as surgical reconstruction (e.g. of hearts), drug therapy (e.g. insulin for a malfunctioning pancreas), mechanical devices (e.g. kidney dialysers) and synthetic prostheses.[53] Cell morbidity is caused by grafting and repetitive surgical operations, while chronic rejection is observed in cases where the use of allografts or synthetic prostheses results in poor integration with the host tissue and failure over time.[54]

9.3.2 Material Requirements

Ideally, a tissue engineered scaffold should attempt to mimic the physical and biological properties of the natural extracellular matrix. The primary roles of the extracellular matrix are to serve as an adhesion substrate for the cells, to provide temporary mechanical support to the newly grown tissue and to guide the development of new tissue with the appropriate function.[55] Therefore, a successful scaffold should possess the following features: interconnecting pores of appropriate scale to favour tissue integration and vascularisation, being made from material with controlled biodegradability or bioresorbability so that the tissue will eventually replace the scaffold, the appropriate surface chemistry to favour cellular attachment, differentiation and proliferation, the appropriate mechanical properties to match the intended site of implantation and handling, ability to be fabricated into a variety of shapes and sizes.[56]

9.3.3 Fabrication Methods

Scaffolds can be produced either using conventional methods or advanced processing techniques.[52] The main conventional methods involve solvent casting and particulate leaching, gas foaming, fibre meshes and fibre bonding, phase separation, melt moulding, emulsion freeze drying, solution casting and freeze drying.[52] In solvent casting, a solution of a polymer such as poly(L-lactic acid) in chloroform is produced while salt particles of specific diameter are added to create a uniform suspension. The solvent is evaporated, leaving a polymer matrix with salt particles embedded throughout.

Then the composite is immersed in water for the salt to leach out and a porous structure is produced.[57]

Mooney et al. described the gas foaming technique with a biodegradable polymer such as poly(D,L-lactic co-glycolic acid) (PLGA).[58] PLGA is saturated with carbon dioxide (CO_2) at high pressures. The gas solubility is then decreased rapidly by bringing the CO_2 pressure back to atmospheric level. As a result there is nucleation and growth of glass bubbles with sizes between 100–500 μm in the polymer.[58] Cima et al. explained the fibre meshes method of producing a scaffold.[59] Textile technology was used to create fibres and then these fibres produced non-woven scaffolds from polymers such as polyglycolic acid (PGA) and PLLA. The main disadvantage of this method was that the scaffolds lacked structural ability and frequently they were deformed when cells attached on them.[59] In an attempt to improve the above scaffolds, a fibre bonding technique was developed to increase the mechanical properties of the scaffolds. The technique involved the formation of a composite material with nonbonded fibres embedded in a matrix, followed by thermal treatment and selective dissolution of the matrix. PGA fibre meshes were bonded using PLLA as a matrix.[57]

In the phase separation method, a biodegradable synthetic polymer is dissolved in molten phenol or naphthalene and biologically active molecules such as alkaline phosphatase are added to the solution.[60] Then a liquid-liquid phase separation is produced by lowering the temperature and, finally, by quenching the temperature, a two-phase solid is formed. Sublimation is used to remove the solvent and to create the pores on the scaffold. The bioactive molecules are incorporated in the structure.[60] Thompson et al. performed the melt moulding technique with PLGA.[61] This technique involves filling a teflon mould with PLGA powder and gelatine microspheres of specific diameter and then heating the mould above the glass-transition temperature of PLGA while applying pressure to the mixture. The PLGA particles bond together, and the mould is removed. Water is used to leach out the gelatine from the mould and the scaffold formed is dried.[61]

The emulsion freeze drying method for PGA scaffolds is another fabrication process for porous scaffolds.[62] This method involves the addition of ultrapure water to a solution of methylene chloride with PGA. The two immiscible layers are then homogenised to form a water-in-oil emulsion that is quenched in liquid nitrogen and freeze-dried to produce the porous structure.[62] Solution casting has to do with dissolving a polymer such as PLGA in chloroform and then precipitate it by the addition of methanol.[63] Demineralised freeze-dried bone can be combined with PLGA and the composite produced is then pressed into a mould and heated to 45–48°C for 24 h to create the scaffold.[63]

The final conventional technique described in this section is freeze drying. Various materials have been processed by this method such as PLGA, collagen, chitin and alginate.[64] In this case, a material such as PLGA is dissolved in glacial acetic acid or benzene and then freeze-dried to yield porous matrices.

The main limitation of the aforementioned techniques is the lack of ability to control the pore size, pore geometry, pore interconnectivity, spatial pore distribution and to construct internal channels within the scaffold.[52] For example the scaffolds produced by solvent casting and particulate leaching cannot guarantee pore interconnectivity and uniform pore size because they both depend on the salt particles and their agglomeration in the composite.[64] The scaffolds produced have a foam structure in which the cells are seeded and expected to grow throughout the scaffold. It was observed that the cells could not migrate deep into the scaffold because of the lack of nutrients and oxygen and also insufficient removal of the waste products.[64] Therefore, the cells are only able to survive close to the surface. The cells present on the surface of the scaffold act as a barrier, preventing the diffusion of the nutrients and oxygen into the scaffold.[64] Martin et al. also reported that in bone tissue engineering cases, the mineralisation of the surface of the scaffold due to the high rates of nutrients and oxygen transfer could also limit the mass transfer of nutrients into the scaffold.[65]

The materials that the scaffold is composed of, too, have their limitations. The degradation products of some of the biodegradable materials could cause adverse effects for the cells attached on the scaffolds and their proliferation. For instance, when lactic acid is released from PLLA into the surrounding environment, the pH is reduced, resulting in faster degradation of the scaffold, thus creating an acidic environment for the cells present in the environment.[64] Moreover, the presence of residual organic solvents in the scaffold could result in risks of toxicity and carcinogenicity for the cells. Furthermore, the surface chemistry of some materials such as polymers is not familiar to the cells. Natural materials, such as collagen, however, are recognised from the cells as they are the major constituent of their extracellular matrix.[64] Collagen itself has the limitation that it might elicit an immune response.

9.4 Bone and Cartilage Engineering Applications

9.4.1 Bone Biology

Bone is a specialised form of connective tissue that comprises the endoskeleton of vertebrates.[66] Its main functions are to protect internal organs, support the body, produce blood cells (haematopoiesis), reserve important minerals in the body such as calcium and phosphorus, and to interact with muscles, tendons, ligaments and joints for the function of movement.[67]

The bone consists of several types of cells and the matrix that surrounds these cells. Osteoprogenitor cells are the precursors cells derived from mesenchymal tissue; osteoblasts are mature osteoprogenitor cells that produce a protein mixture called osteoid (type I collagen) that mineralises to become bone. Osteoblasts also produce hormones such as prostaglandins, matrix proteins and the enzyme, alkaline phosphatase that plays a role in mineralisation of bone. Another cell type found in bone are the osteocytes. These cells are mature osteoblasts that have migrated and became entrapped in the matrix. They are responsible for the calcium homeostasis, maintenance of the matrix and bone formation. The final cell type present in bone is osteoclasts. They are derived from monocytes and are responsible for bone resorption. The matrix is divided into two components: the inorganic matrix, which consists mainly of crystalline mineral salts and calcium in the form of hydroxyapatite, and the organic matrix that includes primarily type I collagen and other glycosaminoglycans, osteocalcin, osteonectin and other proteins. The hydroxyapatite gives bone its rigidity, hardness and stiffness while the collagen part of the organic matrix contributes to the elasticity of this brittle material.[67]

Repeated stress, such as weight-bearing exercises or bone healing, results in the bone thickening at the points of maximum stress. Wolff's law states that 'every change in form and function of a bone, or in its function alone, is followed by certain definite changes in its internal architecture and equally definite secondary alteration in its mathematical laws.[68] Bone is deposited and resorbed in accordance with the stresses and strains placed upon it. This is also known as bone remodelling.

Bone remodelling is a phenomenon that occurs throughout a person's life in order to regulate calcium homeostasis, repair damaged bones from daily stress and also to shape and sculpture the skeleton during growth. It is a process of resorption of mineralised bone by osteoclasts and new bone formation by osteoblasts. The interaction between these two cell types is known as coupling, and it is essential for maintaining a balance between the rates of bone loss and gain.[67]

9.4.2 Bone-Related Clinical Problems

Critical-size bone defects result from bone fracture due to accidents or weakening of bone due to bone diseases such as arthritis, osteoporosis, osteomyelitis, osteosarcoma and avascular necrosis and genetic bone disorders such as osteogenesis imperfecta. Arthritis is divided into different types, but the two most common are osteoarthritis—where the cartilage in the joints breaks down—and rheumatoid arthritis, where the synovial fluid in the joint space destroys the cartilage due to inflammation. In both these cases, the cartilage is lost and the bones rub against each other causing pain and stiffness. Osteoporosis is present when the bone mineral density decreases, osteomyelitis is associated with infection of bone, osteosarcoma is a type of bone cancer, avascular necrosis occurs when the blood supply to the bones is temporarily or permanently lost, and osteogenesis imperfecta involves type I collagen deficiency.[67]

The most common treatments are administration of drugs to increase the bone mass and reduce bone weakening, physiotherapy to strengthen and improve motility without any risk of fracturing the bone, and surgery. Surgical treatment includes the use of pins and plates made of synthetic materials to unite the broken parts of the bone and keep them together by mechanical fixation, while the affected area is immobilised to permit healing of the fracture to occur. Another surgical treatment involves the removal of the diseased/damaged bone and the use of bone substitutes, by autografts (harvested from somewhere else in the body), allografts (from another donor) or by a synthetic material. In the case of autografts, the main risks are availability and cell morbidity while for allografts, there are risks such as chronic rejection and disease transfer.

9.4.3 Bone-Related Applications

Generally, synthetic materials used for medical applications should possess specific properties. They should be made from materials with biocompatible and biodegradable properties, porous in order to allow tissue integration and vascularisation, have surface properties that would favour cell attachment and proliferation and appropriate mechanical properties to match the properties of the intended implantation site.[56]

For bone related applications, the mechanical properties of the implant play a significant role on the biomaterial/tissue interaction and consequently in bone remodelling. The degree to which applied loads are shared between the implant and surrounding bone is a significant determinant of the stress-induced adaptive remodelling of bone tissue. An increased variance in modulus of elasticity and stiffness between the implant and host bone would result in changes of the normal strain distribution in the bone site, which will further cause non-anatomic remodelling of the bone because of atrophy (stress shielding) or hyperplasia (due to abnormally high strains).[69]

The main groups of synthetic materials used for bone applications are metals, ceramics and polymers. The bone related applications are divided into orthopaedic applications and dental applications. Metals are used for load bearing applications such as pins, plates, femoral stems, orthodontic devices, etc., where sufficient mechanical properties are required to stand the load of the surrounding environment. Ceramics are used for wear applications in joint replacements and for bone bonding applications due to their hardness and wear resistance. Polymers are used for their flexibility and stability in low friction articulating surfaces against ceramic components in joint replacements.[69]

The synthetic materials widely used for orthopaedic applications are metals, ceramics, polymers and composites. The following studies give a review on the different nanoscaffolds implemented for bone tissue engineering applications.

Increased osteoblast (bone-forming cells) adhesion on nanograined materials was first reported in 1999.[45,49] For example, alumina with grain sizes ranging from 49–67 nm and titania with grain sizes of 32–56 nm promoted osteoblast adhesion compared to their respective micro-grained materials. Further investigations of these nanoceramics (such as alumina, titania, and hydroxyapatite

(HA)) demonstrated that in vitro osteoblast proliferation and long-term functions (as measured by intracellular and extracellular matrix protein synthesis such as collagen and alkaline phosphatase, as well as calcium-containing mineral deposition) were enhanced on ceramics with grain or fibre sizes less than 100 nm.[50,70] Moreover, osteoblast functions (specifically, viable cell adhesion, proliferation, and calcium deposition) were even further increased on nanofibre structures compared to nanospherical structures of alumina; this was believed to occur because, compared to nanospherical geometries, nanofibres more closely approximate the shape of HA crystals and collagen fibres in bone.[70]

In addition to osteoblast functions, enhanced osteoclast (bone-resorbing cells) functions were also observed on nanophase ceramics. Coordinated functions of osteoblasts and osteoclasts are critical for the formation and maintenance of healthy new bone juxtaposed to an orthopaedic implant.[71] Therefore, the results of promoted functions of osteoblasts, coupled with greater functions of osteoclasts, could assure healthy remodelling of juxtaposed bone formed at implant surfaces composed of nanophase ceramics. Finally, decreased functions of competitive cells such as fibroblasts were observed on nanophase compared to conventional ceramics and polymers.[72,73] Specifically, the ratio of osteoblast to fibroblast adhesion increased from 1:1 on conventional alumina to 3:1 on nanophase alumina.[73] Such observations are found not only on nano-ceramics, but also on nano-metals, nano-polymers, and nano-composites. In the case of Ti alloys (Ti_6Al_4V) and CoCrMo, respective nanophase metals fabricated by traditional powder metallurgy techniques increased osteoblast adhesion, proliferation, collagen synthesis, and calcium deposition.[74]

Different techniques create nanometre surface features on polymers (such as chemical etching, mould casting, e-beam lithography, and polymer demixing[75,76] showing promise for orthopaedic applications. A research group transferred the topography of compacted nanomaterials to poly(latide-co-glycolide) acid (PLGA) using well-established silastic mould-casting techniques.[77] As a result, increased osteoblast adhesion with decreased fibroblast adhesion were observed on PLGA samples which were cast from nanomaterial compact compared to conventional compact moulds. Although increased osteoblast functions and decreased fibroblast functions have been observed on nano compared to conventional compacts,[45] polymers such as PLGA, may contribute more to rehabilitating damaged bone tissue due to their controllable biodegradability.

Natural bone is composed of organic components (mainly type I collagen) and inorganic components (HA). Therefore, there are various studies demonstrating greater osteoblast responses on composites of PLGA combined separately with nanophase alumina, titania, and HA (30/70 ceramic/polymer weight ratio).[78] On the other hand, nanophase ceramic particles tended to agglomerate when added into such polymeric mixtures. Thus, the increase of the dispersion of nanophase titania in PLGA was studied to promote more effectively the nano surface roughness; this resulted in greater osteoblast functions (including adhesion, collagen and alkaline phosphatase synthesis, and calcium deposition) on polymer composites with a greater dispersion of nanoceramics.[79,80]

Moreover, carbon nanofibre/polymer composites possess tailorable electrical and mechanical properties ideal for orthopaedic applications. Promoted responses of osteoblasts have been reported when carbon nanofibres were incorporated into polyurethane (PU).[81] Specifically, three and four times the number of viable osteoblasts adhered on PU when combined with 10 and 25 wt% carbon nanofibres, respectively. Nanostructured biocomposites provide alternatives not yet fully explored for orthopaedic applications. They may be fabricated to possess similar micro- and nano-architecture as that of healthy, physiological bone. Their improved mechanical and biocompatibility properties promise future greater orthopaedic implant efficacy.[45]

9.4.4 Nanomaterials for Cartilage Applications

Once damaged, cartilage does not normally regenerate itself. Mature articular cartilage cannot heal spontaneously owing to its low mitotic activity, which contrasts to the rapid rate of chondrocytic (cartilage synthesising cell) mitosis during normal cartilage growth.[82] Accordingly, surgical strategies for cartilage repair have focused on accessing the regenerative signalling molecules and cells within the subchondral bone marrow.

The downside of these techniques is that they require invasive drilling or abrasion through the overlying articular cartilage into the bone marrow, and thus causes even further cartilage tissue damage before any therapeutic effect is achieved.[45] Moreover, the biomechanical and biochemical properties of the regenerated tissue fail to match that of the uninjured cartilage. It is believed that tissue engineering combined with nanomaterials may give opportunities for cartilage regeneration. One approach for cartilage regeneration is to closely match the composition, properties and microstructure of cartilage. A research team prepared nanostructured PLGA by chemically etching PLGA in 1 N NaOH for 10 min.[83] Results demonstrated that NaOH-treated PLGA 3-D scaffolds enhanced chondrocyte function compared to non-treated, traditional PLGA scaffolds. Specifically, viable chondrocyte numbers, total intracellular protein content, and the amount of extracellular matrix components (such as glycosaminoglycans and collagens) were significantly greater when chondrocytes were cultured on NaOH-treated scaffolds than on non-treated PLGA scaffolds.[83]

In addition to polymeric nanomaterials, nanocomposites and natural nanomaterials have also been studied for cartilage regeneration. Cheng et al. reported that human cartilage cells attached and proliferated well on HA nanocrystals homogeneously dispersed in PLA composites.[84] Another group demonstrated that type II collagen formed nonwoven fibrous scaffolds by electrospinning with fibre diameters from 110nm to 1.8 mm to support chondrocyte growth.[85]

9.5 Cardiac Engineering Applications

Tissue engineering has been widely applied to render living semi-lunar heart valves and arteries, either by combining cells and substrates in vitro, or by implanting cell-free substrates intended for cell infiltration, or repopulation in vivo.[1] The preferable option for a valve or vessel substrate is a xenograft or homograft matrix, depleted of its cellular components. Such bio-derived acellular matrices possess a native-like geometry and structure, with mechanical and functional behaviour comparable to their native counterparts.[1] The use of these acellular substrates for valvular and vascular grafts has been studied extensively in vivo; either implanted as such or pre-seeded with various types of cells in vitro ranged from complete failure to encouraging long-term follow-up.[86–89] The tissue growth and remodelling from these matrices was attributed to the host cells repopulation or on the type and fate of cells seeded prior to implantation.

First, the use of animal derived materials goes along with the risk of transferring zoonoses and second, when using homografts as a bioderived matrix, their availability may be a limiting factor for success. Therefore, natural and synthetic polymers have been explored to serve as substrates for cardiac tissue engineering.

Natural polymers in gel-like substrates are widely used as substrates for the creation of cardiac replacements in vitro.[90–95] Cells, mainly (myo)fibroblasts and smooth muscle cells, encapsulated within such gels, produce and organise endogenous ECM. For example, engineered valvular and vascular grafts based on fibrin gel as a substrate functioned well in sheep.[96] Due to their limited initial mechanical integrity and load bearing properties, tissue engineered valves and vessels created from biological gels generally require prolonged in vitro culture and conditioning to develop an endogenous load bearing structure and to produce grafts that survive under in vivo hemodynamic loading conditions.[97,98]

The most commonly used biodegradable synthetic substrates are PGA, PLA, polyhydroxybutyrates (PHB), PCL or their co-polymers, although elastomeric substrates are also gaining popularity in this area.[99,100] Apart from their tuneable mechanical properties, these degradable and biocompatible materials vary in their rates of hydrolysis and processing possibilities, providing the tissue engineer with a high degree of control to design substrates with tailored structural and mechanical properties.[1]

9.6 Nerve Engineering Applications

It has long been hypothesised that any scaffold material has to have the ability to mimic the ECM. The material must be able to interact with cells in a 3-D environment and facilitate this communication. In native tissues—e.g. nerve—the structural ECM proteins (50 to 500 nm-diameter fibres) are one to two orders of magnitude smaller than the cell itself; this allows the cell to be in direct contact with many ECM fibres, thereby defining its 3D orientation. Therefore, it is of significant importance that research in tissue engineering should also focus on nanotechnology. In native tissues, cells are often regularly oriented, and this cell orientation is of crucial importance for tissue function, especially for neural tissue, where neuronal and axonal growth have a precise directionality.[4] In the case of peripheral nervous tissue regeneration, it is very important to enhance the neurite growth direction so as to connect the two lesioned nerve stumps.[4] The scope of the application of aligned polymer nanofibre in tissue engineering is exactly to control cell orientation.

In electrospinning, self-assembling, and phase separation, the diameter of the nanofibre should be controlled by adjusting parameters including polymer concentration, amino acid sequence, flow rate of the polymer solution, solvent conductivity, and temperature.[4] In a study, they aligned electrospun PCL fibres to test their potential in providing contact guidance to human Schwann cells and to reproduce the formation of bands of Büngner.[101] Schwann cells were observed to elongate and align along the axes of the electrospun fibres and to enhance the rate and extent of neurite elongation.[101] They applied the electrospinning technique, using a rotating plate with a sharpened edge as collector, for the alignment of a PLLA nano-micro fibrous scaffold, and, investigated the suitability of the scaffold for neural stem cells (NSC) culture in terms of their fibre alignment and dimension. It was observed that the aligned nanofibres improved neurite outgrowth, whereas the fibre diameter does not show any significant effect on cell orientation.[101]

A new electrospinning technology to prepare polymer nanofibres with a core-shell structure has been introduced.[102–103] This technique uses a specifically designed plunger with a coaxial opening structure, which contains a core opening and a surrounding annular opening. Different solutions can be introduced into the core opening and the annular opening, and electrospun into a core-shell-structured nanofibre. The most likely application of this technique is to embed drugs, growth factors, or genes into the core of the biodegradable polymer nanofibres, producing polymer nanofibres able to release drugs.[102,103]

Another way/application for the core-shell structure is by setting synthetic polymer as the core material and natural polymer such as collagen as the shell material. As a result, nanofibres with strong mechanical strength and good biocompatible surface can be fabricated. This strategy may solve the problem of poor biocompatibility of synthetic polymer nanofibres.[4] Surface modification of biomaterials in order to improve cell and tissue response has been extensively investigated for the recreation of the native nervous structure. An increasingly employed approach for emulating the ECM involves identifying bioactive motifs present in these molecules and grafting synthetic analogues of these signals onto a material. For example, cells engage with ECM ligands via receptors such as integrins.[4] They are known to bind to several common polypeptide motifs like the arginine-glycine-aspartic acid (RGD) motif. The possibility to include these motifs in the biomaterials structure can promote cell interactions and enhance nervous regeneration.

An important feature for a promising nanobiomaterial is its porosity. Cells live in a complex mixture of pores, ridges, and fibres of ECM at nanometre scales.[104] Mimicking the ECM and including nanoscale structural elements in tissue engineering scaffolds affects cell response at the cell–scaffold interface. To engineer functional tissues and organs successfully, the scaffolds must be designed to facilitate cell distribution and guide tissue regeneration in three dimensions.[105,106]

Koh and colleagues have recently enhanced neurite outgrowth using nanostructured scaffolds coupled with laminin,[107] a well-known component of the ECM that is continuously synthesised after nerve injury and that plays a crucial role in cell migration, differentiation, and axonal growth. Laminin was coupled into PLLA nanofibres using three different methods: covalent binding using water-soluble carbodiimide and N-hydroxysuccimide as the coupling reagents, physical adsorption, and physical blending of laminin together with PLLA solution for electrospinning procedure. The last method, blended electrospinning, proved to be the more efficient technique to introduce laminin on the surface and in the interior of the nanofibres. To investigate glial cell migration and axonal growth on electrospun nanofibres, Schnell and colleagues produced, by electrospinning, a collagen–PCL nanofibres scaffold as a guidance structure.[108] Schwann cell migration, neurite orientation, and process formation of Schwann cells, fibroblasts, and olfactory ensheathing cells were improved on collagen-PCL fibres in comparison with pure PCL fibres, demonstrating that electrospun fibres composed of a collagen and PCL blend represent a suitable substrate for supporting cell proliferation, process outgrowth, and migration, and as such would be a good material for artificial nerve implants.[4]

9.7 Correlation of Cell–Material Interactions at the Nanoscale

In order to comprehend the degree to which nanomaterials interact with various cells, it is important to further understand the mechanisms as to why nanomaterials demonstrate unique biological properties.[45]

Natural tissues are designed and assembled in controlled ways from micro- and nanoscale building blocks. These size scales are very important in describing the hierarchical structures of natural tissues and understanding the relationship between such structures at various levels. Figure 9.1 illustrates the relative scales of proteins, cells, tissues, etc. in biological systems that nanomaterials attempt to mimic.[109]

For example, natural bone is a candidate of a nanostructured composite material. There are three levels of structures in bone: (1) the nanostructure (a few nanometres to a few hundred nanometres) including non-collageneous organicproteins, fibrillar collagen and embedded mineral (HA) crystals; (2) the microstructure (from 1 to 500 mm) including lamellae, osteons, and Haversian systems; and (3) the macrostructure including cancellous and cortical bone.[110]

Unique surface energy of nanomaterials: protein mediated cell interactions. Before cells adhere to a material surface, proteins will adsorb onto the surface within milliseconds to potentially interact with select cell membrane receptors, as shown in Fig. 9.2.[111] Accessibility of cell adhesive domains (such as specific amino acid sequences of adsorbed vitronectin, fibronectin, and laminin) may either enhance or inhibit subsequent cellular adhesion and growth. The type, concentration, conformation, and bioactivity of plasma proteins adsorbed onto materials depend on surface chemistry, hydrophilicity or hydrophobicity, charge, topography, roughness, and energy. For example, maximum vitronectin adsorption was noted on hydrophilic surfaces of high surface roughness.[112]

Nanomaterials have unique surface properties and energetics due to higher surface areas, higher surface roughness, higher amounts of surface defects (including grain boundaries), altered electron distributions, etc.[45] All of these unusual properties inherent for nanomaterials will affect interactions with proteins since all proteins are nanoscale entities. For example, increased surface areas and nanoscale surface features on nanomaterials can provide for more available sites for protein adsorption and, thus, alter the amount of cellular interactions.[45]

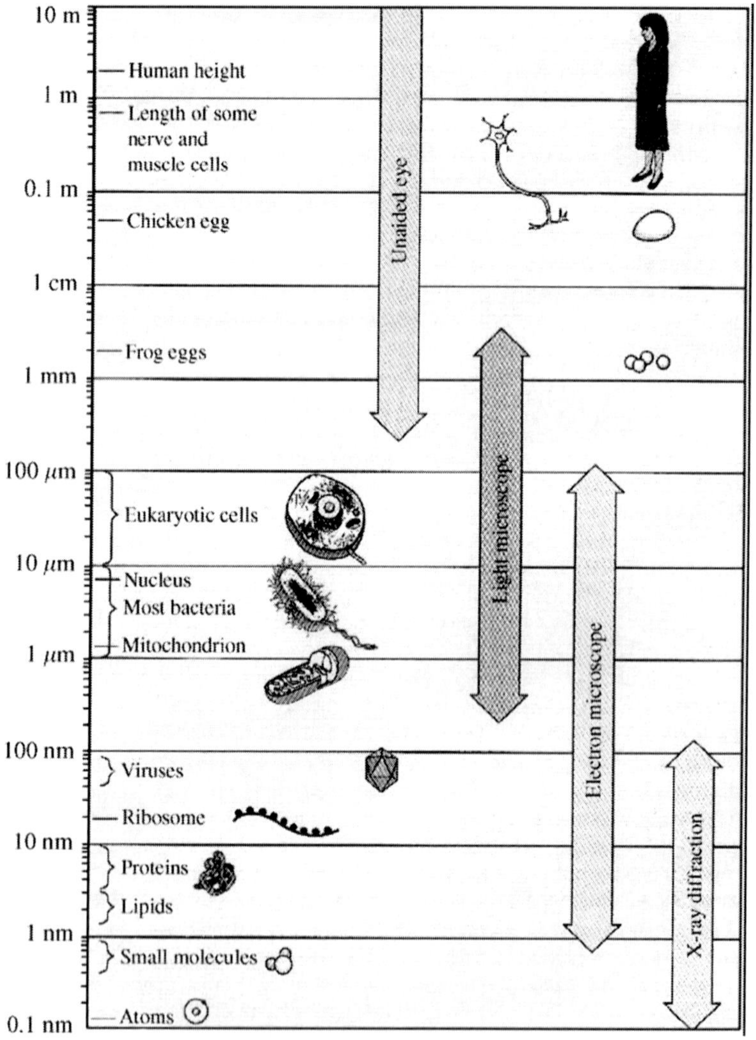

Figure 9.1 Relative scales of proteins, cells, tissues, etc., in biological systems that nanomaterials attempt to mimic.

Specifically, Miller et al. examined fibronectin interactions with nanomaterials with various nanoscale surface features under atomic force microscope and demonstrated for the first time how proteins responded differently to such nanoscale surface features.[113] Specifically, fibronectin (5 mg/mL) adsorbed to PLGA surfaces with 500 nm spherical bumps showed little to no interconnectivity; fibronectin (5 mg/mL) adsorbed to PLGA surfaces with 200 nm spherical bumps showed a higher degree of interconnectivity; fibronectin (5 mg/mL) adsorbed to PLGA surfaces with 100 nm spherical bumps showed well-spread fibronectin molecules with the highest degree of interconnectivity leading to a masking of the underlying PLGA nanometre surface features.

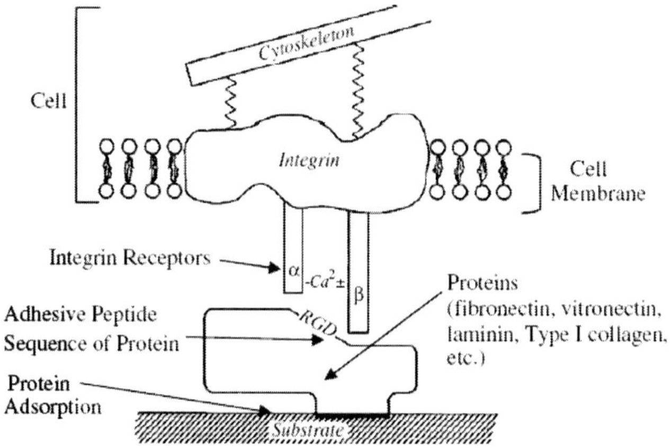

Figure 9.2 Schematic representation of protein-mediated cell adhesion on biomaterial surfaces.

A few studies have addressed the mechanisms of enhanced cellular activity (such as osteoblast, chondrocyte etc.) on nanophase materials. One set of in vitro studies pinpoints grain size in the nanometre regime as the major parameter for enhancing ceramic cytocompatibility. Investigations of the underlying mechanisms revealed that the concentration, conformation, and bioactivity of vitronectin (a protein contained in serum that is known to mediate osteoblast adhesion) was responsible for the select, enhanced adhesion of osteoblasts (a crucial prerequisite for subsequent, anchorage-dependent cell functions) on these novel nanophase ceramic formulations. Vitronectin is a linear protein, 15 nm in length, that preferentially adsorbed to the small pores present in nanophase ceramics (such as 0.98nm pore diameters for nanophase titania compacts).[45]

In addition to experimental evidence, there is also ample theoretical evidence to support the unique mechanical and surface properties of nanomaterials for tissue-engineering applications. For example, mechanical deformation theory indicates that the high-volume fraction of interfacial regions compared to bulk materials leads to increased deformation by grain-boundary sliding and short-range diffusion-healing events as grain size is reduced (thus, increasing ductility for nanocrystalline ceramics). In addition, compared to conventional ceramics, nanophase ceramics possess greater surface roughness, resulting from both decreased grain size and possibly decreased diameter of surface pores. Moreover, nanophase ceramics possess enhanced surface wettability due to greater surface roughness and greater numbers of grain boundaries (or, in other words, material defects) on their surfaces.[45] From this theory, it is clear that the implementation of nanotechnology into tissue engineering does not represent a traditional biomaterial trial-and-error approach, but rather, possesses numerous scientific rationales.

9.8 Conclusions

In this chaptyer, it becomes apparent that tissue engineering and nanotechnology are two fields that are interconnected and complement each other since all the biological tissues are composed of synthetic and complex nanostructures. It is very important to select the correct fabrication methods and the appropriate materials (either metals, polymers, ceramics or a combination of them) in order to create the suitable 3D environment that will mimic the physical and biological properties of the natural extracellular matrix for the regenerated tissue. There are many challenges and hurdles in the scaffold that to a great extent influence the progress in the tissue engineering field.

References

1. Bouten C. V. C., Dankers P. Y. W., Driessen-Mol A., Pedron S., Brizard A. M. A., Baaijens F. P. T. (2011) Substrates for cardiovascular tissue engineering, *Adv Drug Deliv Rev*, 63:221–241.
2. Hlatky M. A., Boothroyd D. B., Melsop K. A., Brooks M. M., Mark D. B., Pitt B., Reeder G. S., Rogers W. J., Ryan T. J., Whitlow P. L., Wiens R. D. (2004)

Medical costs and quality of life 10 to 12 years after randomization to angioplasty or bypass surgery for multivessel coronary artery disease, *Circulation*, 110:1960–1966.

3. Vacanti J. P., Langer R. (1999) Tissue engineering: the design and fabrication of living replacement devices for surgical reconstruction and transplantation, *Lancet*, 254: SI32–SI43.
4. Cunha C., Panseri S., Antonini S. (2011) Emerging nanotechnology approaches in tissue engineering for peripheral nerve regeneration, *Nanomed: Nanotechnol Biol Med*, 7, 50–59.
5. Matthews J. A., Wnek G. E., Simpson D. G., Bowlin G. L. (2002) Electrospinning of collagen nanofibers, *Biomacromolecules*, 3:232–238.
6. Huang Z. M., Zhang Y. Z., Ramakrishna S., Lim C. T. (2004) Electrospinning and mechanical characterization of gelatin nanofibers, *Polymer*, 45:5361–5368.
7. Neal R. A., McClugage S. G., Link M. C., Sefcik L. S., Ogle R. C., Botchwey E. A. (2009) Laminin nanofiber meshes that mimic morphological properties and bioactivity of basement membranes. *Tissue Eng C*, 15:11–21.
8. Wang W., Itoh S., Matsuda A., Ichinose S., Shinomiya K., Hata Y., et al. (2008) Influences of mechanical properties and permeability on chitosan nano/microfiber mesh tubes as a scaffold for nerve regeneration, *J Biomed Mater Res A*, 84:557–566.
9. Wang W., Itoh S., Matsuda A., Aizawa T., Demura M., Ichinose S., et al. (2008) Enhanced nerve regeneration through a bilayered chitosan tube: the effect of introduction of glycine spacer into the CYIGSR sequence, *J Biomed Mater Res A*, 85:919–928.
10. Lee C. H., Singla A., Lee Y. (2001) Biomedical applications of collagen, *Int J Pharm* 221:1–22.
11. Keeley R. D., Nguyen K. D., Stephanides M. J., Padilla J., Rosen J. M. (1991) The artificial nerve graft: a comparison of blended elastomer-hydrogel with polyglycolic acid conduits, *J Reconstr Microsurg*, 7:93–100.
12. Corey J. M., Lin D. Y., Mycek K. B., Chen Q., Samuel S., Feldman E. L., et al. (2007) Aligned electrospun nanofibers specify the direction of dorsal root ganglia neurite growth, *J Biomed Mater Res A*, 83:636–645.
13. Yang F., Murugan R, Wang S, Ramakrishna S.,(2005) Electrospinning of nano/micro scale poly(L-lactic acid) aligned fibers and their potential in neural tissue engineering, *Biomaterials*, 26:2603–2610.
14. Ghasemi-Mobarakeh L., Prabhakaran M. P., Morshed M., Nasr-Esfahani M. H., Ramakrishna S. (2008) Electrospun poly(ε-caprolactone)/gelatin

nanofibrous scaffolds for nerve tissue engineering, *Biomaterials*, 29:4532–4539.

15. Panseri S., Cunha C., Lowery J., Del Carro U., Taraballi F., Amadio S., et al. (2008) Electrospun micro- and nanofiber tubes for functional nervous regeneration in sciatic nerve transections, *BMC Biotechnol*, 8:39.
16. Davis M. E., Motion J. P., Narmoneva D. A., Takahashi T., Hakuno D., Kamm R. D., et al (2005) Injectable self-assembling peptide nanofibers create intramyocardial microenvironments for endothelial cells, *Circulation*, 111:442–450.
17. Ellis-Behnke R. G., Liang Y. X., Tay D. K., Kau P. W., Schneider G. E., Zhang S., et al. (2006) Nano hemostat solution: immediate hemostasis at the nanoscale, *Nanomed Nanotechnol Biol Med*, 2:207–215.
18. Ellis-Behnke R. G., Liang Y. X., You S. W., Tay D. K., Zhang S., So K. F., et al. (2006) Nano neuro knitting: peptide nanofiber scaffold for brain repair and axon regeneration with functional return of vision, *Proc Natl Acad Sci U S A*, 103:5054–5059.
19. Bawa P., Pillay V., Choonara Y. E., du Toit L. C. (2009) Stimuli-responsive polymers and their applications in drug delivery, *Biomed Mater*, 4:22001.
20. Branco M. C., Pochan D. J., Wagner N. J., Schneider J. P. (2009) Macromolecular diffusion and release from self-assembled β-hairpin peptide hydrogels, *Biomaterials*, 30:1339–1347.
21. Kyle S., Aggeli A., Ingham E., McPherson M. J. (2009) Production of selfassembling biomaterials for tissue engineering, *Trends Biotechnol*, 27:423–433.
22. Elbert D. L., Pratt A. B., Lutolf M. P., Halstenberg S., Hubbell J. A. (2001) Protein delivery from materials formed by self-selective conjugate addition reactions, *J Control Release*, 76:11–25.
23. Mann B. K., Gobin A. S., Tsai A. T., Schmedlen R. H., West J. L. (2001) Smooth muscle cell growth in photopolymerized hydrogels with cell adhesive and proteolytically degradable domains: synthetic ECM analogs for tissue engineering, *Biomaterials*, 22:3045–3051.
24. Lyons J., Li C., Ko F. (2004) Melt-electrospinning part I: processing parameters and geometric properties, *Polymer*, 45:7597–7603.
25. Tan S. H., Inai R., Kotaki M., Ramakrishna S. (2005) Systematic parameter study for ultra-fine fiber fabrication via electrospinning process, *Polymer* 46:6128–6134.
26. Gersbach C. A., Byers B. A., Pavlath G. K., Guldberg R. E., Garcia A. J. (2004) Runx2/Cbfa1-genetically engineered skeletal myoblasts mineralize collagen scaffolds in vitro, *Biotechnol Bioeng*, 88:369–378.

27. Shields K. J., Beckman M. J., Bowlin G. L., Wayne J. S. (2004) Mechanical properties and cellular proliferation of electrospun collagen type II,*Tissue Eng,* 10:1510–1517.

28. Bhattarai N., Edmondson D., Veiseh O., Matsen F. A., Zhang M. (2005) Electrospun chitosan-based nanofibers and their cellular compatibility, *Biomaterials*, 26:6176–6184.

29. Geng X., Kwon O. H., Jang J. (2005) Electrospinning of chitosan dissolved in concentrated acetic acid solution, *Biomaterials*, 26:5427–5432.

30. Um I. C., Fang D., Hsiao B. S., Okamoto A., Chu B. (2004) Electro-spinning and electro-blowing of hyaluronic acid, *Biomacromolecules*, 5:1428–1436.

31. Jin H. J., Chen J., Karageorgiou V., Altman G. H., Kaplan D. L. (2004) Human bone marrow stromal cell responses on electrospun silk fibroin mats, *Biomaterials*, 25:1039–1047.

32. Cai J., Peng X., Nelson K. D., Eberhart R., Smith G. M. (2005) Permeable guidance channels containing microfilament scaffolds enhance axon growth and maturation, *J Biomed Mater Res A*, 75:374–386.

33. Hadlock T., Elisseeff J., Langer R., Vacanti J., Cheney M. (1998) A tissue engineered conduit for peripheral nerve repair, *Arch Otolaryngol Head Neck Surg*, 124:1081–1086.

34. Luis A. L., Rodrigues J. M., Geuna S., Amado S., Shirosaki Y., Lee J. M., et al. (2008) Use of PLGA 90:10 scaffolds enriched with in vitro-differentiated neural cells for repairing rat sciatic nerve defects, *Tissue Eng Part A*, 14:979–993.

35. Evans G. R., Brandt K., Niederbichler A. D., Chauvin P., Herrman S., Bogle M., et al. (2000) Clinical long-term in vivo evaluation of poly(L-lactic acid) porous conduits for peripheral nerve regeneration, *J Biomater Sci Polym Ed*, 11:869–878.

36. Fairman R., Akerfeldt K. S. (2005) Peptides as novel smart materials, *Curr Opin Struct Biol*, 15:453–463.

37. Taraballi F., Campione M., Sassella A., Vescovi A., Paleari A., Hwang W. et al. (2009) Effect of functionalization on the self-assembling propensity of β-sheet forming peptides, *Soft Matter*, 5:660–668.

38. Taraballi F., Natalello A., Campione M., Villa O., Doglia S. M., Paleari A., et al. (2010) Glycine-spacers influence functional motifs exposure and self assembling propensity of functionalized substrates tailored for neural stem cell cultures, *Frontiers Neuroeng*, 3:1.

39. Gelain F., Bottai D., Vescovi A., Zhang S. (2006) Designer self-assembling peptide nanofiber scaffolds for adult mouse neural stem cell 3-dimensional cultures, *PLoS ONE*, e119:1.

40. Schense J. C., Bloch J., Aebischer P., Hubbell J. A. (2000) Enzymatic incorporation of bioactive peptides into fibrin matrices enhances neurite extension, *Nat Biotechnol*, 18:415–419.

41. Ma P. X., Zhang R. (1999) Synthetic nano-scale fibrous extracellular matrix, *J Biomed Mater Res*, 46:60–72.

42. Ma P. X., Zhang R. (2001) Microtubular architecture of biodegradable polymer scaffolds, *J Biochem Mater Res Part A*, 56:469–477.

43. Hua F. J., Kim G. E., Lee J. D., Son Y. K., Lee D. S. (2002) Macroporous poly(L-lactide) scaffold 1. Preparation of a macroporous scaffold by liquid-liquid phase separation of a PLLA-dioxane-water system, *J Biomed Mater Res*, 63:161–167.

44. Yang F., Murugan R., Ramakrishna S., Wang X., Ma Y. X., Wang S. (2004) Fabrication of nano-structured porous PLLA scaffold intended for nerve tissue engineering, *Biomaterials*, 25:1891–1900.

45. Liu H., Webster T. J. (2007) Nanomedicine for implants: A review of studies and necessary experimental tools, *Biomaterials*, 28:354–369.

46. Qin X. Y., Kim J. G., Lee J. S. (1999) Synthesis and magnetic properties of nanostructured g-Ni–Fe alloys, *Nanostruct Mater*, 11(2):259–270.

47. Ferrari M (2005) Cancer nanotechnology: opportunities and challenges, *Nat Rev Cancer*, 5(3):161–171.

48. Vasir J. K., Reddy M. K., Labhasetwar V. D. (2005) Nanosystems in drug targeting: opportunities and challenges, *Curr Nanosci*, 1(1):47–64.

49. Webster T. J., Siegel R. W., Bizios R. (1999) Osteoblast adhesion on nanophase ceramics, *Biomaterials*, 20(13):1221–1227.

50. Webster T. J., Ergun C., Doremus R. H. (2000) Enhanced functions of osteoblasts on nanophase ceramics, *Biomaterials*, 21(17):1803–1810.

51. Langer, R., Vacanti, J. (1993) Tissue engineering, *Science*, 260:920–926.

52. Yeong, W.-Y., Chua, C.-K., Leong, K.-F., Chandrasekaran, M. (2004) Review-Rapid Prototyping in tissue engineering: challenges and potential, *Trends Biotechnol*, 22(12):643–652.

53. Chua, C. K., Leong, K. F., Cheah, C. M., Chua, S. W. (2003) Development of a tissue engineering scaffold structure library for rapid prototyping. Part 1: investigation and classification, *Int J Adv Manuf Technol*, 21:291–301.

54. Hollister, S. J. (2005) Porous scaffold design for tissue engineering, *Nat Mater*, 4: 518–524.

55. Kim, B. S., Mooney, D. J. (1998a) Development of biocompatible synthetic extracellular matrices for tissue engineering, *Trends Biotechnol*, 16(5):224–230.

56. Hutmacher, D. W. (2001a) Scaffold design and fabrication technologies for engineering tissues-state of the art and future perspectives, *J Biomater Sci Polym Ed*, 12:107–124.

57. Mikos, A. G., Thorsen, A. J., Czerwonka, L. A., Bao, Y., Langer, R. (1994) Preparation and characterisation of poly(L-lactic acid) foams, *Polymer*, 35:1068–1077.

58. Mooney, D. J., Baldwin, D. F., Suh, N. P., Vacanti, J. P., Langer, R. (1996) Novel approach to fabricate porous sponges of poly(D,L-lactic co-glycolic acid) without the use of organic solvents, *Biomaterials*, 17(14):1417–1422.

59. Cima, G., Vacanti, J. P., Vacanti, C., Inger, D., Mooney, D., Langer, R. (1991) Tissue engineering by cell transplantation using degradable polymer substrates, *J Biomech Eng Trans Am Soc Mech Eng*, 113: 143–151.

60. Lo, H., Ponticiello, M. S., Leong, K. W. (1995) Fabrication of controlled release biodegradable foams by phase separation, *Tissue Eng*, 1:15–28.

61. Thompson, R. C., Yaszemski, M. J., Powers, J. M., Mikos, A. G. (1995) Fabrication of biodegradable polymer scaffolds to engineering trabecular bone, *J Biomater Sci Polym Ed*, 7:23–28.

62. Whang, K., Thomas, C. K., Nuber, G., Healy, K. E. (1995) A novel method to fabricate bioabsorbable scaffolds, *Polymer*, 36:837.

63. Schmitz, J. P., Hollinger, J. O. (1988) A preliminary study of the osteogenic potential of a biodegradable alloplastic-osteoconductive alloimplant, *Clin Orthop*, 237:245–255.

64. Sachlos, E., Czernuszka, J. T. (2003) Making tissue engineering scaffolds work. Review on the application of solid freeform fabrication technology to the production of tissue engineering scaffolds, *Eur Cells Mater*, 5:29–40.

65. Martin, I., Padea, R. F., Vunjak-Novakovic, G., Freed, E. (1998) In vitro differentiation of chick embryo bone marrow stromal cells into cartilaginous and bone-like tissues, *J Orthop Res*, 16:181–189.

66. Marks, S. C., Odgren, P. R. (2002) Structure and development of the skeleton, in *Principles of Bone Biology*, Academic press, 3–15.

67. Tortora, G.J., Derrickson, B., (2009), The Skeletal System: Bone Tissue. In: Principles of Anatomy and Physiology, 12th Ed., Wiley; Chapter 6, 176–179, 182–187.

68. Wolff, J. (1892) *Das Gesetz der Transformation der Knochen*, Berlin, A. Hirschwald, published with support from the Royal Academy of Sciences in Berlin English Translation by P. Maquet and R. Furlong, Berlin.

69. Katz, J. L. (1996) Orthopaedic applications, in *Biomaterials Science: An Introduction to Materials in Medicine* (ed. Ratner, B. D., Hoffman, A. S., Schoen, F. J., Lemons, J. E.) Academic Press, Chapter 7, Part 7.7, pp. 335–356.

70. Gutwein LG, Tepper F, Webster TJ. (2004) Increased osteoblast function on nanofibered alumina, *26th Annual American Ceramic Society Meeting*, Cocoa Beach, FL.

71. Webster TJ, Ergun CD, Siegel RW, Bizios R. (2001) Enhanced functions of osteoclast-like cells on nanophase ceramics, *Biomaterials*, 22(11):1327–1333.

72. Webster T. J., Siegel R. W., Bizios R. (2001) Enhanced surface and mechanical properties of nanophase ceramics to achieve orthopaedic/dental implant efficacy, *Key Eng Mater*, 192–195:321–324.

73. Vance R. J., Miller D. C., Thapa A., Haberstroh K. M., Webster T. J. (2004) Decreased fibroblast cell density on chemically degraded poly-lacticco-glycolic acid, polyurethane, and polycaprolactone, *Biomaterials*, 25(11):2095–2103.

74. Webster T. J., Ejiofor J. U. (2004) Increased osteoblast adhesion on nanophase metals: Ti, Ti_6Al_4V, and CoCrMo, *Biomaterials*, 25(19): 4731–4739.

75. Thapa A., Miller D. C., Webster T. J., et al. (2003) Nano-structured polymers enhance bladder smooth muscle cell function, *Biomaterials*, 24:2915–2926.

76. Schift H., Heyderman L. J., Padeste C., Gobrecht J. (2002) Chemical nanopatterning using hot embossing lithography, *Microelectron Eng*, 61–62:423–428.

77. Palin E., Liu H., Webster T. J. (2005) Mimicking the nanofeatures of bone increases bone-forming cell adhesion and proliferation, *Nanotechnology*, 16:1828–1835.

78. Kay S., Thapa A., Haberstroh K. M., Webster T. J. (2002) Nanostructured polymer/nanophase ceramic composites enhance osteoblast and chondrocyte adhesion, *Tissue Eng*, 8(5):753–761.

79. Liu H., Slamovich E. B., Webster T. J. (2005) Improved dispersion of nanophase titania in PLGA enhances osteoblast adhesion, *Ceramic Transactions, vol. 159, Ceramic Nanomaterials and Nanotechnology*

III—Proceedings of the 106th Annual Meeting of the American Ceramic Society, pp. 247–55.

80. Liu H., Slamovich E. B., Webster T. J. (2005) Increased osteoblast functions on nanophase titania dispersed in poly-lactic-co-glycolic acid composites, *Nanotechnology*, 16(7):S601–S608.
81. Price R. L., Waid M. C., Haberstroh K. M., Webster T. J. (2003) Selective bone cell adhesion on formulations containing carbon nanofibers, *Biomaterials*, 24(11):1877–1887.
82. Tamai N., Myoui A., Hirao M. (2005) A new biotechnology for articular cartilage repair: subchondral implantation of a composite of interconnected porous hydroxyapatite, synthetic polymer (PLAPEG), and bone morphogenetic protein-2 (rhBMP-2), *Osteo Arth Cartilage*, 13:405–417.
83. Park G. E., Pattison M. A., Park K. (2005) Accelerated chondrocyte functions on NaOH-treated PLGA scaffolds, *Biomaterials*, 26: 3075–3082.
84. Cheng L, Zhang SM, Chen PP. (2006) Fabrication and characterization of nano-hydroxyapatite/poly (D,L-lactide) composite porous scaffolds for human cartilage tissue engineering. *Key Eng Mater*, 309–311 II, *Bioceramics*, 18:943–946.
85. Matthews JA, Boland ED, Wnek GE, et al. (2003) Electrospinning of collagen type II: a feasibility study, *J Bioactive Compatible Polym*, 18(2):125–134.
86. Simon P., Kasimir M. T., Seebacher G., Weigel G., Ullrich R., Salzer-Muhar U., Rieder E., Wolner E. (2003) Early failure of the tissue-engineered porcine heart valve SYNERGRAFT in pediatric patients, *Eur J Cardiothorac Surg*, 23:1002–1006.
87. W. Erdbrugger, W. Konertz, P. M. Dohmen, S. Posner, H. Ellerbrok, O. E. Brodde, H. Robenek, D. Modersohn, A. Pruss, S. Holinski, M. Stein-Konertz, G. Pauli (2006) Decellularized xenogenic heart valves reveal remodeling and growth potential in vivo, *Tissue Eng.* 12:2059–2068.
88. P. M. Dohmen, W. Konertz (2006) Results with decellularized xenografts, *Circ Res*, 99:e10.
89. A. Bayrak, M. Tyralla, L. Ladhoff, M. Schleicher, U. A. Stock, H. D. Volk, M. Seifert (2010) Human immune responses to porcine xenogeneic matrices and their extracellular matrix constituents in vitro, *Biomaterials*, 31:3793–3803.
90. D. Seliktar, R. A. Black, R. P. Vito, R. M. Nerem (2000) Dynamic mechanical conditioning of collagen-gel blood vessel constructs induces remodeling in vitro, *Ann Biomed Eng*, 28:351–362.

91. S. Jockenhoevel, K. Chalabi, J. S. Sachweh, H. V. Groesdonk, L. Demircan, M. Grossmann, G. Zund, B. J. Messmer (2001) Tissue engineering: complete autologous valve conduit—a new moulding technique, *Thorac Cardiovasc Surg*, 49:287–290.

92. C. L. Cummings, D. Gawlitta, R. M. Nerem, J. P. Stegemann (2004) Properties of engineered vascular constructs made from collagen, fibrin and collagen–fibrin mixtures, *Biomaterials*, 25:3699–3706.

93. T. Aper, A. Schmidt, M. Duchrow, H. P. Bruch (2007) Autologous blood vessels engineered from peripheral blood samples, *Eur J Vasc Endovasc Surg*, 33:33–39.

94. T. C. Flanagan, C. Cornelissen, S. Koch, B. Tschoeke, J. S. Sadweh, T. Schmitz-Rode, S. Jockenhoevel (2007) The in vitro development of autologous fibrin-based tissueengineered heart valves through optimised dynamic conditioning, *Biomaterials*, 28:3388–3397.

95. P. S. Robinson, S. L. Johnson, M. C. Evans, V. H. Barocas, R. T. Tranquillo (2008) Functional tissue-engineered valves from cell-remodeled fibrin with commissural alignment of cell-produced collagen, *Tissue Eng A*, 14:83–95.

96. J. Y. Liu, D. D. Swartz, H. F. Peng, S. F. Gugino, J. A. Russell, S. T. Andreadis (2007) Functional tissue-engineered blood vessels from bone marrow progenitor cells, *Cardiovasc Res*, 75:618–628.

97. T. N. Huynh, R. T. Tranquillo (2010) Fusion of concentrically layered tubular tissue constructs increases burst strength, *Ann. Biomed. Eng.* 38:2226–2236.

98. Z. H. Syedain, L. A. Meier, J. W. Bjork, A. Lee, R. T. Tranquillo (2010) Implantable arterial grafts from human fibroblasts and fibrin using a multi-graft pulsed flow-stretch bioreactor with noninvasive strength monitoring, *Biomaterials*, 32:714–722.

99. J. A. Stella, J. Liao, Y. Hong, D. W. Merryman, W. R. Wagner, M. S. Sacks (2008) Tissue-to cellular level deformation coupling in cell micro-integrated elastomeric scaffolds, *Biomaterials*, 29:3228–3236.

100. V. L. Sales, G. C. Engelmayr, J. A. Johnson, J. Gao, Y. Wang, M. S. Sacks, J. E. Mayer (2007) Protein precoating of elastomeric tissue-engineering scaffolds increased cellularity, enhanced extracellular matrix protein production, and differentially regulated the phenotypes of circulating endothelial progenitor cells, *Circulation*, 116:I55–I63.

101. Chew S. Y., Mi R., Hoke A., Leong K. W. (2008) The effect of the alignment of electrospun fibrous scaffolds on Schwann cell maturation, *Biomaterials*, 29:653–661.

102. Sun Z. C., Zussman E., Yarin A. L., Wendorff J. H., Greiner A. (2003) Compound core–shell polymer nanofibers by co-electrospinning. *Adv Mater*, 15:1929–1932.

103. Zhang Y., Huang Z. M., Xu X., Lim C. T., Ramakrishna S. (2004) Preparation of core–shell structured PCL-r-gelatin bi-component nanofibers by coaxial electrospinning, *Chem Mater*, 16:3406–3409.

104. Desai T. A. (2000) Micro- and nanoscale structures for tissue engineering constructs, *Med Eng Phys*, 22:595–606.

105. Flemming R. G., Murphy C. J., Abrams G. A., Goodman S. L., Nealey P. F. (1999) Effects of synthetic micro- and nano-structured surfaces on cell behavior, *Biomaterials*, 20:573–588.

106. Liu X., Ma P. X. (2004) Polymeric scaffolds for bone tissue engineering, *Ann Biomed Eng*, 32:477–486.

107. Koh H. S., Yong T., Chan C. K., Ramakrishna S. (2008) Enhancement of neurite outgrowth using nano-structured scaffolds coupled with laminin, *Biomaterials*, 29:3574–3582.

108. Schnell E., Klinkhammer K., Balzer S., Brook G., Klee D., Dalton P., et al. (2007) Guidance of glial cell migration and axonal growth on electrospun nanofibers of poly-ε-caprolactone and a collagen/poly-ε-caprolactone blend, *Biomaterials*, 28:3012–3025.

109. http://personalpages.manchester.ac.uk/staff/R.Ulijn/rein_lectures.html.

110. Rho J. Y., Kuhn-Spearing L., Zioupos P. (1998) Mechanical properties and the hierarchical structure of bone, *Med Eng Phys*, 20:92–102.

111. Webster T. J. (2001) Nanophase ceramics: the future orthopedic and dental implant material. In: *Advances in Chemical Engineering Volume 27: Nanostructured Materials* (ed. Ying J. Y.). San Diego, CA: Academic Press; pp. 126–160.

112. Lopes M. A., Monteiro F. J., Santos J. D., et al. (1999) Hydrophobicity, surface tension, and zeta potential measurements of glass-reinforced hydroxyapatite composites. *J Biomed Mater Res*, 45(4):370–375.

113. Miller D. C. (2006) *Nanostructured polymers for vascular grafts*, PhD thesis, Purdue University, West Lafayette, IN.

Chapter 10

Biocompatible 2D and 3D Polymeric Scaffolds for Medical Devices

Masaru Tanaka
Biomaterials Science and Tissue Engineering Group,
Department of Biochemical Engineering
and Department of Polymer Science and Engineering,
Graduate School of Science and Engineering,
Yamagata University, Jonan 4-3-16, Yonezawa, Japan
tanaka@yz.yamagata-u.ac.jp

Polymeric biomaterials have a significant impact on an ageing society. Biocompatible surfaces have emerged during the past decades to promise extraordinary breakthroughs in a wide range of diagnostic and therapeutic endeavours. Understanding and controlling the interfacial interactions of polymeric biomaterials with biological entities, such as proteins and biological cells are of paramount importance towards their successful implementation in biomedical applications. Here, we highlight recent developments of flat (2D) and three-dimensional (3D) polymeric biomaterials for medical devices and tissue engineering.

Horizons in Clinical Nanomedicine
Edited by Varvara Karagkiozaki and Stergios Logothetidis

ISBN 978-981-4411-56-1 (Hardcover), 978-981-4411-57-8 (eBook)
www.panstanford.com

10.1 Introduction: An Explanation of the Design and Research on Polymeric Biomaterials

The variety of polymeric biomaterials with distinct chemical structures, and with the precise control of the molecular architecture and morphology, rationalise the numerous uses of polymers in academia and industry over the past few decades. For example, biocompatible polymers are used as artificial organs and drug delivery systems [1, 2]. The nature and biological fate of the polymers, however, depend on their molecular architecture, self-organisation and biocompatibility.

In biomedical applications, there are continuous efforts to enhance methods, materials, and devices. The recent development of novel biomaterials and their application to medical problems have dramatically improved the treatment of many diseases [3–5]. Although various types of materials in medicine have been used widely, many biomaterials lack the desired functional properties to interface with biological systems and have not been engineered for optimum performance. Therefore, there is an increasing demand to develop novel materials to address such problems in the nano-biomedicine arena. Self-organised 2D and 3D biocompatible polymers are a class of new generation of biomaterials that have demonstrated great potential for biomedical devices, scaffolds for tissue engineering, drug delivery and biosensors and actuators.

In this chapter, we describe the recent development of polymeric biomaterials for various applications in biomedical devices and tissue engineering. Here, we present various synthetic and physical strategies for the preparation of biomaterials scaffolds, which include characteristic properties of biomaterials, biocompatibility, and 3D patterning of polymer based biomaterials in a self-organised manner. In addition, we describe the applications of polymer-based biomaterials in tissue engineering and medical devices. The advantages and drawbacks in the devices application are also discussed.

10.2 Biocompatible Polymers

Polymeric materials for medical devices that may come in contact with human blood should have the capacity to resist protein adsorption and blood cell adhesion, thus triggering the organism's defence mechanism [6]. Some biocompatible polymer surfaces have been developed, and they fall into the following three categories: (i) hydrophilic surfaces [7], (ii) surfaces with micro-phase-separated domains [8], and (iii) biomembrane-like surfaces [9], including zwitterionic groups [10–13]. Physicochemical properties including surface charge, wettability, surface free energy, stiffness, topography, and the presence of specific chemical functionalities and surface bound water appear to bear an instrumental role in the biological response induced by the polymers [14–18]. New generation polymer: Poly(2-methoxyethyl acrylate) (PMEA) (Fig. 10.1) shows excellent bio and blood compatibility, and has been approved for medical use by the FDA [19]. For instance, PMEA coated circuits exhibit significantly reduced blood cell activation when used in cardiopulmonary bypass and catheters. It has been maintained that PMEA's compatibility with platelet, white blood cells, complement and coagulation systems has been dictated by the presence of the *intermediate water* (see Section 10.4) [20–22].

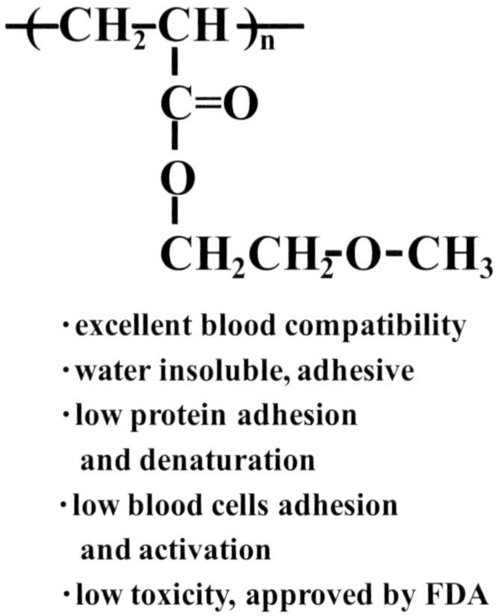

• excellent blood compatibility
• water insoluble, adhesive
• low protein adhesion and denaturation
• low blood cells adhesion and activation
• low toxicity, approved by FDA

Figure 10.1 Chemical structure and properties of novel biocompatible polymer: poly(2-methoxyethyl acrylate) (PMEA).

It should be noted that the word 'biocompatible' is used in general as the term evaluating properties of materials which do not cause adverse effects when the materials come into contact with living organisms, such as proteins, biological cells and tissues. This chapter primarily deals with 'biocompatibility' of polymer materials against various biological elements in the blood flow system.

10.3 Protein Adsorption on Polymer Surfaces

One of the important properties on biocompatibility is the amount of proteins absorbed on the polymer surface. Well-known methods to determine the amount of adsorption are infrared spectroscopy (IR), ultraviolet spectroscopy (UV), X-ray photoelectron spectroscopy (XPS), radioisotope-labelled immunoassay (RI) and circular dichroism (CD) [23–29]. In situ studies on adsorption properties, radioisotope-labelling and fluorescent-labelling techniques have been reported. One of the methods which have been successfully applied to get information on adsorbed proteins is total internal reflection fluorescence spectroscopy (TIRF). The high sensitivity of fluorescence spectroscopy enables the quantification of small amounts of the adsorbed proteins, including competitive adsorption, interfacial conformation changes, and the surface mobility of the adsorbed proteins. Surface plasmon resonance has been applied to in situ detection of adsorbed protein. Also, as an effective and easy method to analyse the in situ biomolecular interaction, quartz crystal microbalance (QCM) has been recommended [30, 31]. QCM is known to be a very sensitive mass-measuring device in air and in aqueous solution. The resonance frequency of the QCM electrode decreases linearly with the increase in the mass of the electrode due to the adsorption of some compounds, and the sensitivity at the nanogram level. Several researchers have reported the interaction between polymeric biomaterial and protein by using this method.

Protein adsorption behaviours on various kinds of polymer surfaces have been extensively investigated. It is reported that the important factor expressing the biocompatibility is not the amount of adsorbed proteins on the surfaces, but the structure or orientation of the adsorbed proteins [25]. A very interesting objective is whether the protein adsorption is reversible or not, and much research on it has been carried out. There are many reports that insisted on the irreversible adsorption of proteins on the polymer surface [26], whereas, some researchers reported the reversible adsorption [32].

Thus, it is important to analyse the kinetics during adsorption of protein on a polymer surface in addition to the adsorption amount of the protein when the biocompatibility of polymers is discussed [27].

In order to clarify the reasons for excellent biocompatibility of PMEA, the amount and kinetics in the early stage of plasma protein adsorbed onto PMEA and the secondary structure of the protein were investigated. The amount of protein adsorbed onto PMEA was very small, and the quantity was similar to that adsorbed onto poly(2-hydroxyethyl methacrylate) (PHEMA) [30, 31]. Circular dichroism (CD) spectroscopy revealed a significant conformation change in proteins adsorbed onto PHEMA, whereas the conformational change of the proteins adsorbed on PMEA was very small [31]. Using the quartz crystal microbalance (QCM) measurement, we investigated the adsorption/desorption behaviour of proteins on the PMEA surface in terms of their binding constants and association and dissociation rates. The CD and QCM results suggested that the excellent biocompatibility of PMEA is related to the low denaturation and the high dissociation rates of the proteins attached onto PMEA [31]. The schematic representation of proposed adsorption of BSA, Fibrinogen and IgG on PMEA, PHEMA and polypropylene (PP) is presented in Fig. 10.2.

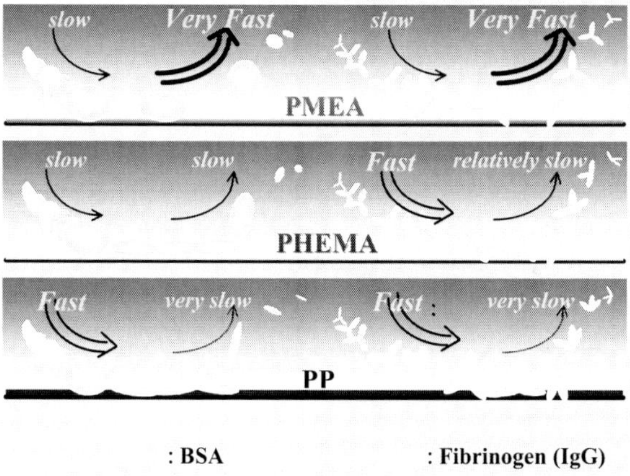

Figure 10.2 Schematic representation of the assumed adsorption state of BSA and Fibrinogen or IgG. The low denaturation and the high dissociation rates of the proteins adsorbed onto PMEA compared to PHEMA and Polypropylene (PP).

The adhesion force between PMEA and fibrinogen as well as PMEA and bovine serum albumin (BSA) were measured by atomic force microscopy. The PMEA surface showed almost no adhesion to native protein molecules [31]. The denaturation of the adsorbed protein could lead to platelet activation and subsequent thrombus formation. In other words, when the protein molecule which adsorbs onto a polymer surface retains its native conformation, platelets cannot adhere to the surface. If we were able to develop a polymer surface that does not denature proteins, that surface would be biocompatible.

10.4 Water Structure and Dynamics of Polymer

Although considerable theoretical and experimental efforts have been devoted to clarifying this issue in the past few decades, the factors responsible for biocompatibility of polymers have not been elucidated [33]. Water molecules serve as a medium for adhesion, and play a role in cell morphology and other cellular functions. Water is thought to be a fundamental factor in the biological response induced by artificial materials. Many researchers have insisted that water structure on a polymer surface is one of the key factors for its biocompatibility. However, the proposed structures and/or the functions of water are different in many cases, and there is little consistency among structures. Detailed studies on the dynamics and structure of hydrated polymers are required to clarify the mechanism underlying the biocompatibility of polymers.

The hydrated water in polymer could be classified into three types: *free water* (or *freezing water*), *freezing-bound water* (or *intermediate water*), and *non-freezing water* (or *non-freezing-bound water*) (Fig. 10.3). The hydrated PMEA possessed a unique water structure, observed as cold crystallization of water in differential scanning calorimetry (DSC) (Fig. 10.4). Cold crystallization is interpreted as ice formation at low temperatures below 0°C, an attribute of *intermediate water* in PMEA. The presence of three types of water in PMEA is supported by the results of attenuated total reflection infrared (ATR-IR) spectroscopy [34] and NMR [35]. The *intermediate water* molecules interact weakly with the methoxy group of PMEA. While investigating the main factor responsible for the excellent biocompatibility of PMEA, it is important to reveal the *intermediate water* structure on the polymer surface.

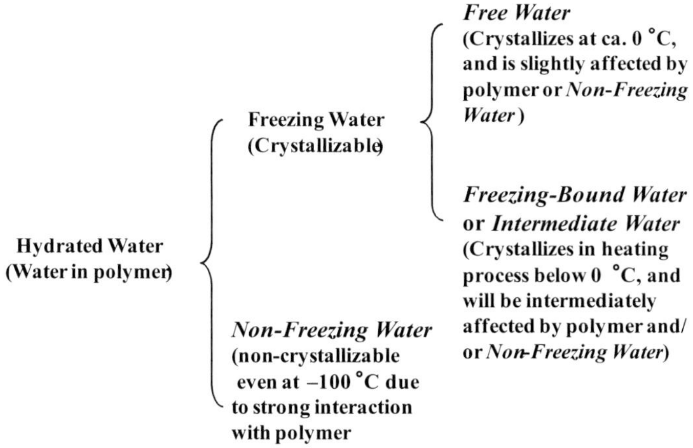

Figure 10.3 Classification of water in hydrated polymer. The hydrated water in a polymer can be classified into three types—*non-freezing water*, *freezing-bound water* (*intermediate water*) and *free water*—on the basis of the equilibrium water content and the enthalpy changes due to the phase transition observed by differential scanning calorimetric (DSC) analysis.

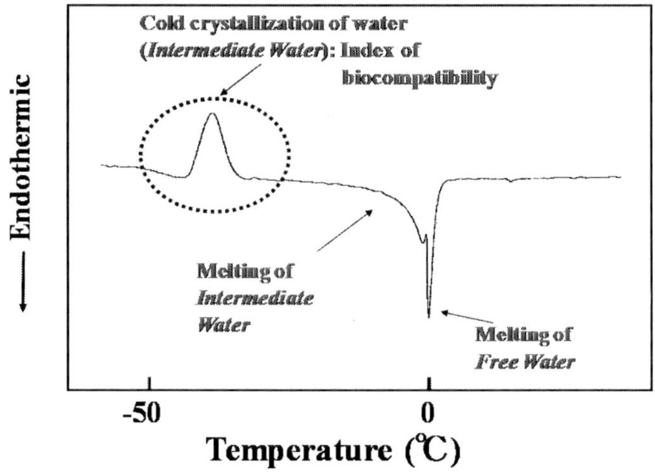

Figure 10.4 Differential scanning calorimetric (DSC) heating curve of a biocompatible polymer: PMEA–water system.

When the polymer surface comes in contact with blood, it first absorbs water, and a specific water structure is formed on the surface. If the resulting structure formed is the first layer, then the layers will be in the following order: polymer surface→*non-freezing water*→(*intermediate water*)→*free water*→*bulk water* (Fig. 10.5). *Free water* is unlikely to activate the system, and is unable to shield the polymer surface or *non-freezing* Water on the polymer surface; this is because *free water* exchanges with *bulk water* very freely, resulting in a structure similar to *bulk water*. Since *intermediate water* is weakly (loosely) bound to the polymer molecule or *non-freezing water* (tightly bound water), this layer forms a rather stable structure compared to that of *free water*. Thus, when the *intermediate water* layer becomes adequately thick, it prevents the cell or the protein from directly contacting the polymer surface or *non-freezing water*.

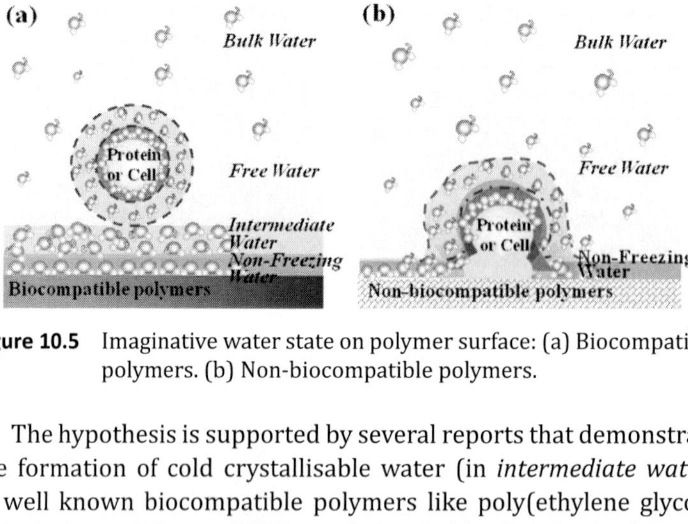

Figure 10.5 Imaginative water state on polymer surface: (a) Biocompatible polymers. (b) Non-biocompatible polymers.

The hypothesis is supported by several reports that demonstrate the formation of cold crystallisable water (in *intermediate water*) in well known biocompatible polymers like poly(ethylene glycol), polyvinylpyrrolidone (PVP), poly(methylvinyl ether) (PMVE), poly(2-methacryloyloxyethyl phosphorylcholine) (PMPC), poly(tetrahydrofulfuryl acrylate) (PTHFA), poly(2-(2-ethoxyethoxy) ethyl acrylate) and other biocompatible polymers, gelatin, albumin, cytochrome C, and various polysaccharides, including hyaluronan, alginate, and gum [36–41]. On the other hand, cold crystallization of water was not observed in hydrated PMEA analogous polymers, which do not show excellent biocompatibility. Based on these findings, the *intermediate water*, which prevents the biocomponents from directly contacting the polymer surface, or *non-freezing water* on the polymer surface, must play an important role in the excellent biocompatibility of PMEA. Here, we proposed the "*intermediate water*" concept; the water exhibited both clear peak for cold crystallization in DSC chart and a strong peak at 3400 cm^{-1} in a time-resolved IR spectrum; the localised hydration structure consisting of the three hydrated water in PMEA.

10.5 3D Polymeric Scaffolds for Tissue Engineering

Tissue engineering aims to replace, repair, or regenerate tissue or organ function and to create artificial tissues and organs for transplantation. Autografts and allografts represent current strategies for surgical intervention and subsequent tissue repair, but each possesses limitations, such as donor-site morbidity with the use of autograft and the risk of disease transmission with the use of allograft. It is mandatory for any optimal tissue engineering scaffold to act as a temporary 3D support for cell adhesion, proliferation and extracellular matrix (ECM) deposition. The 3D scaffolds fabricated from polymers play critical roles in tissue engineering as temporary ECMs and have, therefore, been widely used [42]. Strict requirements for scaffolds are biocompatibility, a design closely resembling the natural extracellular structure, an appropriate surface chemistry to promote cell adhesion, proliferation and differentiation, and a sufficient mechanical strength to withstand in vivo stresses and physiological loading. The biocompatible 3D scaffolds of appropriate architectures facilitate cell adhesion, proliferation, differentiation, and eventual tissue regeneration in a natural manner. Cultures of cells on 3D scaffolds may create tissues suitable either for applications in reconstructive surgery or that can serve as novel in vitro model systems. It is well known that 3D nano/micro patterns fabricated on the surfaces of such scaffolds significantly influence the morphology, proliferation, differentiation, and function of various cells [43, 44]. Finally, the degradation of the ideal scaffold should proceed in a controlled way, still keeping a sufficient structural integrity until the newly grown tissue has replaced the scaffold's supporting functions. In this chapter, we describe top-down and bottom-up fabrication methods.

10.6 Methods of 3D Scaffold Fabrication

10.6.1 Top-Down Fabrication

A recent advancement of polymer patterning surfaces [45–47] has attracted attention in tissue engineering. Polymeric biomaterials can be used as integral components in micro- and nano- devices using various approaches that have been developed. Here, we introduce the recent advances in patterning of polymeric biomaterials using direct-write lithography, nano/micro imprinting, inkjet printing, nano/microcontact printing, robotic deposition and their application in tissue engineering.

Direct-write lithography is widely used in research because it is inexpensive and offers flexibility in pattern generation. Currently, electron-beam lithography is the most commonly used direct-write technique. 3D printing is the first rapid prototyping devices developed for tissue engineering applications and regenerative medicine. Rapid prototyping techniques can process a wide number of biomaterials in a custom-made shape and with the desired mechanical properties for specific applications. Products of this technology have finely controllable porosity, pore size and shape, and have a completely interconnected pore network allowing better cell migration and nutrient perfusion than those provided by the 3D scaffolds fabricated with conventional techniques. The following section focuses on the recent advancement in lithography and 3D printing, and self-organisation techniques for porous scaffold fabrication in bio-related and medicinal research, particularly in the study of cells and tissue engineering applications.

10.6.2 Bottom-Up Fabrication

A number of highly useful fabrication methods, including phase separation [48], porogen leaching [49], gas foaming [50], fibre meshing [51], supercritical fluid processing [52], laser sintering [53], and three-dimensional printing [54], lithography techniques [55], more recently self-organized honeycomb porous structures [56] have been developed and used to fabricate tissue engineering scaffolds from polymers.

As mentioned before, many techniques have been developed and used to fabricate tissue engineering and medical devices scaffolds from polymeric materials. Each of the above mentioned techniques is unique and requires specific process conditions. These fabrication techniques, however, require a large amount of energy and involve numerous steps. In addition, the starting materials used in these techniques have limited availability. As a result, there is a high demand for suitable alternative methods that can be used to prepare scaffolds from functional polymers [57, 58].

Biology is full of examples of so-called bottom-up fabrication—self organisation of organic and inorganic components into hierarchical, sophisticated structures under mild ambient conditions. For example, established interference patterns found on butterfly wings show the same self-cleaning properties as do lotus leaves and photonic crystals [59]. Gecko feet, research on which has been a major topic of interest, have nearly 500,000 keratinous hairs or setae and can generate a strong adhesive force [60, 61]. Inspired by the organisation exhibited in biological structures, it is expected the self-organisation of polymers into hierarchical and sophisticated structures to be a suitable alternative to the conventional techniques of nano/micro fabrication of medical devices scaffolds. Self-fabrication of polymers requires physiological conditions, in contrast to conventional nano/micro fabrication techniques, which require harsh conditions [56, 57]. This self-organisation method requires neither high temperatures nor harsh chemical conditions and can, therefore, be used with a variety of polymers.

The self-organisation technique can be used with various polymers because of its physical generality. Regular structures have been formed during the casting of polymer solutions on solid surfaces; for example, self-organised honeycomb-patterned polymer films (honeycomb films) with highly regular porous structures can be prepared using various polymers under different conditions [62–69]. The micrographs (Fig. 10.6) show a highly regular hexagonal pore arrangement (honeycomb patterned structure) as well as a uniform pore structure. Using SEM, the honeycomb film had a top layer and a bottom layer and laterally interconnected structures with side pores (Fig. 10.6). The tilted and side-view images of the honeycomb film reveal two hexagonal lattices connected at the vertices of the hexagons by vertical columns (Fig. 10.6).

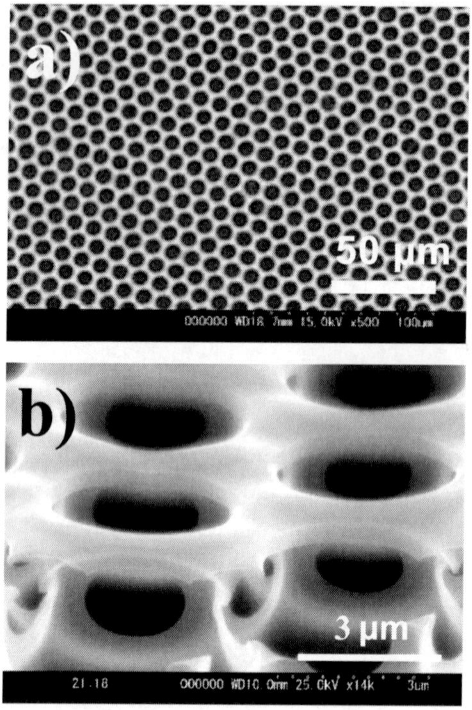

Figure 10.6 Scanning electron micrographs of the structure of honeycomb film. (a) Top view. (b) Tilted view.

This double-layered self-organised structure reflects the morphology of the template, which is a self-organised and hexagonally packed array of water droplets. The use of a water-immiscible solvent for preparing the polymer solution resulted in condensation of water droplets after the solution was cast onto the substrate; this was due to the evaporative cooling of humid air used in the experiment. Self-packed, mono-dispersed water droplets that formed on the surface of the polymer solution acted as a temporary template for the pores. In general, the condensed water droplets are not stable and eventually start to coalesce. In order to prepare a highly regular honeycomb film, it is necessary to stabilise the water droplets. The amphiphilic polymer prevents the fusion of the water droplets during pattern formation. The amphiphilic polymer acts as a surfactant and contributes to the stabilisation of the water droplets at the interface of the polymer solution and water. As a result, fusion of the water droplets is prevented by the intervening amphiphilic polymer layer. Most polymers dissolved in a water-immiscible solvent can thus be made to form honeycomb films through the addition of amphiphilic compounds. Various experimental factors affect the structure of the pores and distribution. A uniform pore size can be achieved by changing the casting conditions [63]. The casting volume controls the pore size of the fabricated honeycomb films. The rims of the films widened as the pore size increased. The porosity of each film was found to be approximately 50%.

It also reported the formation of pincushion structures on a glass substrate after peeling off the top layer of the polymeric honeycomb films using adhesive tape [70]. These polymeric scaffolds are ideal for biomedical applications such as tissue engineering and drug delivery. For such applications, it would be advantageous to create nano-and micropatterned structures that can gradually degrade and be resorbed by the body. The mechanical properties and biodegradability of tissue engineering and medical devices scaffolds should resemble those of healthy tissues during tissue regeneration. The fabrication of hexagonal arrays of biodegradable polymer pincushions could be used as novel tissue engineering scaffolds. As a precursor to these polymer pincushion arrays, highly regular honeycomb films were prepared on the glass substrate by a simple casting technique using essential biodegradable, and biocompatible polymers such as poly (ε-caprolactone) (PCL), poly(L-lactide) (PLA), poly(D,L-lactide-coglycolide) (PLGA), and poly(3-hydroxybutyrate) (PHB) [71]. The use of these polymers in certain clinical practices has been approved by the U.S. Food and Drug Administration; hence, they can be used to fabricate honeycomb films by a self-organisation technique.

When the top layer of the honeycomb film was peeled off using adhesive tape under ambient conditions, arrays of polymer pincushions were formed on both sides of the glass substrate and on the adhesive tape. Pincushions (Fig. 10.7a) on both sides of the glass substrate and on the adhesive tape were generated. Each air hole is surrounded by six pincushions with a diameter of approximately 0.1–0.5 μm. Figure 10.7b shows the peeling Interface of the honeycomb film. For reference, only half of the honeycomb film was formed into a pincushion pattern. The contact angle of the typical pincushion was approximately 150, regardless of the pore size, indicating that its hydrophobicity is greater than that of the honeycomb film, whose contact angle was c.a. 100. Figure 10.78shows the

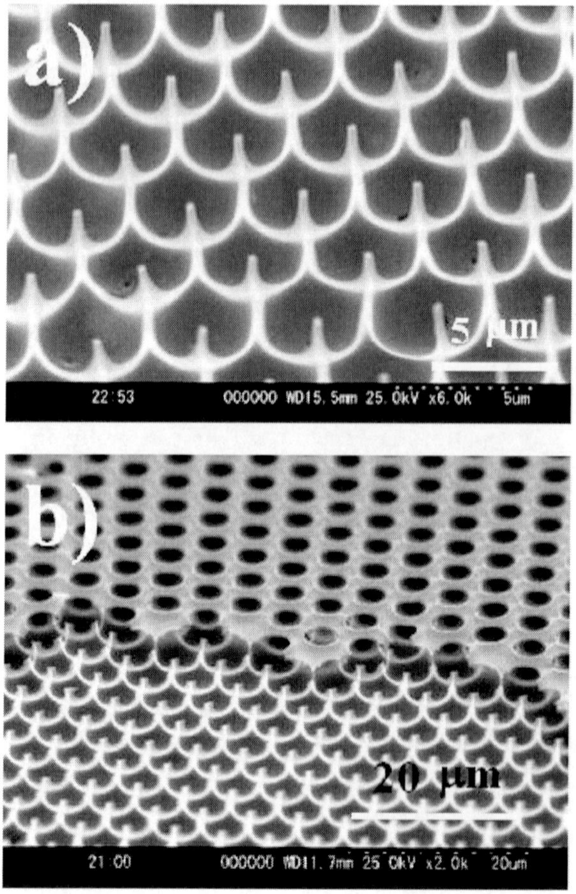

Figure 10.7 Scanning electron micrographs of the surface topography of arrays of polymer pincushions (a) and the peeling interface (b).

tilt-angle SEM images of the pincushion structures obtained from various honeycomb films. The morphologies of these pincushions, which were prepared under the same peeling conditions, can be categorised into two types: vertically aligned and hairly aligned. The film remaining after the top layer of the honeycomb film had been peeled off was able to be removed from the glass substrate as a free-standing film. In contrast, the pincushions formed on the tape were found to adhere strongly to the tape and were flexible. The heights, widths, and distances of separation of the pincushions were dependent on the type of polymer used and the pore size of the original honeycomb film. PCL pincushions showed (Fig. 10.8) an elongated hair-like morphology compared to other polymer pincushions, including the nonbiodegradable polymer controls.

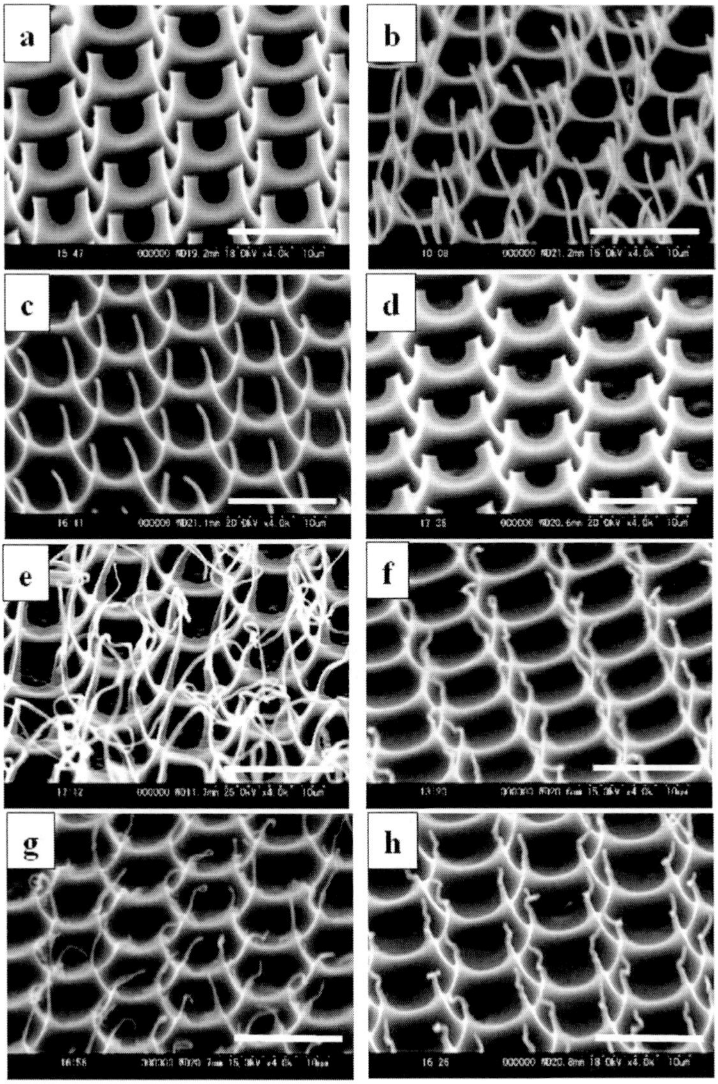

Figure 10.8 Tilted angled SEM images of the polymer pincushions (a) PS, (b) PMMA, (c) PC, (d) PTHFMA, (e) PCL, (f) PLA, (g) PLGA, (h) PHB (scale bar = 10 μm).

The fabrication is performed under physiological conditions, in contrast to conventional industrial processing techniques that require harsh conditions; this makes the fabrication method simpler and less expensive than conventional techniques. It has been reported that the nano structured surfaces utilised for long-term maintenance of stem cell phenotype and multipotency [72] can have a marked influence on cellular function. These pincushion surfaces may improve cell adhesion and positively affect the flow of nutrients and wastes. In particular, biologically inspired hairy pincushions made from polymers may generate significant adhesive force. This fast and inexpensive method of fabricating polymeric biomaterials can be used to produce nano- and micro topographies for cell support scaffolds for tissue engineering, drug delivery, and for use as tissue adhesives for medical applications.

10.7 Control of Cell Adhesion and Functions

Most tissue-derived cells are anchorage-dependent and require attachment to a solid surface for viability and growth. For this reason, cell adhesion to a surface is critical because adhesion precedes other events, such as cell spreading, cell migration, and differentiated cell functions. Thus, the physical and chemical properties of culture substrates are a significant factor in determining the cellular response. The honeycomb films exerted a strong influence on cell morphology, proliferation, cytoskeleton, focal adhesion, and ECM production profiles [73–80]. For example, hepatocytes formed spheroids, and specific activities, such as albumin secretion and urea synthesis, were upregulated by changing the pore size of the films [74]. The proliferation of endothelial cells (ECs) was the highest on the honeycomb films (pore size: 5 μm) [73]. Honeycomb films with a pore size of 3 μm promoted the proliferation of neural stem cells (NSCs), while preventing the differentiation of NSCs into neurons [76]. In addition, the pore size of the honeycomb pattern also affects mesenteric-visceral adipocytes function and a honeycomb film with a pore size of 20 μm had the highest cell functions.

On the other hand, growth of cancer cells on honeycomb films was less than that of cells on control flat films. In 27 of 58 cell lines, more than 50% inhibition was observed [77]. The topography of the honeycomb film in contact with cancer cells has a potential anticancer effect. Thus, the effects of honeycomb films on cellular phenotypes are considered to be dependent on the cell lineage, that is, whether they are ECs, NSCs and other normal, cancer and stem cells. These successes were achieved without the use of growth factors. In these results, the honeycomb films with different pore sizes could regulate the cell adhesion, morphologies, and functions without growth factors (Fig. 10.9).

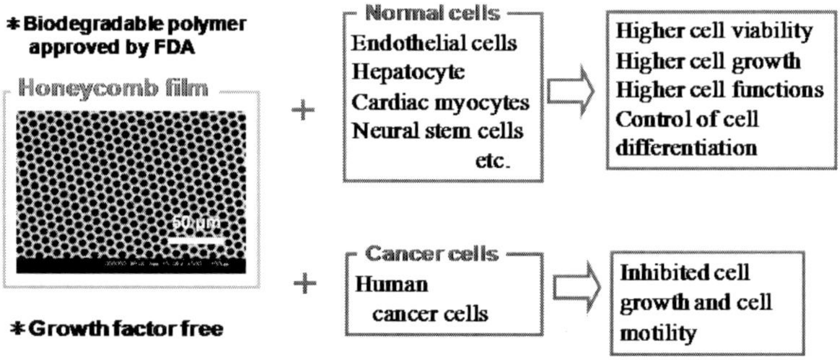

Figure 10.9 Cell behaviour on the honeycomb films. Cell morphologies and specific functions were greatly influenced by topography of the film.

To promote the appropriate use of new or emerging 3D fabrication technologies, this chapter describes an update on the technical specifications, efficacy, safety, and financial considerations regarding the use of stents in the vascular and bile duct. Stents are devices used to maintain or restore the lumen of hollow organs, vessels, and ducts. Stents consist of woven, knitted, or laser cut metal mesh cylinders that exert self-expansive forces until they reach their maximum fixed diameter. They are generally packaged in a compressed form and constrained on a delivery device. Current metallic stents used to treat disease have suboptimal biocompatibility, reducing their clinical efficacy. Stents available for application in the alimentary tract include self-expanding metal stents for oesophageal, gastroduodenal, and colonic malignant obstruction and a self-expanding plastic (polymer) stent for benign or malignant vascular, bile duct, enteral, oesophageal, duodenal, and colonic strictures.

Cancer and vascular disease remains one of the leading causes of premature death in virtually all countries of the world. To prevent tumour ingrowth, the interstices between the metal mesh of stent should be wholly or partially covered by a biocompatible film. Recently, in a vertically open-pored honeycomb film, support for the tubule was given by a metallic tubular mesh, and this was commercialised as a bile duct stent (Fig. 10.10). This has great potential to improve stent efficacy.

Figure 10.10 Novel stent covered with a vertically open-pored honeycomb film.

Furthermore, co-culture systems such as ECs and smooth muscle cells on the inner and outer sides of the tubular honeycomb film are expected to be developed for novel artificial vessels.

10.8 Conclusions and Perspectives

2D and nano- and micro-patterned 3D structures of tissue engineering scaffolds made of biocompatible polymers profoundly influence cell behaviour. We proposed the "*intermediate water*" concept, but it needs the quantitative description of biocompatibility driven by novel approaches. To achieve cost-effective 3D patterning and patterns in multiple length scales, a combination of different patterning techniques will be necessary. The use of both photolithography and printing methods together would allow the fabrication of low-cost patterns with enhanced performance. Recently, self-organisation techniques have shown a very promising approach to fabricate 3D honeycomb structures in a cost effective manner, with exiting in vitro and in vivo results. Such well-ordered, biologically inspired structures could find application in biomedical, photonic, and electronic materials.

Acknowledgements

The author gratefully acknowledges the financial support from the Funding Program for Next Generation World-Leading Researchers (NEXT Program, Japan).

References

1. Hoffman, A. S. (2002) Hydrogels for biomedical applications, *Adv Drug Deliv Rev*, 43, 3–13.
2. Shi, D. (2006) *Introduction to Biomaterials*, World Scientific, Tsinghua University Press, pp. 143–210.
3. Liu, W.G., and Griffith, M., and Li, F. (2008) Alginate microsphere-collagen composite hydrogels for ocular drug delivery and implantation, *J Mater Sci: Mater Med*, 19, 3365–3371.
4. Yang, F., Wang, Y., Zhang, Z., Hsu, B., Jabs, E.W., and Elisseeff, J.H. (2008) The study of abnormal bone development in the Apert syndrome Fgfr2$^{+/S252W}$ mouse using a 3D hydrogel culture model, *Bone*, 43, 55–63.
5. Eljarrat-Binstock, E., Orucov, F., Frucht-Pery, J., Pe'er, J., and Domb, A.J. (2008) Methylprednisolone delivery to the back of the eye using hydrogel iontophoresis, *J Ocular Pharm Ther*, 24, 344–350.
6. Severian, D. (2002) *Polymeric Biomaterials*, Mercel Dekker, New York.
7. Peppas, N. A. (1987) *Hydrogel in Medicine and Pharmacy*, Vol. 2, CRC Press, Boca Raton, FL.
8. Okano, T., Nishiyama, S., Shinohara, I., Akaike, T., Sakurai, Y., Kataoka, K., and Tsuruta, T. (1981) Effect of hydrophilic and hydrophobic microdomains on mode of interaction between block copolymer and blob platelets, *J Biomed Mater Res*,15, 393–403.
9. Ishihara, K., Nomura, H., Mihara, T., Kurita, K., Iwasaki, Y., and Nakabayashi, N. (1998) Why do phospholipid polymers reduce protein adsorption? *J Biomed Mater Res*, 39, 323–330.
10. Holmlin, R.E., Chen, X., Chapman, R.G., Takayama, S., and Whitesides, G.M. (2001) Zwitterionic SAMs that resist nonspecific adsorption of protein from aqueous buffer, *Langmuir*, 17, 2841–2850.
11. Kitano, H., Tada, S., Mori, T., Takaha, K., Gemmei-Ide, M., Tanaka, M., Fukuda, M., and Yokoyama, Y. (2005) Correlation between the structure of water in the vicinity of carboxybetaine polymers and their blood-compatibility, *Langmuir*, 21, 11932–11940.
12. Zhang, Z., Chen, S., Chang, Y., and Jiang, S. (2006) Surface grafted sulfobetaine polymers via atom transfer radical polymerization as superlow fouling coatings, *J Phys Chem B*, 110, 10799–10804.
13. Tada, S., Inaba, C., Mizukami, K., Fujishita, S., Gemmei-Ide, M., Kitano, H., Mochizuki, A., Tanaka, M., and Matsunaga, T. (2009) Anti-biofouling properties of polymers with a carboxybetaine moiety, *Macromol Biosci*, 9, 63–70.

14. Stevens, M. M., and George, J. H. (2005) Exploring and engineering the cell surface interface, *Science*, 310, 1135–1138.

15. Discher, D.E., Moony, D.J., and Zandsrta, P.W. (2009) Growth factors, matrices, and forces combine and control stem cells, *Science*, 324, 1673–1677.

16. Mitragotri, S., and Lahann, J. (2009) Physical approaches to biomaterial design, *Nat Mater*, 8, 15–23.

17. Place, E. S., Evans, N. D., and Stevens, M. M. (2009) Complexity in biomaterials for tissue engineering, *Nat Mater*, 8, 457–470.

18. Nell, A. E., Mädler, L., Velegol, D., Xia, T., Hoek, E. M. V., Somasundaran, P., Klaessig, F., Castranova, V., and Thompson, M. (2009) Understanding biophysicochemical interactions at the nano-bio interface, *Nat Mater*, 8, 543–557.

19. Tanaka, M., Motomura, T., Kawada, M., Anzai, T., Kasori, Y., Shiroya, T., Shimura, K., Onishi, M., and Mochizuki, A. (2000) Blood compatible aspects of Poly(2-methoxyethylacrylate) (PMEA)—Relationship between protein adsorption and platelet adhesion on PMEA surface, *Biomaterials*, 21, 1471–1481.

20. Tanaka, M., Motomura, T., Ishii, N., Shimura, K., Onishi, M., Mochizuki, A., and Hatakeyama, T. (2000) Cold crystallization of water in hydrated poly(2-methoxyethylacrylate) (PMEA), *Polym Int*, 49, 1709–1713.

21. Tanaka, M., Mochizuki, A., Ishii, N., Motomura, T., and Hatakeyama, T. (2002) Study on blood compatibility of poly(2-methoxyethylacrylate). Relationship between water structure and platelet compatibility in poly(2-methoxyethylacrylate-co-2- hydroxyethylmethacrylate), *Biomacromolecules*, 3, 36–41.

22. Tanaka, M., and Mochizuki, A. (2004) Effect of water structure on blood compatibility, Thermal analysis of water in poly(meth)acrylate, *J Biomed Mater Res*, 68A, 684–695.

23. Greenfield, N.J., and Fasman, G.D. (1969) Computed circular dichroism spectra for the evaluation of protein conformation, *Biochemistry*, 8, 4108–4112.

24. Hunter, J. B., and Hunter, S. M. (1987) Quantification of proteins in the low nanogram range by staining with the colloidal gold stain aurodye, *Anal Biochem*, 164, 430–433.

25. Soderquist, M. E., and Walton, A.G. (1980) Structural changes in protein adsorbed on polymer surfaces, *J Colloid Interface Sci*, 75, 386–397.

26. Castillo, E.J, Koenig, J. L., Anderson, J.M., and Lo, J. (1984) Characterization of protein adsorption on soft contact lenses. I. Conformational changes of adsorbed human serum albumin, *Biomaterials*, 5, 319–325.

27. Brash, J.L., and Horbett, T.A. (ed. Brash, J. L., and Horbett, T.A.) (1987) Proteins at interfaces: physicochemical and biochemical studies, vol. 343, ACS Symposium Series, Washington, DC. pp. 1–33.
28. Lenk, T. J., Ratner, B.D., Gendreau, R.,M., and Chittur, K.,K. (1989) IR spectral changes of bovine serum albumin upon surface adsorption, *J Biomed Mater Res*, 23, 549–569.
29. Horbett, T.A., Brash, J.,L.(eds) (1995) *Proteins at Interfaces. II. Fundamental and Applications*, vol. 602, ACS Symposium Series, Washington, DC.
30. Tanaka, M., Mochizuki, A., Motomura, T., Shimura, K., Onishi, M., and Okahata, Y. (2001) In situ studies on protein adsorption onto a poly(2-methoxyethyl acrylate) surface by a quartz crystal microbalance, *Colloids Surf A: Physicochem Eng Aspects*, 193, 145–152.
31. Tanaka, M., Mochizuki, A., Shiroya, T., Motomura, T., Shimura, K., Onishi, M., and Okahata, Y. (2002) Study on kinetics of early stage protein adsorption and desorption on poly(2- methoxyethyl acrylate) (PMEA) surface, *Colloids Surf A: Physicochem Eng Aspects*, 203, 195–204.
32. Minton, A. P. (2001) Effects of excluded surface area and adsorbate clustering on surface adsorption of proteins. II. Kinetic models, *Biophys J*, 80, 1641–1648.
33. Ratner, B. D. (2000) Blood compatibility—a prespective, *J Biomater Sci Polym Ed*, 11, 1107–1119.
34. Morita, S., Tanaka, M., and Ozaki, Y. (2007) Time-resolved in-situ ATR-IR observations of the process of water into a poly(2-methoxyethyl acrylate) (PMEA) film, *Langmuir*, 23, 3750–3761.
35. Miwa, Y., Ishida, H., Saito, H., Tanaka, M., and Mochizuki, A. (2009) Network structures and dynamics of dry and swollen poly(acrylate) s. Characterization of high-and low-frequency-motions as revealed by suppressed or recovered intensities (SRI) analysis of 13 C NMR, *Polymer*, 50, 6091–6099.
36. Harris, M. (1992) *Poly(ethylene glycol) Chemistry, Biotechnical and Biomedical Applications*, Plenum Press, New York.
37. Wolfgang, G.G., and Hatakeyama, H. (1992) *Viscoelasticity of Biomaterials*, ACS Symposium Series 489, American Chemical Society, Washington, DC.
38. Hatakeyma, T., Kasuga, H., Tanaka, M., and Hatakeyama, H. (2007) Cold crystallization of poly (ethylene glcyol)-water systems, *Thermochimica Act*, 465, 59–66.

39. Mochizuki, A., Hatakeyama, T., Tomono, Y., and Tanaka, M. (2009) Water structure and blood compatibility of poly(tetrahydrofulfuryl acrylate), *J Biomater Sci Polym Ed*, 20, 591–603.

40. Hatakayama, T., Tanaka, M., and Hatakayama, H. (2010) Studies on bound water restrained by poly(2-methacryloyloxyethyl phosphorylcholine) (PMPC): comparison of the polysaccharides-water systems, *Acta Biomater*, 6, 2077–2082.

41. Tanaka, M., and Mochizuki, A. (2010) Clalification of blood compatibility mechanism by controlling water structure, *J Biomater Sci Polym Ed*, 21, 1849–1863.

42. Hollister, S.J. (2005) Porous scaffold design for tissue engineering, *Nat Mater*, 4, 518–524.

43. Curtis, A.S.G., and Wilkinson, C. D. W. (1997) Topographical control of cells, *Biomaterials*, 18, 1573–1583.

44. Dalby, M. J., Gadegaard, N., Tare, R., Andar, A., Riehle, M.O., Herzyk, P., Wilkinson, C. D. W, and Oreffo, R. O. C. (2007) The control of human mesenchymal cell differentiation using nanoscale symmetry and disorder, *Nat Mater*, 6, 997–1003.

45. Théry, M., Racine, V., Pépin, A., Piel, M., Chen, Y., Sibarita, J. B., and Bornens, M. (2005) The extracellular matrix guides the orientation of the cell division axis, *Nature Cell Biol*, 7, 947–953.

46. Théry M, Racine V, Piel M, Pépin A, Dimitrov A, Chen Y, Sibarita, J. B., and Bornens, M. (2006) Anisotropy of cell adhesive microenvironment governs cell internal organization and orientation of polarity, *Proc Natl Acad Sci USA*, 103, 19771–19776.

47. Seol, Y. J., Kang, T. Y., and Cho, D. W. (2012) Solid freeform fabrication technology applied to tissue engineering with various biomaterials, *Soft Matter*, 8, 1730–1735.

48. Asefnejad, A., Khorasani, M. T., Behnamghader, A., Farsadzadeh, B., and Bonakdar, S. (2011) Manufacturing of biodegradable polyurethane scaffolds based on polycaprolactone using a phase separation method: physical properties and in vitro assay, *Int J Nanomed*, 6, 2375–2384.

49. Allaf, R. M., and Rivero, I. V. (2011) Fabrication and characterization of interconnected porous biodegradable poly(ε-caprolactone) load bearing scaffolds, *J Mater Sci: Mater Med*, 22(8), 1843–1853.

50. Wang, C. Z., Ho, M, L., Chen, W.C., Chiu, C. C., Hung, Y. L., Wang, C. K., and Wu, S. C. (2011) Characterization and enhancement of chondrogenesis in porous hyaluronic acid-modified scaffolds made of PLGA(75/25) blended with PEI-grafted PLGA(50/50), *Mater Sci Eng C: Mater Biolog Appl*, 31(7), 1343–1351.

51. Saraf, A., Baggett, L. S., Raphael, R. M., Kasper, F. K., and Mikos, A. G. (2010) Regulated non-viral gene delivery from coaxial electrospun fiber mesh scaffolds, *J Control Release*, 143, 95–103.

52. Reverchon, E., Cardea, S., and Rapuano, C. (2008) A new supercritical fluid-based process to produce scaffolds for tissue replacement, *J Supercrit Fluids*, 45(3), 365–373.

53. Duan, B., Wang, M., Zhou, W. Y., Cheung, W. L., Li, Z. Y., and Lu, W. W. (2010) Three-dimensional nanocomposite scaffolds fabricated via selective laser sintering for bone tissue engineering, *Acta Biomater*, 6(12), 4495–4505.

54. Khan, F., and Ahmad, S.R. Fabrication of 3D Scaffolds and Organ Printing for Tissue Regeneration, In *Handbook of Biomaterials and Stem Cells in Regenerative Medicine* (ed. Ramalingam, M., Ramakrishna, S., and Best, S.), Taylor & Francis (in press).

55. Gates, B. D., Xu, Q., Stewart, M., Ryan, D., Willson, C.G., and Whitesides, G. M. (2005) New approaches to nanofabrication: molding, printing, and other techniques, *Chem Rev*, 105, 1171–1196.

56. Widawski, G., Rawiso, M., and Francois, B. (1994) Self-organized honeycomb morphology of star-polymer polystyrene films, *Nature*, 369, 387–389.

57. Nie, Z., and Kumacheva, E. (2008) Patterning surfaces with functional polymers, *Nat Mater*, 7, 277–290.

58. Campo, A., and Arzt, E. (2008) Fabrication approaches for generating complex micro- and nanopatterns on polymeric surfaces, *Chem Rev*, 108, 911–945.

59. Vukusic, P., and Sambles, J. R. (2003) Photonic structures in biology, *Nature*, 424, 852–855.

60. Autumn, K., Liang, Y.A., Hsieh, S.T., Sesch, W., Chan, W. P., Kenny, T. W., Fearing, R., and Full, R. J. (2000) Adhesive force of a single gecko foot-hair, *Nature*, 405, 681–685.

61. Mahdavi, A., Ferreira, L., Sundback, C., Nicho, J.W., Chan, E. P., and Carter, D. J. D., et al. (2008) A biodegradable and biocompatible gecko-inspired tissue adhesive, *Proc Natl Acad Sci USA*, 105, 2307–2312.

62. McMillan, J. R., Akiyama, M., Tanaka, M., Yamamoto, S., Goto, M., Abe, R., Sawamura, D., Shimomura, M., and Shimizu, H. (2007) Small diameter porous biodegradable membranes enhance adhesion and growth of human cultured epidermal keratinocyte and dermal fibroblast cells, *Tissue Eng*, 13, 789–798.

63. Tanaka, M., Takebayashi, M., Miyama, M., Nishida, J., and Shimomura, M. (2004) Design of novel biointerfaces (II). Fabrication of self-organized porous polymer film with highly uniform pores, *Biomed Mater Eng*, 14, 439–445.

64. Fukuhira, Y., Kitazono, E., Hayashi, T., Kaneko, H., Tanaka, M., Shimomura, M., and Sumi, Y. (2006) Biodegradable honeycomb-patterned film composed of poly(lactic acid) and dioleoylphosphatidylethanolamine, *Biomaterials*, 27, 1797–1802.

65. Maruyama, N., Koito, T., Sawadaishi, T., Karthaus, O., Ijiro, K., Nishi, N., and Shimomura, M. (1998) Mesoscopic patterns of molecular aggregates on solid substrates, *Thin Solid Films*, 327–329, 854–856.

66. Stenzel-Rosenbaum, M. H., Davis, T. P., Fane, A. G., and Chen, V. (2001) Porous polymer films and honeycomb structures made by the self-organization of well-defined macromolecular structures created by living radical polymerization techniques, *Angew Chem Int Ed*, 40, 3428–3432.

67. Yabu, H., Tanaka, M., Ijiro, K., and Shimomura, M. (2003) Preparation of honeycomb patterned polyimide films by self-organization, *Langmuir*, 19, 6297–6300.

68. Tanaka, M., Yoshizawa, K., Tsuruma, A., Sunami, H., Yamamoto, S., and Shimomura, M. (2008) Formation of hydroxyapatite on self-organized honeycomb-patterned polymer film, *Colloids Surfaces A: Physicochem Eng Aspects*, 313–314, 515–519.

69. Fukuhira, Y., Ito, M., Kaneko, H., Sumi, Y., Tanaka, M., Yamamoto, S., and Shimomura, M. (2008) Prevention of postoperative adhesions by honeycomb-patterned poly(lactide) film in a rat experimental model, *J Biomed Mater Res Part B: Appl Biomater*, 86B, 353–359.

70. Yabu, H., Takebayashi, M., Tanaka, M., and Shimomura, M. (2005) Superhydrophobic and lipophobic properties of self-organized honeycomb and pincushion structures, *Langmuir*, 21, 3235–3237.

71. Tanaka, M., Takebayashi, M., and Shimomura, M. (2009) Fabrication of ordered arrays of biodegradable polymer pincushions using self-organized honeycomb-patterned film, *Macromol Symp*, 279, 175–182.

72. McMurray, R. J., Gadegaard, N., Tsimbouri, P. M., Burgess, K. V., McNamara, L. E., Tare, R., Murawski, K., Kingham, E., Oreffo, R. O. C., and Dalby, M. J. (2011) Nanoscale surfaces for the long-term maintenance of mesenchymal stem cell phenotype and multipotency, *Nat Mater*, 10(8), 637–644.

73. Tanaka, M., Takayama, A., Ito, E., Sunami, H., Yamamoto, S., and Shimomura, M. (2007) Effect of pore size of self-organized honeycomb-

patterned polymer films on spreading, focal adhesion, proliferation, and function of endothelial cells, *J Nanosci Nanotech*, 7, 763–772.

74. Tanaka, M., Nishikawa, K., Okubo, H., Kamachi, H., Kawai, T., Matsushita, M., Todo, S., and Shimomura, M. (2006) Control of hepatocyte adhesion and function on self-organized honeycomb-patterned polymer film, *Colloids Surfaces A: Physicochem Eng Aspects*, 284–285, 464–469.
75. Shimomura, M., Nishikawa, T., Mochizuki, A., and Tanaka, M. (2001) *Honeycomb structure and its preparation method for cell culture scaffold*, Patent application JP2001, 157574.
76. Tsuruma, A., Tanaka, M., Yamamoto, S., and Shimomura, M. (2008) Control of neural stem cell differentiation on honeycomb films, *Colloids Surfaces A: Physicochem Eng Aspects*, 313–314, 536–540.
77. Tanaka, M. (2011) Design of Novel 2D and 3D Bio-interfaces using self-organization to control cell behavior, *Nanotechnologies: Emerging Applications in Biomedicine, Biochimica et Biophysica Acta (BBA) - General Subjects*, 1810, 251–258.
78. Yamamoto, S., Tanaka, M., Sunami, H., Yamashita, S., Morita, Y., and Shimomura, M. (2007) Effects of honeycomb-patterned surface topography on the adhesion and signal transduction of porcine aortic endothelial cells, *Langmuir*, 23, 8114–8120.
79. Ishihata, H., Tanaka, M., Iwama, N., Ara, M., Shimonishi, M., Nagamine, M., Murakami, N., Kanaya, S., Nemoto, E., Shimauchi, H., and Shimomura, M. (2010) Proliferation of periodontal ligament cells on biodegradable honeycomb film scaffold with unified micropore organization, *J Biomech Sci Eng*, Special Issue on Micro Nanobiotech for Cells, 5, 252–261.
80. Sato, T., Tanaka, M., Yamamoto, S., Ito, E., Shimizu, K., Igarashi, Y., Shimomura, M., and Inokuchi, J. (2010) Effect of honeycomb-patterned surface topography on the function of mesenteric adipocytes, *J Biomater Sci Polym Ed*, 21, 1947–1956.

Chapter 11

Regenerative Dentistry: Stem Cells Meet Nanotechnology

Lucía Jiménez-Rojo, Zoraide Granchi, Anna Woloszyk, Anna Filatova, Pierfrancesco Pagella, and Thimios A. Mitsiadis
Institute of Oral Biology, ZZM, Faculty of Medicine, University of Zurich, Plattenstrasse 11, Zurich, CH-8032, Switzerland
thimios.mitsiadis@zzm.uzh.ch

11.1 Introduction

Humans develop primary (deciduous) and secondary (permanent) single-rowed dentitions. Teeth form as a result of sequential and reciprocal interactions between oral epithelium and neural crest-derived mesenchyme.[1] During early stages of tooth development, and due to the interactions with the underlying mesenchyme, oral epithelium thickens and grows to form the enamel organ. The enamel organ is composed of four different epithelial cell populations, which are the stellate reticulum, stratum intermedium, inner enamel epithelium, and outer enamel epithelium. At more advanced developmental stages, cells from the dental papilla mesenchyme that face the inner enamel epithelium polarise and differentiate into the

Horizons in Clinical Nanomedicine
Edited by Varvara Karagkiozaki and Stergios Logothetidis

ISBN 978-981-4411-56-1 (Hardcover), 978-981-4411-57-8 (eBook)
www.panstanford.com

dentin-secreting odontoblasts. It has been shown that the presence of dentin is essential for the differentiation of inner dental epithelial cells into ameloblasts.[2] Ameloblasts start to produce and secrete specific enamel matrix proteins and completely disappear soon after tooth eruption. During that time, tooth root develops, accompanied by cementum deposition and periodontium formation.[3]

Human teeth have a very limited capacity to regenerate upon injury or disease. Sophisticated dental materials have been successfully used the last years for tooth repair after carious lesion and injury, while dental implants are still used for tooth replacement.[4] Recent efforts focus on partial or whole tooth regeneration for the treatment of common dental diseases and tooth loss.[4]

11.2 Dental Stem Cells

Stem cells are characterised by their potential to self-replicate and their capacity to give rise to more differentiated cells.[5] Adult stem cells are undifferentiated pluripotent cells found in various tissues of the human body from both epithelial and mesenchymal origin, including skin,[6] bone marrow,[7] adipose tissue,[8] periosteum,[9] cartilage,[10] and blood.[11] Due to their ability to give rise to every cell type, adult stem cells are responsible for tissue/organ homeostasis and repair. During the last decades, adult stem cells from different tissues have been isolated, characterised, and tested for their potential applications in regenerative medicine.

Since teeth are formed by a continuous crosstalk between oral epithelial and neural crest-derived mesenchymal cells, regenerating or building a brand new tooth requires the association of mesenchymal (that will form odontoblasts, cementoblasts, osteoblasts, and fibroblasts) and epithelial (that will form ameloblasts) stem cells. To date, mesenchymal stem cells (MSCs) have been isolated from different locations within adult or postnatal dental tissues. In contrast to dental mesenchymal cells, most of dental epithelial cells disappear shortly after tooth eruption. Thus, identifying epithelial stem cells (EpSCs) residing in human adult teeth constitutes a major challenge. The current knowledge on dental EpSCs has been obtained mainly from animal models such as rodents, in which adult EpSCs have been already described as the main source for the renewal of the epithelium in their continuously growing incisors.

11.2.1 Dental Mesenchymal Stem Cells

MSCs were first isolated from bone marrow in 1970[7] and have been found to express markers such as STRO-1, CD146, and CD44,[12] MSCs have the potential to differentiate into mesodermal lineages, which give rise to connective tissues, such as bone, cartilage, and adipose tissue.[13–15]

Dental mesenchymal stem cells (DMSCs) were first identified in dental pulp of human permanent teeth.[16] DMSCs have also been isolated from dental pulp from exfoliated deciduous teeth,[17] the apical part of dental papilla,[18] the dental follicle,[19] and the periodontal ligament[20,21] (Fig. 11.1).

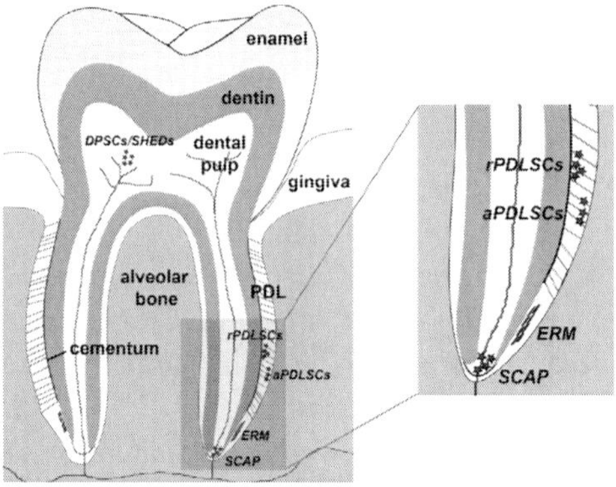

Figure 11.1 Stem cell niches in adult human teeth. Insert: DMSCs residing in the apex of the dental root. Abbreviations: aPDLSCs, alveolar periodontal ligament stem cells; DPSCs, dental pulp stem cells; ERM, epithelial cell rests of Mallassez; PDL, periodontal ligament; rPDLSCs, root periodontal ligament stem cells; SCAP, stem cells from the apical papilla; SHEDs, stem cells from human exfoliated deciduous teeth.

11.2.1.1 Dental pulp stem cells

The most common source of DMSCs is the dental pulp of third molars, since these teeth are often extracted.[22] Dental pulp stem cells (DPSCs) were first isolated in 2000.[16]

In order to isolate the dental pulp, the tooth has to be sectioned at the cementum-enamel junction. Single cell suspensions can be expanded in vitro. Due to the lack of specific dental stem cell markers, commonly used identification procedures rely on morphology, selective adherence properties, proliferation and differentiation potential, and the ability to tissue repair the cells.[5] Purification of the isolated cells can be obtained after labelling of cells with fluorescent antibodies followed by fluorescence-activated cell sorting (FACS). Typical MSCs markers such as STRO-1, CD146 (MUC-18), and CD44 are commonly used for the identification of dental stem cell populations.[23]

DPSCs are able to form odontogenic,[16,17,23–26] adipogenic,[23,27] chondrogenic,[27] osteogenic,[28] myogenic,[29] and neurogenic[23,30] lineages in vitro. Moreover, in vivo transplantation of DPSCs mixed with hydroxyapatite/tricalcium phosphate resulted in ectopic pulp-dentin tissue formation.[16,31] Recently, the first clinical trial of autologous DPSC transplantation in combination with a collagen sponge scaffold was successfully performed in humans for bone reconstruction after the extraction of third molar teeth.[32] A study in rats with acute myocardial infarction revealed another important feature of DPSCs. Even though DPSCs were not found to differentiate into cardiac or smooth muscle cells in vivo, they still were able to repair the infarcted myocardium by increasing the number of vessels and reducing the infarct size. A possible explanation could be the ability of DPSCs to secrete proangiogenic and antiapoptotic factors.[33] The myogenic potential of human DPSCs has been also assessed in dystrophic dogs, an animal model of muscular dystrophy. Transplanted human DPSCs were able to engraft in the muscles of the dogs, thus improving their clinical image[29] (Table 11.1).

11.2.1.2 Stem cells from human exfoliated deciduous teeth

Exfoliated human deciduous teeth contain a population of multipotent stem cells (SHEDs) that can be isolated using the same procedure as for DPSCs. Similar to DPSCs, SHEDs express the surface molecules STRO-1 and CD146,[17] but also several neural and glial markers such as nestin, βIII tubulin, GAD, NeuN, GFAP, NFM, and CNPase,[17] possibly due to their neural-crest origin.[44] When compared to DPSCs, SHEDs show higher proliferation rate but have less capacity to form dentin-pulp complexes in vivo.[17]

In various differentiation assays in vitro, SHEDs have been shown to be odontogenic, osteogenic, adipogenic, and neurogenic,[17] but also myogenic, and chondrogenic.[29] In vivo, SHEDs are able to induce bone and dentin formation.[17] Moreover, the transplantation of SHEDs seeded in biodegradable scaffolds prepared with human tooth slices into immunodefficient mice resulted in the formation of a dental pulp-like tissue[38] (Table 11.1).

Table 11.1 Properties of human dental mesenchymal stem cells

		Differentiation potential		Cell
Cell source	**Abbr.**	**In vitro**	**In vivo**	**markers**
Dental pulp	DPSC	Odontogenic,[16,17,23–26] Adipogenic,[25,27] chondrogenic,[27] osteogenic,[28,34] Myogenic,[29] neurogenic[23,30]	Dentin-pulp-like complex,[16,31] bone,[32,35] heart,[33] muscles,[29] teeth[36,37]	STRO-1, CD146, CD44
Exfoliated deciduous teeth	SHED	Neurogenic,[17] adipogenic,[17] osteogenic,[17] odontogenic,[17] myogenic,[29] chondrogenic[29]	Bone,[17] dentin,[17] teeth[38]	STRO-1, CD146
Root apical papilla	SCAP	Odontogenic,[18,39] adipogenic[18,39]	Dentin[18]	STRO-1, CD146, CD44, CD24
Periodontal ligament	PDLSC	Adipogenic,[40] osteogenic,[40] chondrogenic,[20,40] cementogenic[20]	cementum/PDL-like tissues[20]	STRO-1, CD146, CD44
Dental follicle	DFSC	Cementogenic,[41] osteogenic[41]	Cementum,[42] PDL[43]	STRO-1, CD44, BMPR-IA, BMPR-IB, BMPR-II

11.2.1.3 Stem cells from the apical part of the papilla

Stem cells from the apical part of the papilla (SCAPs) can be extracted from the root apical papillae, a soft tissue at the apexes of developing permanent teeth.[18,39] Compared to DPSCs, SCAPs have a higher proliferative rate and telomerase activity, increased tissue regeneration potential, as well as enhanced migration capacity in a scratch assay.[18]

As DPSCs, SCAPs do also express the three MSCs surface markers. Interestingly, CD24 has been shown to be a SCAPs-specific marker,[18] since DPSCs were found to be CD24-negative.[18,22]

Besides the in vitro potential of SCAPs to generate odontogenic and adipogenic lineages,[18] they were also shown to form dentin in vivo, when transplanted on a hydroxyapatite/tricalcium phosphate carrier into the incisor socket of a mini-pig[18] (Table 11.1).

11.2.1.4 Periodontal ligament stem cells

The PDL consists of connective tissue fibres that are located between the tooth root cementum and the alveolar bone (Fig. 11.1). PDL contributes to tooth stability, homeostasis, nutrition, and repair of damaged tissues.[45,46] Periodontal ligament stem cells (PDLSCs) or periodontal ligament stem cells from the root surface (rPDLSCs) were first isolated from third molars by separating PDL from the root surface.[20] As DPSCs and SCAPs, rPDLSCs do express the MSCs specific cell-surface markers STRO-1, CD146, and CD44.

rPDLSCs can be differentiated into adipogenic and osteogenic cells under defined culture conditions in vitro.[40] In vivo transplantation experiments showed that PDLSCs can give rise to cementum/PDL tissues when injected into immunocompromised mice and rats, therefore contributing to the regeneration of the periodontium[20] (Table 11.1).

Recently, another periodontal stem cell population has been evidenced close to the alveolar bone, the alveolar periodontal ligament stem cells (aPDLSCs). aPDLSCs differ from the previously described rPDLSCs.[21] Both stem cell populations express the same surface markers, but aPDLSCs exhibit much higher osteogenic and adipogenic potential and are more efficient in repairing the periodontal defects when compared to the rPDLSCs.[21]

11.2.1.5 Stem cells from the dental follicle

The dental follicle is a mesenchymal tissue that contains progenitors for the PDL, alveolar bone, and cementum. Its biological function is the coordination of tooth eruption. Stem cells from the dental follicle (DFSCs) have been isolated by plastic adherence from extracted dental tissues.[19] Apart from STRO-1 and CD44, DFSCs express the BMP receptors BMPR-IA, BMPR-IB and BMPR-II.[41]

DFSCs demonstrate both cementogenic and osteogenic potentials in vitro.[41] Interestingly, bovine DFSCs transplanted into immunodeficient mice (SCID mice) were able to form cementum.[41] Similar results were observed in a different study which showed that a PDL-like tissue could be created after transplantation of mouse DFSCs into SCID mice[43] (Table 11.1).

11.2.2 Dental Epithelial Stem Cells

11.2.2.1 Epithelial stem cells in rodent incisors

Rodents follow a very abrasive diet and, consequently, they need to continuously renew their teeth. Rat and mice incisors, vole molars, rabbit and sloth teeth are some examples of continuously growing teeth in mammals. Thus, MSCs and EpSCs are supporting this continuous growth that takes place throughout the entire life of the animal. Given the limited knowledge of EpSCs in human teeth, many studies on dental EpSCs have been performed in rodent incisors.

Rodent incisors have an elongated cylindrical shape where two distinct epithelia (i.e., lingual and labial side epithelia) surround the dental pulp mesenchyme[47] (Fig. 11.2). The lingual side of the incisor represents the molar root analogue. The epithelium of the lingual side is composed by two layers of cells (i.e., outer and inner enamel epithelia) that will never differentiate into ameloblasts. Enamel is deposited only in the labial side of the incisor, which is thus equivalent to the molar crown. The epithelium of the labial side is composed by four layers of cells (i.e., inner enamel epithelium, outer enamel epithelium, stellate reticulum, stratum intermedium) (Fig. 11.3). In this side, cells from the inner enamel epithelium show a differentiation gradient from the posterior part (i.e., cervical loop area) towards the anterior part of the incisor, becoming progressively preameloblasts, secretory ameloblasts, and post-secretory ameloblasts.[1,48]

EpSCs that sustain the continuous generation of enamel-producing ameloblasts in rodents have been identified for the first time in a putative stem cell niche located in the cervical loop (CL) of the mouse incisor49. Fluorescent dye DiI (1,1-dioctadecyl-6,6-di (4-sulfophenyl)-3,3,3′,3′-tetramethylindocarbocyanine) has been used to label stem cells from the CL and trace their fate.

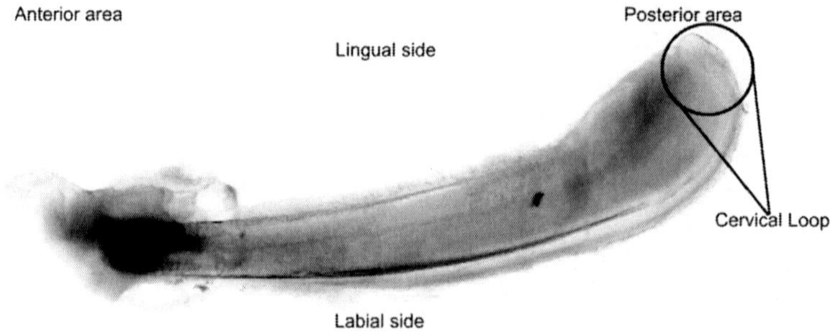

Figure 11.2 Freshly isolated postnatal mouse incisor in which the main areas are indicated.

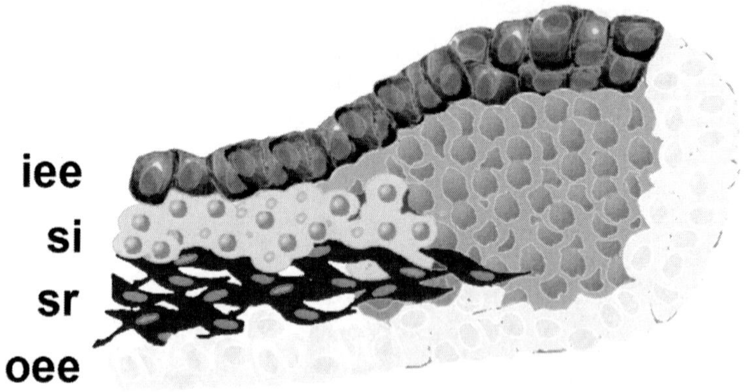

Figure 11.3 Schematic representation of the cervical loop area (labial side of the mouse incisor) showing the different epithelial cell populations. Abbreviations: iee, inner enamel epithelium; oee, outer enamel epithelium; si, stratum intermedium; sr, stellate reticulum.

Adult stem cells from different tissues have been identified by label-retention assays using BrdU (5-bromo-2′-deoxyuridine).[49] BrdU is integrated into the newly synthesised genome of actively cycling cells, and it is diluted out after the labelled cells undergo a number of divisions. Short exposure to BrdU, results in labelling of actively cycling cells, while longer exposure labels also the rarely dividing cells. Experiments of "pulse and chase", are used to analyse the kinetics of rarely cycling cells. After the chase period, those cells (label-retaining cells or LRCs) would be the only ones that retain the labelling while all the faster cycling cells dilute their labelling during the chase period. The CL has been shown to contain cells that retain the labelling after a long chase period. Further studies have supported the idea of stem cells residing in the CL area,[50] identifying some of the key regulatory molecules involved in stem cell and transit amplifying cells maintenance and proliferation.[51,52]

11.2.2.2 Epithelial stem cells in human teeth

As mentioned above, cells from the inner enamel epithelium differentiate into ameloblasts and produce the enamel, which is the hardest tissue in the human body. Soon after tooth eruption, however, ameloblasts and their precursors disappear. This makes enamel regeneration impossible after a carious lesion or injury.

One of the possible sources of dental EpSCs in humans is the wisdom tooth (i.e., third molars) that develops postnatally. In the third molars, the enamel organ forms around the 72nd month of the human life. Several studies report on the isolation of EpSCs from third molars of newborn or juvenile animals.[53–55] These epithelial progenitor cells were amplified and associated in vitro with MSCs, isolated from the same tooth, on biomaterials such as collagen sponges and synthetic polymers, in order to generate new teeth.

Surprisingly, epithelial cells can be also isolated from human deciduous dental pulp.[56] In these experiments, human deciduous pulp tissue was cultured in a serum-free medium. Rounded cells with epithelial appearance during the culturing process showed colony-forming proliferation and had cuboidal or polygonal shape. These cells expressed some of the epithelial markers (i.e., E-cadherin and pan-cytokeratin) and markers of EpSCs (i.e., ABCG2, Bmi-1, ΔNp63, p75). These cells, however, did not express muscle, mesenchymal, dental pulp stem cell, hematopoietic, and endothelial markers.[56] Although their functional role is not yet elucidated, these results suggest that EpSCs may exist in human deciduous teeth.

After the development of the crown, the outer dental epithelium and the inner dental epithelium fuse to form the Hertwig's epithelial root sheath (HERS). Reciprocal interactions between HERS and mesenchyme lead to root formation. Upon dentin mineralisation, HERS disintegrates into strands of epithelial cells, which are called epithelial cell rest of Malassez (ERM).[57,58] ERM cells express EpSCs markers such as ABCG2, Bmi-1, ΔNp63, EpCAM, and p75, as well as embryonic stem cell markers such as Oct-4, Nanog and SSEA-4. The transcription factors Oct-4 and Nanog are essential for pluripotency and self-renewal of embryonic stem cells. To date, ERM seem to be the most promising source of dental EpSCs.

11.3 Regenerative Dentistry

Despite the considerable progress made in dental treatment and tooth decay prevention, tooth loss is still a frequent problem. Tooth loss can result from numerous pathologies, such as periodontal and carious diseases, fractures, injuries, and genetic alterations. [4,59]

Dental caries is a very common disease, and it consists essentially in an infection of the mineral tissues, which eventually reaches the dental pulp causing inflammation and potential tooth loss. The dental pulp has a key function in providing nutrients, oxygen and nerve supply to the tooth, and it fosters cells of the immune system that tackle infections; moreover, it produces reaction or reparative dentin in response to external stimuli and injury.[59–61] In clinical practice, damaged pulp is usually amputated and substituted, after disinfection of the pulp cavity, with an artificial material.[59] Although the tooth is saved in its normal position, it is not vital anymore and it cannot fulfill to the full extent its role.[59] We could imagine that in the near future dentists will focus on pulp regeneration for the treatment of these clinical cases.

Periodontitis is perhaps the most common infectious disease in humans. Periodontitis is triggered by microorganisms that are present in the oral cavity, attach on the teeth, and finally cause inflammation and damage on the periodontal tissues.[59] Periodontal tissues include the gingiva, alveolar bone, cementum and periodontal ligament (PDL), and they are specifically tailored to support the tooth in its functional position. Severe damage of the periodontal tissues often results in tooth loss.[59] So far, the approaches have focused on the removal of the inflamed tissue, or on providing appropriate conditions for spontaneous wound healing, exploiting the natural, limited regenerative capacity of the periodontium.[62] These approaches, however, do not ensure a predictable outcome of periodontal regeneration and often result in healing with epithelial lining rather than new tissue formation.[62]

When traumatic events or severe infections cause damage to the pulp and/or to the periodontal tissues, a limited repair occurs that prevents bacterial invasion. In the adult pulp, cell division and secretory activity of the odontoblasts are reduced, but they can be reactivated following injury.[63] However, in severe tooth lesions the spontaneous regeneration processes of periodontium or dental pulp are often insufficient, resulting in tooth loss. Moreover, damaged enamel cannot be naturally regenerated, since ameloblasts are not present anymore in humans after tooth eruption.

These pathologies, together with ageing, are commonly responsible for tooth loss, and have a substantial effect in the quality of life of the patient. So far, dental implants have been the only solution for the replacement of missing teeth.

Although the development of new, biocompatible dental materials has improved the quality of treatment, there are several limitations in functionality and longevity of the implants. This is mainly due to their inability to interact properly with the surrounding environment, as natural teeth do, and the strong dependence on the quantity and the quality of the surrounding bone.[64] The metal/bone interface does not ensure complete integration of the implant, thus reducing its performance and long-term stability. Another major problem is given by mechanical stress. During mastication, teeth are the vehicle of high mechanical stresses, which in normal conditions are absorbed and modulated by the PDL, which mediates the interaction between the tooth and the surrounding alveolar bone. PDL is not formed around dental implants, making the alveolar bone vulnerable in excessive forces during mastication.[4,64].

To overcome these problems, new ideas and approaches have emerged recently from the fields of stem cell biology and tissue engineering.

11.3.1 Approaches for Tooth Regeneration

Damaged teeth may require different therapeutic approaches, depending on the degree and type of injury. In many cases, regeneration of a whole tooth would not be necessary, while partial dental tissue regeneration may be preferable. Actual therapies are mainly oriented towards specific dental tissue repair, where the damaged tissue is substituted with an appropriate "non-biological" material. Regeneration, on the other hand, involves the replacement of the injured tissue by the same tissue and the restoration of its biological function. It is thus obvious that tissue regeneration is highly preferable for the treatment of any damaged tissue or organ.

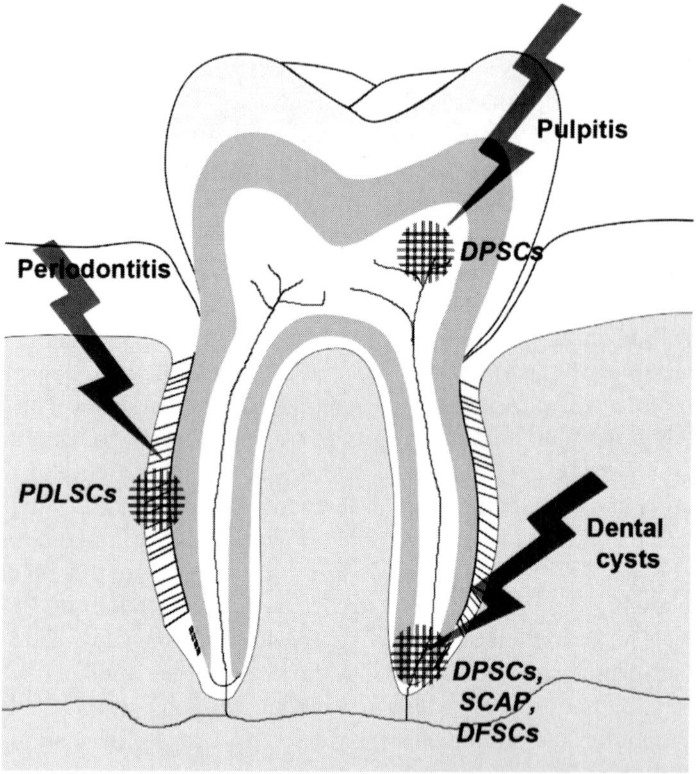

Figure 11.4 Most common diseases affecting tooth, and potential stem cell-based therapies. DPSCs injected into the dental pulp have been proved to regenerate both pulp and dentin tissues. Similarly, PDLSCs are able to regenerate the periodontium and DPSCs, SCAPs or DFSCs can refill emptied root canals.

11.3.1.1 Regeneration of the pulp/dentin complex

In carious lesions, the regeneration of the dentin-pulp interface represents the ideal solution that allows dentinal bridge formation at the pathological site, and protection of infections and other external stimuli. Recently, growth factors have been used to stimulate and increase the natural regenerative response of the dental pulp.[65] This approach, however, would require pulp exposure and limited inflammation.

The proper regeneration of the dentin-pulp complex requires the revascularisation and reinnervation of the pulp, and new dentin deposition. When DPSCs are placed in contact with dentin in vitro, they can differentiate into odontoblasts, endothelial cells, and neurons. In mice, transplantation of DPSCs after pulpotomy can regenerate both pulp and dentin tissues in vivo.[16,66] Another in vivo study in dogs demonstrated that DPSCs combined with calcium hydroxyapatite were able to form dentin.

DPSCs and SCAPs isolated from human third molars, seeded on a poly-D,L-lactide/glycole scaffold, and transplanted into the empty root canal space of mouse teeth, were able to refill the empty space with a newly formed vascularised pulp.[65,67] A continuous layer of mineralised tissue resembling dentin was deposited in the existing dental walls of the canals.[65,67]

Although these principles and results prove that DPSCs can regenerate dental pulp, further studies are clearly required to investigate their potential clinical applications.

11.3.1.2 Regeneration of periodontal tissues

Periodontal tissues have a clear regenerative capacity due to the presence of multipotent progenitor cells.[20,59] Extensive studies in animal models have depleted the various components of the periodontal tissues and analysed their potential to form PDL, cementum, and alveolar bone.[68,69] Slowly cycling progenitor cells were identified in mouse and human teeth. When these human progenitor cells were transplanted into immunocompromised mice, they were able to contribute to periodontium regeneration, thus indicating their huge potential for future cell-based therapies in dental clinic.[20,59]

Equally important for the development of stem cell-based therapies in dentistry is the use of signalling molecules. Several molecules involved in periodontal development are under investigation, and in some rare cases are already in use in clinical practice. Among these molecules, the most promising is the PDGF family of growth factors, either used alone or in combination with FGFs, IGFs, EGFs. It has been shown that PDGF molecules are able to stimulate periodontal healing and regeneration.[70,71] BMP molecules are also capable of inducing formation of new alveolar bone and cementum.[72] Quite frequently, however, the use of BMPs may have undesirable effects by inducing tooth ankylosis.

Ameloblasts secrete amelogenins, which are the main enamel matrix proteins. Although they are frequently used in dental clinics for the stimulation of periodontal tissues regeneration, and they are able to stimulate periodontal regeneration, their mechanism of action is still unclear.[73]

Current studies focus on the identification of the accurate population of cells, suitable signalling molecules, and desirable scaffold materials that will be used as carriers for specific cell types. In addition, it is important to note that periodontal regeneration would occur in conditions of strong inflammation and infection, issues that should be carefully taken in consideration.

11.3.1.3 Regeneration of enamel

Regarding regeneration of dental epithelium, it has been shown that ERM derived from porcine mandible can differentiate into ameloblasts after co-culture with dental pulp cells in vitro. These ameloblast-like cells were positive for CK14 and amelogenin. Moreover, after transplantation of ERM cells combined with primary dental pulp cells, an enamel-like tissue was produced in the implant. Histological analysis revealed that appropriate stages of amelogenesis from initiation to maturation were present in all implants. Thick enamel-dentin structures were clearly recognised, and ameloblast-like cells expressed CK14 and amelogenin, 8 weeks post-transplantation.[74]

11.3.1.4 Regeneration of an entire tooth

In case of tooth loss, whole tooth regeneration would be the ideal therapeutic approach. Functional teeth can be experimentally bioengineered in mice by re-association of dissociated dental cells. In fact, in mice, recombination of embryonic dental epithelium and mesenchyme can contribute to the formation of a tooth. Numerous attempts have been made in order to form teeth in vivo, resulting in the differentiation of the transplanted cells into functional odontoblasts and ameloblasts.[4]

There are two main approaches in constructing a brand-new tooth. The first consists in associating epithelial and mesenchymal stem cells in vitro, where they can form a tooth germ. This tooth germ could then be transplanted into the alveolar bone, where it will finally develop, erupt and become a functional tooth.

The second approach consists of implanting into the jaw tooth-shaped polymeric scaffolds that are filled with both epithelial and mesenchymal stem cells. Ideally, this scaffold should be biodegradable and reproduce the three-dimensional structure required for the transplanted cells. This will support their differentiation and finally give rise to a functional tooth.[4]

Bioengineered teeth formed with human cells have been produced so far in ectopic sites and they are still missing some essential elements such as a complete root and periodontal tissue, which would allow correct anchoring into the alveolar bone.[4] Experiments in mice, however, showed that it is also possible to obtain such tissues and structures with the above-mentioned approaches. In one recent study, epithelial and mesenchymal mouse cells were subsequently seeded into a collagen drop, serving as a scaffold, and implanted into the tooth cavity of adult mice. With this technique, the presence of all dental tissues such as odontoblasts, ameloblasts, dental pulp, blood vessels, crown, periodontal ligament, root, and alveolar bone could be observed.[75] The implantation of these tooth germs in the mandible allowed their development, eruption and maturation, and their full integration into the recipient alveolar bone. Such results have not been observed using human cells. In fact, a major issue in whole tooth regeneration is the availability of odontogenic stem cells. Mouse embryonic cells were usually used in these studies, and in these experiments, both the mesenchymal and the epithelial components were odontogenic and were able to induce an odontogenic fate. In humans, various populations of adult dental mesenchymal stem cells are under study, while adult dental epithelial stem cells have not been identified so far. As mentioned above, a potential source of dental EpSCs is constituted by the ERM, which are able to differentiate into enamel-secreting ameloblasts; these cells, however, lack odontogenic inductive properties. Alternatively, epithelial stem cells of non-dental origin may be used, in the presence of odontogenic stimuli.

Experiments on human cells are mainly focused on root and periodontal regeneration, rather than on whole tooth regeneration. Using different scaffolds, it has been possible to induce differentiation of PDLSCs or DPSCs into the various cell types composing the root and/or the periodontal tissues in vitro and after subcutaneous transplantation in immunocompromised mice.[76,77] Recently, a complete dentin-pulp complex with neurons was obtained by transplanting subcutaneously a sheet of human dental follicle cells, obtained by non-erupted wisdom teeth, placed into a human treated-dentin matrix (TDM), obtained by human premolars[76]; this structure contained dentin, pulp, cementum, and PDL tissues. Although a root was successfully constructed, several issues are still to be solved. This approach still has to be tested in the alveolar bone, and has to be determined whether the regenerated root is strong enough to support a dental crown.[76]

11.3.1.5 Challenges of dental tissue regeneration

So far, the general bases of tooth regeneration have been discussed, meaning the principles and the approaches underlying the bare generation of a tooth and its components. Further issues have to be solved, however, before even successfully regenerated teeth and tissues could be introduced in clinical practice. More precisely, the issues of timing, aesthetics, and dimension have to be clearly addressed before regenerated teeth can be implanted in patients.

Enamel formation is obtained through a series of complex dynamic and programmed events that include tightly regulated production and secretion of proteins, cellular movements, and matrix maturation.[78]

The first great challenge in enamel regeneration in humans is constituted by the search for the appropriate epithelial precursors. As previously mentioned, adult human teeth lack populations of dental epithelia stem or precursor cells. Thus, it is necessary to identify non-dental epithelial stem cells that, under the effect of the appropriate odontogenic stimuli, might differentiate into enamel-producing ameloblasts. Enamel owes its peculiarities to the complex mechanisms and dynamics underlying its formation; therefore, stem or precursors cells destined to form enamel in regenerated teeth have to undergo the same stages and the same cellular movements in order to properly form ordered enamel.

An additional and key issue in enamel regeneration is constituted by the colour of the enamel. Although apparently negligible, regenerated teeth should fully meet patients' expectations, including aesthetic requests. Studies in mice demonstrated that enamel colour depends on its ion content, and most particularly the iron.[79,80] Apart from this, little is known about how ion content is regulated in enamel, and what is the effect of other ions and molecules on enamel colour; even less is known about the determination of enamel colour in humans.

Another major issue to solve in human enamel regeneration is time. Enamel formation is regulated by complex interactions of clock genes,[4,81] that influence timing of many other developmental and physiological processes in other organs. The relative expression of these genes seems to be correlated with the differentiation stage of the various tissues,[81] particularly odontoblasts and ameloblasts, although no causal links between clock genes expression and enamel development have been demonstrated so far.[81] In humans, the whole process of enamel formation takes more than five years.[81] Such a span of time is clearly unacceptable in clinical practice; thus, the entire process should also be accelerated before it could be applied on human patients.

Finally, a key issue in whole tooth regeneration consists of tooth shape and size, both in crown and root counterparts. Regarding the size, recent work on mice demonstrated that it is possible to control the crown length and the root extension of bioengineered teeth.[82] In these experiments, the same approach described above[82] for whole tooth regeneration was used; the growing tooth, however, was locked into a ring-shaped size-control device. The resulting bioengineered tooth had the desired width and root length, and it successfully integrated in the mouse alveolar bone.[82]

More problematic is the issue of tooth shape. The majority of studies so far concentrated on molars. From studies on mice, we know that even development of maxillary and mandibular molars is regulated by different sets of genes.[83] While mice have only molars and incisors, human dentition is also composed of other types of teeth (premolars and canines). During development, tooth number and location are mainly determined by the oral epithelium; in contrast, tooth type and shape are determined by the mesenchyme. It is extremely important to understand deeply how reciprocal signals between the various components of the teeth determine their type and shape, in order to obtain the desired tooth morphology.

11.4 Stem Cells Meet Nanotechnology

The combination of stem cells with novel nanotechnology platforms holds great promise for applications in the biomedical arena. The development of either improved or innovative nanostructured materials (i.e., on the scale of 1–1000 nm) could be useful in manipulating stem cells that will be used for tissue repair in the clinics. In the dental field, nanomaterials can also be used for extracellular drug delivery, intracellular RNA interference (RNAi), DNA and protein delivery, scaffolds and implants, and finally as diagnostic tools.[84,85] It is important, however, to measure the toxicity of these materials as well as their interference with the self-renewal potential of stem cells. We will briefly develop the benefits of nanotechnology in the continuously growing field of stem cells.

11.4.1 Follow-Up of Stem Cells after Transplantation

11.4.1.1 Magnetic nanoparticles

Stem cell-based therapies for tissue repair and/or regeneration necessitate thorough testing, firstly in animals and finally in humans. For the evaluation of the therapeutic efficacy of the transplanted stem cells or progenitor cells, it is important to track their survival, migration, fate and regenerative impact in vivo. Transplanted stem cells can be assessed for a long-term period using non-invasive imaging techniques.[86,87] For example, stem cells can be tracked in vivo after their transplantation due to their initial labelling with a fluorescent dye, or to their transfection with the LacZ gene or the green fluorescence protein (GFP). Labelled stem cells might be tracked with cell imaging systems either in vivo or in vitro. Furthermore, magnetic nanoparticles, such as the superparamagnetic iron oxide (SPIO) nanoparticles, have been used for studying the fate of transplanted stem cells.[88,89] Their visualisation requires complex imaging systems, such as magnetic resonance imaging (MRI).[86,90] Often, MRI detection needs clusters of numerous labelled cells.[91] SPIO nanoparticles have an overall diameter of 60–150 nm, are composed of biodegradable iron that might be recycled by cells, and are coated with dextran or carboxydextran to prevent their aggregation and ensure aqueous solubility.[92,93] They can either attach to the stem cell surface or be internalised by phagocytosis or endocytosis. Internalisation of SPIO nanoparticles by stem cells, such as human mesenchymal stem cells (hMSCs), occurs generally through endocytosis,[94] and does not affect hMSCs viability, growth and differentiation into fat, cartilage or bone cells.[95] Several membrane receptor binding or coating agents (*e.g.* protamine sulphate) have been used to facilitate the internalisation of SPIO nanoparticles into stem cell.[89,96]

SPIO nanoparticles might yield important information about stem cell migration and differentiation in the context of dental pathology (e.g., periodontitis, injury, pulpitis) and repair.[5,59,97] For example, we could monitor kinetics and the fate of SPIO-labeled stem cells injected into the periodontal space or pulp chamber after dental injury. These approaches are necessary to evaluate the therapeutic effects of stem cells when exposed to a specific microenvironment before any clinical application.

11.4.1.2 Quantum dots

Quantum dots (Qdots) are light-emitting nanomaterials composed of 2–10 nm in diameter nanocrystals and are used for long-term labelling of stem cells.[98,99] Their detection relies on optical imaging, which is less sophisticated and complex than the MRI system. Qdots present a superior photostability and durable fluorescence intensity when compared to organic dyes (e.g., DiI) and fluorescence proteins (e.g., GFP).[100] Qdots have been used to monitor the dynamics of various stem cell components (e.g., cell membrane) in real time.[101] These studies bring valuable new information concerning participation and clustering of multiple cell-surface molecules involved in migration and differentiation of stem cells over time in vivo. This knowledge could be used for the design of novel and more efficient engineered scaffolds for homing stem cells before transplantation.

Qdots are internalised by stem cells by endocytosis, a process that is often facilitated by the use of specific peptides (e.g., RGD, phospholipids, and cholera toxin).[102–104] Once internalised, Qdots are trafficked via endosomes to the perinuclear region of the cells.[105] Qdots that cannot be used by the cells are degraded by lysosomes and peroxisomes.[88] This oxidative degradation of Qdots could lead to mitochondria dysfunction and ultimately cell death.[106] Although conflicting results exist concerning cytotoxicity of Qdots,[103,105] most of the studies using Qdots-labeled hMSCs have demonstrated no interference of Qdots in hMSCs differentiation into fat, cartilage, and bone cells.[102,103] The variability of size, surface coating, and chemical composition of the Qdots, however, could influence stem cell behaviour and fate.[106,107]

11.4.2 Gene, Protein and Drug Intracellular Delivery

One of the most attractive concepts in manipulating the fate of stem cells—thus directing their differentiation into specific cell populations—is the use of nanomaterials for intracellular gene delivery (e.g., RNAi, DNA).[108,109] Generally, viral (e.g., adenoviruses, lentiviruses, retroviruses) and non-viral vectors (e.g., lipids, polymers) can be used for cellular transfection and/or nucleofection, thus offering durable gene expression within stem cells.[110–113] Non-viral carriers have a number of advantages over viral vectors, since they exhibit low-risk immunogenicity and insertional mutagenesis, controllable toxicity, and great gene-carrying capacity.[112] Many efforts for the improvement of non-viral vectors are focused on cationic polymers that interact with negatively charged DNA or RNAi. Polymers, including poly(L-lysine)-palmitic acid, poly(L-lysine), and polyethylenimine, condense the genetic material into particles of 200–300 nm in diameter, protect them from enzymes, and facilitate cellular entrance.[111,114] These complexes of polymers with genetic material (called "polyplexes") have a transfection efficiency that is equivalent to the adenoviral vectors.[114]

Nanoparticles, carbon nanotubes and silicon nanowire arrays have also been used for gene delivery.[115,116] The apatite particles coated with E-cadherin and fibronectin, ensure high gene delivery capacity in stem cells.

Polymeric biodegradable nanoparticles of 100–300 nm in diameter could also serve as platforms to incorporate and deliver proteins and chemicals within stem cells. It has been shown that after internalisation, these nanoparticles accumulate in the perinuclear region and have a minimal effect on the viability and proliferation of stem cells, but a high impact on their differentiation.[88]

The cytotoxicity of "polyplexes", nanoparticles, and nanotubes has been evaluated in stem cells, and the results showed that in general the toxicity correlates with the chemistry, concentration, size, shape and coating of the nanomaterials. [88,114,115]

11.4.3 Nanobiomimetics

11.4.3.1 Nanotechnology for the design of artificial stem cell niches

Nanotechnology can be used to create artificial microenvironments that will direct stem cells or progenitor cells towards a precise fate and function. A big challenge is to engineer materials that resemble the structural complexity of stem cell niches, which represent specific anatomic locations homing stem cells and prevent them from exiting the mitotic cycle.[50] Extracellular matrix (ECM) molecules such as collagen, fibronectin, laminin, and proteoglycans represent the non-cellular components of the niches and are important for the creation of a particular microenvironment (e.g., tooth, bone, peritoneum, heart). ECM provides nanoscale structures, such as the 15–300 nm in diameter collagen fibrils, that allow cell adhesion (via integrins) and immobilisation of signalling molecules, thus influencing the fate and behaviour (i.e., proliferation, migration, differentiation) of stem cells.[117] The concentration, size, spacing, surface chemistry and shape (e.g., ridges, grooves, pores, pits) of the artificial nanostructures (e.g., nanotubes, nanolines) are important parameters for the development of cell adhesion sites that monitor stem cell behaviour.[118–120] For example, it has been shown that surface irregularity (e.g., nanoline grating) and diverse surface chemistries (e.g., silica, poly[methyl methacrylate]) are capable of enhancing adhesion, alignment, growth, and osteogenic differentiation of hMSCs.[118,119] These findings indicate that such nanomaterials have the potential to be successfully used in dentistry for the generation of new nanotextured "osteogenic coating" dental implants, leading to direct bone/material contact. The big variety of adult stem cell populations that exist in the human body indicates that their differentiation potential and response to nanoscale materials may be different. There are not yet methodical comparative studies, however, that will allow the assessment—under the same conditions—of nanomaterials on the various stem cell lines. The lack of this crucial information delays the application of stem cell-based therapies in clinics.

11.4.3.2 Design of nanofiber scaffolds

Injection of stem cells into the injured or pathological tissue limits their spreading and, in addition, does not ensure their good engraftment.[121] Injected cells could die due to the absence of trophic factors, oxygen, or lack of a suitable ECM for their adhesion. This can be avoided by placing stem cells in biocompatible/biodegradable nanofibre scaffolds that recreate a temporary fibrous three-dimensional (3D) network of ECM. Stem cells are anchored to the nanofibres of the scaffolds, and then transplanted to the lesion site. This will improve stem cell survival, migration and differentiation potentials, and finally, their 3D organisation.[121] Stem cells (e.g., hMSCs) cultured on nanofibre scaffolds exhibit high viability and lower mobility, and differ in morphology when compared to cells cultured on conventional substrates (e.g., polystyrene).[122,123] Nanofibres with controlled diameter (e.g., 300–1000 nm) are composed of either natural polymers, such as collagen and silk, or synthetic polymers including poly(lactic acid) and poly(amide).[122,124,125] The 3D organisation, surface, and chemistry of the nanofibre scaffolds result in stem cell self-renewal, migration, and differentiation. Nanofibre scaffolds have high porosity and specific surfaces that offer an ideal environment for stem cell homing. The design of tissue-specific artificial niches offers new perspectives to stem cell-based applications for the treatment of peculiar anatomic sites (e.g., long bones, alveolar bone, teeth, periodontium, heart, muscles). In vivo transplantation of stem cells anchored to nanofibre-based scaffolds is a technique successfully used in regenerative medicine.[126,127] Transplanted biodegradable scaffolds act as temporary niches that guide, by controlling stem cell behaviour, the formation of a new specific ECM for tissue repair.

11.5 Conclusions

Numerous studies have shown the ability of various stem cell populations to form dental tissues. Stem cell-based therapies, however, are not yet applicable in dental clinics. There is a need first to develop accurate techniques that will allow monitoring the fate and behaviour of transplanted stem cells at the injured dental site. Nanotechnology, combined with stem cell biology, might provide new, non-invasive, techniques for dental tissues repair. Nanotechnology offers a plethora of exciting perspectives to regenerative dentistry, since it has a great potential to improve the actual stem cell-based therapies for diagnosis, prevention, and treatment of various diseases.

Acknowledgements

This work was supported by the 3100A0-118332 SNSF grant (T. A. M., Z. G.), 31003A_135633 SNSF grant (T. A. M., A. F.) and funds from the University of Zurich (L. J.-R., A. W., A. F., P. P, T. A. M.).

References

1. Mitsiadis TA, Graf D (2009) Cell fate determination during tooth development and regeneration, *Birth Defects Res C Embryo Today*, 87(3), 199–211.
2. Jimenez-Rojo L, Ibarretxe G, Aurrekoetxea M, de Vega S, Nakamura T, Yamada Y, Unda F (2010) Epiprofin/Sp6: a new player in the regulation of tooth development, *Histol Histopathol*, 25(12), 1621–1630.
3. Wright T (2007) The molecular control of and clinical variations in root formation, *Cells Tissues Organs*, 186(1), 86–93.
4. Mitsiadis TA,Papagerakis P (2011) Regenerated teeth: the future of tooth replacement? *Regen Med*, 6(2), 135–139.
5. Mitsiadis TA, Feki A, Papaccio G, Caton J (2011) Dental pulp stem cells, niches, and notch signaling in tooth injury, *Adv Dent Res*, 23(3), 275–279.
6. Toma JG, Akhavan M, Fernandes KJ, Barnabe-Heider F, Sadikot A, Kaplan DR, Miller FD (2001) Isolation of multipotent adult stem cells from the dermis of mammalian skin, *Nat Cell Biol*, 3(9), 778–784.
7. Friedenstein AJ, Chailakhjan RK, Lalykina KS (1970) The development of fibroblast colonies in monolayer cultures of guinea-pig bone marrow and spleen cells, *Cell Tissue Kinet*, 3(4), 393–403.
8. Zuk PA, Zhu M, Ashjian P, De Ugarte DA, Huang JI, Mizuno H, Alfonso ZC, Fraser JK, Benhaim P, Hedrick MH (2002) Human adipose tissue is a source of multipotent stem cells, *Mol Biol Cell*, 13(12), 4279–4295.
9. De Bari C, Dell'Accio F, Luyten FP (2001) Human periosteum-derived cells maintain phenotypic stability and chondrogenic potential throughout expansion regardless of donor age, *Arthritis Rheum*, 44(1), 85–95.
10. Alsalameh S, Amin R, Gemba T, Lotz M (2004) Identification of mesenchymal progenitor cells in normal and osteoarthritic human articular cartilage, *Arthritis Rheum*, 50(5), 1522–1532.
11. Zvaifler NJ, Marinova-Mutafchieva L, Adams G, Edwards CJ, Moss J, Burger JA, Maini RN (2000) Mesenchymal precursor cells in the blood of normal individuals, *Arthritis Res*, 2(6), 477–488.
12. Pittenger MF, Mackay AM, Beck SC, Jaiswal RK, Douglas R, Mosca JD, Moorman MA, Simonetti DW, Craig S, Marshak DR (1999) Multilineage potential of adult human mesenchymal stem cells, *Science*, 284(5411), 143–147.

13. Caplan AI (1991) Mesenchymal stem cells, *J Orthop Res*, 9(5), 641–650.
14. Prockop DJ (1997) Marrow stromal cells as stem cells for nonhematopoietic tissues, *Science*, 276(5309), 71–74.
15. Tuan RS, Boland G, Tuli R (2003) Adult mesenchymal stem cells and cell-based tissue engineering, *Arthritis Res Ther*, **5**(1), 32–45.
16. Gronthos S, Mankani M, Brahim J, Robey PG, Shi S (2000) Postnatal human dental pulp stem cells (DPSCs) in vitro and in vivo, *Proc Natl Acad Sci U S A*, 97(25), 13625–13630.
17. Miura M, Gronthos S, Zhao M, Lu B, Fisher LW, Robey PG, Shi S (2003) SHED: stem cells from human exfoliated deciduous teeth, *Proc Natl Acad Sci U S A*, 100(10), 5807–5812.
18. Sonoyama W, Liu Y, Fang D, Yamaza T, Seo BM, Zhang C, Liu H, Gronthos S, Wang CY, Wang S, Shi S (2006) Mesenchymal stem cell-mediated functional tooth regeneration in swine, *PLoS One*, 1, e79.
19. Morsczeck C, Gotz W, Schierholz J, Zeilhofer F, Kuhn U, Mohl C, Sippel C, Hoffmann KH (2005) Isolation of precursor cells (PCs) from human dental follicle of wisdom teeth, *Matrix Biol*, 24(2), 155–165.
20. Seo BM, Miura M, Gronthos S, Bartold PM, Batouli S, Brahim J, Young M, Robey PG, Wang CY, Shi S (2004) Investigation of multipotent postnatal stem cells from human periodontal ligament, *Lancet*, 364(9429), 149–155.
21. Wang L, Shen H, Zheng W, Tang L, Yang Z, Gao Y, Yang Q, Wang C, Duan Y, Jin Y (2011) Characterization of stem cells from alveolar periodontal ligament, *Tissue Eng Part A*, 17(7–8), 1015–1026.
22. Huang GT, Gronthos S, Shi S (2009) Mesenchymal stem cells derived from dental tissues vs. those from other sources: their biology and role in regenerative medicine, *J Dent Res*, 88(9), 792–806.
23. Gronthos S, Brahim J, Li W, Fisher LW, Cherman N, Boyde A, DenBesten P, Robey PG, Shi S (2002) Stem cell properties of human dental pulp stem cells, *J Dent Res*, 81(8), 531–535.
24. About I, Bottero MJ, de Denato P, Camps J, Franquin JC, Mitsiadis TA (2000) Human dentin production in vitro, *Exp Cell Res*, 258(1), 33–41.
25. Alliot-Licht B, Bluteau G, Magne D, Lopez-Cazaux S, Lieubeau B, Daculsi G, Guicheux J (2005) Dexamethasone stimulates differentiation of odontoblast-like cells in human dental pulp cultures, *Cell Tissue Res*, 321(3), 391–400.
26. Tecles O, Laurent P, Zygouritsas S, Burger AS, Camps J, Dejou J, About I (2005) Activation of human dental pulp progenitor/stem cells in response to odontoblast injury, *Arch Oral Biol*, 50(2), 103–108.

27. Waddington RJ, Youde SJ, Lee CP, Sloan AJ (2009) Isolation of distinct progenitor stem cell populations from dental pulp, *Cells Tissues Organs*, 189(1–4), 268–274.

28. de Mendonca Costa A, Bueno DF, Martins MT, Kerkis I, Kerkis A, Fanganiello RD, Cerruti H, Alonso N, Passos-Bueno MR (2008) Reconstruction of large cranial defects in nonimmunosuppressed experimental design with human dental pulp stem cells, *J Craniofac Surg*, 19(1), 204–210.

29. Kerkis I, Ambrosio CE, Kerkis A, Martins DS, Zucconi E, Fonseca SA, Cabral RM, Maranduba CM, Gaiad TP, Morini AC, Vieira NM, Brolio MP, Sant'Anna OA, Miglino MA, Zatz M (2008) Early transplantation of human immature dental pulp stem cells from baby teeth to golden retriever muscular dystrophy (GRMD) dogs: Local or systemic?, *J Transl Med*, 6, 35.

30. Nosrat IV, Widenfalk J, Olson L, Nosrat CA (2001) Dental pulp cells produce neurotrophic factors, interact with trigeminal neurons in vitro, and rescue motoneurons after spinal cord injury, *Dev Biol*, 238(1), 120–132.

31. Batouli S, Miura M, Brahim J, Tsutsui TW, Fisher LW, Gronthos S, Robey PG, Shi S (2003) Comparison of stem-cell-mediated osteogenesis and dentinogenesis, *J Dent Res*, 82(12), 976–981.

32. d'Aquino R, De Rosa A, Lanza V, Tirino V, Laino L, Graziano A, Desiderio V, Laino G, Papaccio G (2009) Human mandible bone defect repair by the grafting of dental pulp stem/progenitor cells and collagen sponge biocomplexes, *Eur Cell Mater*, 18, 75–83.

33. Gandia C, Arminan A, Garcia-Verdugo JM, Lledo E, Ruiz A, Minana MD, Sanchez-Torrijos J, Paya R, Mirabet V, Carbonell-Uberos F, Llop M, Montero JA, Sepulveda P (2008) Human dental pulp stem cells improve left ventricular function, induce angiogenesis, and reduce infarct size in rats with acute myocardial infarction, *Stem Cells*, 26(3), 638–645.

34. Graziano A, d'Aquino R, Cusella-De Angelis MG, De Francesco F, Giordano A, Laino G, Piattelli A, Traini T, De Rosa A, Papaccio G (2008) Scaffold's surface geometry significantly affects human stem cell bone tissue engineering, *J Cell Physiol*, 214(1), 166–172.

35. Graziano A, d'Aquino R, Laino G, Papaccio G (2008) Dental pulp stem cells: a promising tool for bone regeneration, *Stem Cell Rev*, 4(1), 21–26.

36. Onyekwelu O, Seppala M, Zoupa M, Cobourne MT (2007) Tooth development: 2. Regenerating teeth in the laboratory, *Dent Update*, 34(1), 20–22, 25–26, 29.

37. Nedel F, Andre Dde A, de Oliveira IO, Cordeiro MM, Casagrande L, Tarquinio SB, Nor JE, Demarco FF (2009) Stem cells: therapeutic potential in dentistry, *J Contemp Dent Pract*, 10(4), 90–96.
38. Cordeiro MM, Dong Z, Kaneko T, Zhang Z, Miyazawa M, Shi S, Smith AJ, Nor JE (2008) Dental pulp tissue engineering with stem cells from exfoliated deciduous teeth, *J Endod*, 34(8), 962–969.
39. Sonoyama W, Liu Y, Yamaza T, Tuan RS, Wang S, Shi S, Huang GT (2008) Characterization of the apical papilla and its residing stem cells from human immature permanent teeth: a pilot study, *J Endod*, 34(2), 166–171.
40. Gay IC, Chen S, MacDougall M (2007) Isolation and characterization of multipotent human periodontal ligament stem cells, *Orthod Craniofac Res*, 10(3), 149–160.
41. Kemoun P, Laurencin-Dalicieux S, Rue J, Farges JC, Gennero I, Conte-Auriol F, Briand-Mesange F, Gadelorge M, Arzate H, Narayanan AS, Brunel G, Salles JP (2007) Human dental follicle cells acquire cementoblast features under stimulation by BMP-2/-7 and enamel matrix derivatives (EMD) in vitro, *Cell Tissue Res*, 329(2), 283–294.
42. Handa K, Saito M, Tsunoda A, Yamauchi M, Hattori S, Sato S, Toyoda M, Teranaka T, Narayanan AS (2002) Progenitor cells from dental follicle are able to form cementum matrix in vivo, *Connect Tissue Res*, 43(2–3), 406–408.
43. Yokoi T, Saito M, Kiyono T, Iseki S, Kosaka K, Nishida E, Tsubakimoto T, Harada H, Eto K, Noguchi T, Teranaka T (2006) Establishment of immortalized dental follicle cells for generating periodontal ligament in vivo, *Cell Tissue Res*, 327(2), 301–311.
44. Chai Y, Jiang X, Ito Y, Bringas P, Jr., Han J, Rowitch DH, Soriano P, McMahon AP, Sucov HM (2000) Fate of the mammalian cranial neural crest during tooth and mandibular morphogenesis, *Development*, 127(8), 1671–1679.
45. Bartold PM, McCulloch CA, Narayanan AS, Pitaru S (2000) Tissue engineering: a new paradigm for periodontal regeneration based on molecular and cell biology, Periodontol 2000, 24, 253–269.
46. Shimono M, Ishikawa T, Ishikawa H, Matsuzaki H, Hashimoto S, Muramatsu T, Shima K, Matsuzaka K, Inoue T (2003) Regulatory mechanisms of periodontal regeneration, *Microsc Res Tech*, 60(5), 491–502.
47. Thesleff I, Wang XP, Suomalainen M (2007) Regulation of epithelial stem cells in tooth regeneration, *C R Biol*, 330(6–7), 561–564.

48. Smith CE,Warshawsky H (1975) Cellular renewal in the enamel organ and the odontoblast layer of the rat incisor as followed by radioautography using 3H-thymidine, *Anat Rec*, 183(4), 523–561.
49. Harada H, Kettunen P, Jung HS, Mustonen T, Wang YA, Thesleff I (1999) Localization of putative stem cells in dental epithelium and their association with Notch and FGF signaling, *J Cell Biol*, 147(1), 105–120.
50. Mitsiadis TA, Barrandon O, Rochat A, Barrandon Y, De Bari C (2007) Stem cell niches in mammals, *Exp Cell Res*, 313(16), 3377–3385.
51. Harada H, Toyono T, Toyoshima K, Yamasaki M, Itoh N, Kato S, Sekine K, Ohuchi H (2002) FGF10 maintains stem cell compartment in developing mouse incisors, *Development*, 129(6), 1533–1541.
52. Jimenez-Rojo L, Granchi Z, Graf D, Mitsiadis TA (2012) Stem cell fate determination during development and regeneration of ectodermal organs, *Front Physiol*, 3, 107.
53. Honda MJ, Sumita Y, Kagami H, Ueda M (2005) Histological and immunohistochemical studies of tissue engineered odontogenesis, *Arch Histol Cytol*, 68(2), 89–101.
54. Honda MJ, Shinohara Y, Hata KI, Ueda M (2007) Subcultured odontogenic epithelial cells in combination with dental mesenchymal cells produce enamel-dentin-like complex structures, *Cell Transplant*, 16(8), 833–847.
55. Young CS, Terada S, Vacanti JP, Honda M, Bartlett JD, Yelick PC (2002) Tissue engineering of complex tooth structures on biodegradable polymer scaffolds, *J Dent Res*, 81(10), 695–700.
56. Nam H,Lee G (2009) Identification of novel epithelial stem cell-like cells in human deciduous dental pulp, *Biochem Biophys Res Commun*, 386(1), 135–139.
57. Mallasez L (1885) Sur l'existence damas epitheliaux autour de la racine des dents chez l'homme adulte et a l'etat normal, *Arch Physiol*, 5, 129–148.
58. Diekwisch TG (2001) The developmental biology of cementum, *Int J Dev Biol*, 45(5–6), 695–706.
59. Caton J, Bostanci N, Remboutsika E, De Bari C, Mitsiadis TA (2011) Future dentistry: cell therapy meets tooth and periodontal repair and regeneration, *J Cell Mol Med*, 15(5), 1054–1065.
60. Bergenholtz G (2000) Evidence for bacterial causation of adverse pulpal responses in resin-based dental restorations, *Crit Rev Oral Biol Med*, 11(4), 467–480.

61. Bjorndal L (2001) Presence or absence of tertiary dentinogenesis in relation to caries progression, *Adv Dent Res*, 15, 80–83.

62. Caton J,Zander HA (1976) Osseous repair of an infrabony pocket without new attachment of connective tissue, *J Clin Periodontol*, 3(1), 54–58.

63. Mitsiadis TA, De Bari C, About I (2008) Apoptosis in developmental and repair-related human tooth remodeling: a view from the inside, *Exp Cell Res*, 314(4), 869–877.

64. Taba M, Jr., Jin Q, Sugai JV, Giannobile WV (2005) Current concepts in periodontal bioengineering, *Orthod Craniofac Res*, 8(4), 292–302.

65. Huang GT, Yamaza T, Shea LD, Djouad F, Kuhn NZ, Tuan RS, Shi S (2010) Stem/progenitor cell-mediated de novo regeneration of dental pulp with newly deposited continuous layer of dentin in an in vivo model, *Tissue Eng Part A*, 16(2), 605–615.

66. Huang GT, Shagramanova K, Chan SW (2006) Formation of odontoblast-like cells from cultured human dental pulp cells on dentin in vitro, *J Endod*, 32(11), 1066–1073.

67. Volponi AA, Pang Y, Sharpe PT (2010) Stem cell-based biological tooth repair and regeneration, *Trends Cell Biol*, 20(12), 715–722.

68. Karring T, Nyman S, Lindhe J (1980) Healing following implantation of periodontitis affected roots into bone tissue, *J Clin Periodontol*, 7(2), 96–105.

69. Nyman S, Karring T, Lindhe J, Planten S (1980) Healing following implantation of periodontitis-affected roots into gingival connective tissue, *J Clin Periodontol*, 7(5), 394–401.

70. Howell TH, Fiorellini JP, Paquette DW, Offenbacher S, Giannobile WV, Lynch SE (1997) A phase I/II clinical trial to evaluate a combination of recombinant human platelet-derived growth factor-BB and recombinant human insulin-like growth factor-I in patients with periodontal disease, *J Periodontol*, 68(12), 1186–1193.

71. Lynch SE, Williams RC, Polson AM, Howell TH, Reddy MS, Zappa UE, Antoniades HN (1989) A combination of platelet-derived and insulin-like growth factors enhances periodontal regeneration, *J Clin Periodontol*, 16(8), 545–548.

72. Selvig KA, Sorensen RG, Wozney JM, Wikesjo UM (2002) Bone repair following recombinant human bone morphogenetic protein-2 stimulated periodontal regeneration, *J Periodontol*, 73(9), 1020–1029.

73. Veis A, Tompkins K, Alvares K, Wei K, Wang L, Wang XS, Brownell AG, Jengh SM, Healy KE (2000) Specific amelogenin gene splice products

have signaling effects on cells in culture and in implants in vivo, *J Biol Chem*, 275(52), 41263–41272.

74. Shinmura Y, Tsuchiya S, Hata K, Honda MJ (2008) Quiescent epithelial cell rests of Malassez can differentiate into ameloblast-like cells, *J Cell Physiol*, 217(3), 728–738.
75. Ikeda E, Morita R, Nakao K, Ishida K, Nakamura T, Takano-Yamamoto T, Ogawa M, Mizuno M, Kasugai S, Tsuji T (2009) Fully functional bioengineered tooth replacement as an organ replacement therapy, *Proc Natl Acad Sci U S A*, 106(32), 13475–13480.
76. Washio K, Iwata T, Mizutani M, Ando T, Yamato M, Okano T, Ishikawa I (2010) Assessment of cell sheets derived from human periodontal ligament cells: a pre-clinical study, *Cell Tissue Res*, 341(3), 397–404.
77. Yang B, Chen G, Li J, Zou Q, Xie D, Chen Y, Wang H, Zheng X, Long J, Tang W, Guo W, Tian W (2012) Tooth root regeneration using dental follicle cell sheets in combination with a dentin matrix - based scaffold, *Biomaterials*, 33(8), 2449–2461.
78. Smith CE (1998) Cellular and chemical events during enamel maturation, *Crit Rev Oral Biol Med*, 9(2), 128–161.
79. Lu Y, Papagerakis P, Yamakoshi Y, Hu JC, Bartlett JD, Simmer JP (2008) Functions of KLK4 and MMP-20 in dental enamel formation, *Biol Chem*, 389(6), 695–700.
80. Yanagawa T, Itoh K, Uwayama J, Shibata Y, Yamaguchi A, Sano T, Ishii T, Yoshida H, Yamamoto M (2004) Nrf2 deficiency causes tooth decolourization due to iron transport disorder in enamel organ, *Genes Cells*, 9(7), 641–651.
81. Zheng L, Papagerakis S, Schnell SD, Hoogerwerf WA, Papagerakis P (2011) Expression of clock proteins in developing tooth, *Gene Expr Patterns*, 11(3–4), 202–206.
82. Oshima M, Mizuno M, Imamura A, Ogawa M, Yasukawa M, Yamazaki H, Morita R, Ikeda E, Nakao K, Takano-Yamamoto T, Kasugai S, Saito M, Tsuji T (2011) Functional tooth regeneration using a bioengineered tooth unit as a mature organ replacement regenerative therapy, *PLoS One*, 6(7), e21531.
83. Sun HH, Jin T, Yu Q, Chen FM (2011) Biological approaches toward dental pulp regeneration by tissue engineering, *J Tissue Eng Regen Med*, 5(4), e1–16.
84. Moghimi SM, Hunter AC, Murray JC (2005) Nanomedicine: current status and future prospects, *FASEB J*, 19(3), 311–330.

85. Moghimi SM, Hunter AC, Andresen TL (2012) Factors controlling nanoparticle pharmacokinetics: an integrated analysis and perspective, *Annu Rev Pharmacol Toxicol*, 52, 481–503.

86. Arbab AS, Janic B, Haller J, Pawelczyk E, Liu W, Frank JA (2009) In vivo cellular imaging for translational medical research, *Curr Med Imaging Rev*, 5(1), 19–38.

87. Gera A, Steinberg GK, Guzman R (2010) In vivo neural stem cell imaging: current modalities and future directions, *Regen Med*, 5(1), 73–86.

88. Ferreira L, Karp JM, Nobre L, Langer R (2008) New opportunities: the use of nanotechnologies to manipulate and track stem cells, *Cell Stem Cell*, 3(2), 136–146.

89. Thu MS, Bryant LH, Coppola T, Jordan EK, Budde MD, Lewis BK, Chaudhry A, Ren J, Varma NR, Arbab AS, Frank JA (2012) Self-assembling nanocomplexes by combining ferumoxytol, heparin and protamine for cell tracking by magnetic resonance imaging, *Nat Med*, 18(3), 463–467.

90. Lewin M, Carlesso N, Tung CH, Tang XW, Cory D, Scadden DT, Weissleder R (2000) Tat peptide-derivatized magnetic nanoparticles allow in vivo tracking and recovery of progenitor cells, *Nat Biotechnol*, 18(4), 410–414.

91. Guzman R, Uchida N, Bliss TM, He D, Christopherson KK, Stellwagen D, Capela A, Greve J, Malenka RC, Moseley ME, Palmer TD, Steinberg GK (2007) Long-term monitoring of transplanted human neural stem cells in developmental and pathological contexts with MRI, *Proc Natl Acad Sci U S A*, 104(24), 10211–10216.

92. Reimer P,Balzer T (2003) Ferucarbotran (Resovist): a new clinically approved RES-specific contrast agent for contrast-enhanced MRI of the liver: properties, clinical development, and applications, *Eur Radiol*, 13(6), 1266–1276.

93. Wang YX, Hussain SM, Krestin GP (2001) Superparamagnetic iron oxide contrast agents: physicochemical characteristics and applications in MR imaging, *Eur Radiol*, 11(11), 2319–2331.

94. Hsiao JK, Tai MF, Chu HH, Chen ST, Li H, Lai DM, Hsieh ST, Wang JL, Liu HM (2007) Magnetic nanoparticle labeling of mesenchymal stem cells without transfection agent: cellular behavior and capability of detection with clinical 1.5 T magnetic resonance at the single cell level, *Magn Reson Med*, 58(4), 717–724.

95. Song YS,Ku JH (2007) Monitoring transplanted human mesenchymal stem cells in rat and rabbit bladders using molecular magnetic resonance imaging, *Neurourol Urodyn*, 26(4), 584–593.

96. Arbab AS, Yocum GT, Kalish H, Jordan EK, Anderson SA, Khakoo AY, Read EJ, Frank JA (2004) Efficient magnetic cell labeling with protamine sulfate complexed to ferumoxides for cellular MRI, *Blood*, 104(4), 1217–1223.
97. Mitsiadis TA,Rahiotis C (2004) Parallels between tooth development and repair: conserved molecular mechanisms following carious and dental injury, *J Dent Res*, 83(12), 896–902.
98. Bruchez MP (2011) Quantum dots find their stride in single molecule tracking, *Curr Opin Chem Biol*, 15(6), 775–780.
99. Lin S, Xie X, Patel MR, Yang YH, Li Z, Cao F, Gheysens O, Zhang Y, Gambhir SS, Rao JH, Wu JC (2007) Quantum dot imaging for embryonic stem cells, *BMC Biotechnol*, 7, 67.
100. Alivisatos P (2004) The use of nanocrystals in biological detection, *Nat Biotechnol*, 22(1), 47–52.
101. Chen H, Titushkin I, Stroscio M, Cho M (2007) Altered membrane dynamics of quantum dot-conjugated integrins during osteogenic differentiation of human bone marrow derived progenitor cells, *Biophys J*, 92(4), 1399–1408.
102. Chakraborty SK, Fitzpatrick JA, Phillippi JA, Andreko S, Waggoner AS, Bruchez MP, Ballou B (2007) Cholera toxin B conjugated quantum dots for live cell labeling, *Nano Lett*, 7(9), 2618–2626.
103. Shah BS,Mao JJ (2011) Labeling of mesenchymal stem cells with bioconjugated quantum dots, *Methods Mol Biol*, 680, 61–75.
104. Slotkin JR, Chakrabarti L, Dai HN, Carney RS, Hirata T, Bregman BS, Gallicano GI, Corbin JG, Haydar TF (2007) In vivo quantum dot labeling of mammalian stem and progenitor cells, *Dev Dyn*, 236(12), 3393–3401.
105. Hsieh SC, Wang FF, Lin CS, Chen YJ, Hung SC, Wang YJ (2006) The inhibition of osteogenesis with human bone marrow mesenchymal stem cells by CdSe/ZnS quantum dot labels, *Biomaterials*, 27(8), 1656–1664.
106. Maysinger D, Lovric J, Eisenberg A, Savic R (2007) Fate of micelles and quantum dots in cells, *Eur J Pharm Biopharm*, 65(3), 270–281.
107. Ruoslahti E, Bhatia SN, Sailor MJ (2010) Targeting of drugs and nanoparticles to tumors, *J Cell Biol*, 188(6), 759–768.
108. Derfus AM, Chen AA, Min DH, Ruoslahti E, Bhatia SN (2007) Targeted quantum dot conjugates for siRNA delivery, *Bioconjug Chem*, 18(5), 1391–1396.

109. Harris TJ, Green JJ, Fung PW, Langer R, Anderson DG, Bhatia SN (2010) Tissue-specific gene delivery via nanoparticle coating, *Biomaterials*, 31(5), 998–1006.

110. Clements MO, Godfrey A, Crossley J, Wilson SJ, Takeuchi Y, Boshoff C (2006) Lentiviral manipulation of gene expression in human adult and embryonic stem cells, *Tissue Eng*, 12(7), 1741–1751.

111. Clements BA, Incani V, Kucharski C, Lavasanifar A, Ritchie B, Uludag H (2007) A comparative evaluation of poly-L-lysine-palmitic acid and Lipofectamine 2000 for plasmid delivery to bone marrow stromal cells, *Biomaterials*, 28(31), 4693–4704.

112. Glover DJ, Lipps HJ, Jans DA (2005) Towards safe, non-viral therapeutic gene expression in humans, *Nat Rev Genet*, 6(4), 299–310.

113. Gropp M,Reubinoff B (2006) Lentiviral vector-mediated gene delivery into human embryonic stem cells, *Methods Enzymol*, 420, 64–81.

114. Aliabadi HM, Landry B, Sun C, Tang T, Uludag H (2012) Supramolecular assemblies in functional siRNA delivery: where do we stand? *Biomaterials*, 33(8), 2546–2569.

115. Bianco A, Kostarelos K, Prato M (2011) Making carbon nanotubes biocompatible and biodegradable, *Chem Commun (Camb)*, 47(37), 10182–10188.

116. Kostarelos K (2010) Carbon nanotubes: fibrillar pharmacology, *Nat Mater*, 9(10), 793–795.

117. Kraehenbuehl TP, Langer R, Ferreira LS (2011) Three-dimensional biomaterials for the study of human pluripotent stem cells, *Nat Methods*, 8(9), 731–736.

118. Dalby MJ, Andar A, Nag A, Affrossman S, Tare R, McFarlane S, Oreffo RO (2008) Genomic expression of mesenchymal stem cells to altered nanoscale topographies, *J R Soc Interface*, 5(26), 1055–1065.

119. Dickinson LE, Kusuma S, Gerecht S (2011) Reconstructing the differentiation niche of embryonic stem cells using biomaterials, *Macromol Biosci*, 11(1), 36–49.

120. Lipski AM, Pino CJ, Haselton FR, Chen IW, Shastri VP (2008) The effect of silica nanoparticle-modified surfaces on cell morphology, cytoskeletal organization and function, *Biomaterials*, 29(28), 3836–3846.

121. Mooney DJ,Vandenburgh H (2008) Cell delivery mechanisms for tissue repair, *Cell Stem Cell*, 2(3), 205–213.

122. Kuo SW, Lin HI, Hui-Chun Ho J, Shih YR, Chen HF, Yen TJ, Lee OK (2012) Regulation of the fate of human mesenchymal stem cells by mechanical

and stereo-topographical cues provided by silicon nanowires, *Biomaterials*, 33(20), 5013–5022.

123. Shih YR, Chen CN, Tsai SW, Wang YJ, Lee OK (2006) Growth of mesenchymal stem cells on electrospun type I collagen nanofibers, *Stem Cells*, 24(11), 2391–2397.

124. Dzenis Y (2008) Materials science. Structural nanocomposites, *Science*, 319(5862), 419–420.

125. Murugan R,Ramakrishna S (2007) Design strategies of tissue engineering scaffolds with controlled fiber orientation, *Tissue Eng*, 13(8), 1845–1866.

126. Hashi CK, Zhu Y, Yang GY, Young WL, Hsiao BS, Wang K, Chu B, Li S (2007) Antithrombogenic property of bone marrow mesenchymal stem cells in nanofibrous vascular grafts, *Proc Natl Acad Sci U S A*, 104(29), 11915–11920.

127. Hashi CK, Derugin N, Janairo RR, Lee R, Schultz D, Lotz J, Li S (2010) Antithrombogenic modification of small-diameter microfibrous vascular grafts, *Arterioscler Thromb Vasc Biol*, 30(8), 1621–1627.

Chapter 12

Toxicity and Genotoxicity of Metal and Metal Oxide Nanomaterials: A General Introduction

Mercedes Rey,[a] **David Sanz,**[b] **and Sergio E. Moya**[b]
[a]*Biodonostia, Paseo Doctor Begiristain, s/n, 20014 Donostia, Spain*
[b]*CIC biomaGUNE, Paseo Miramón 182 C, 20006 San Sebastian, Spain*
smoya@cicbiomagune.es

12.1 Introduction

As a consequence of the fast development of nanotechnology in the last few years, nanomaterials, and more specifically nanoparticles (NPs), are commonly used and will be progressively more applied in fields such as electronics, cosmetics and medicine. Nanomaterials have unique physicochemical properties, such as high conductivity, strength, durability, and chemical reactivity that come from their nanostructure, which render them particularly interesting for industrial applications They are also very promising for biomedical applications, being able to act as drug carriers, and therefore potentially relevant for cancer therapies or as carriers for therapeutics.

Horizons in Clinical Nanomedicine
Edited by Varvara Karagkiozaki and Stergios Logothetidis

ISBN 978-981-4411-56-1 (Hardcover), 978-981-4411-57-8 (eBook)
www.panstanford.com

We are thus directly and repeatedly exposed to such materials in our daily life, through inhalation, ingestion, dermal contact and injection. Upon exposure, NPs may be translocated into the body via the skin, the GI tract, the upper respiratory tract or the lung by crossing *epithelial* barriers. For medical purposes, NPs may also be administered parenterally. The small size of NPs (they must be at least in one dimension less than 100 nm in size) facilitates their uptake into cells as well as transcytosis across epithelial cells into blood and lymph circulation, to reach different sites, such as the central nervous system.[1]

The growing use of nanomaterials has raised public concern about their potential risks to human health, since the safety of these compounds has not been fully assessed, partly because nanomaterials have been considered as safe as common larger sized materials, which are not absorbed by the body. Could Nanotechnology be dangerous? Indeed, many nanomaterials are based on molecules, which are, per se, toxic; others are non-degradable and if humans are exposed to these materials, they can potentially accumulate in organs with unknown effects. Others, because of their surface properties or quantum effects, can have a catalytic activity that may interfere with biological processes. In some cases, the nano form may facilitate the uptake of molecules or ions at the cellular or body level that would otherwise not be possible, with toxicological consequences.

In particular, NPs are able to influence the immune system of the host, and immune organs have been shown to be the main sites for the deposition of some NPs following systemic exposure.[2] Once inside the cell, NPs accumulate in compartments like endosomes and lysosomes. Before any cyto- or genotoxic event takes place, NPs are likely to induce immune responses, involving the activation of biochemical (e.g., complement cascade) and cellular components of the immune system. It is a well-known fact that macrophages play a key role in this process,[3] but the exact events that occur in the interaction between the NPs and the immune cells are still largely unknown, and results are often contradictory, mainly due to a lack of standardisation, both in methods and in reagents.[4,5] In this regard, efforts have been made in order to standardise protocols, for assessing NP interaction with the immune system and its potential effects on their biodistribution.[6,7]

NP toxicity is generally described in terms of oxidative stress, inflammation, adjuvant and procoagulant effects, and interaction with biomolecules that might lead to unwanted toxic effects in the body.[8,9] In this regard, it has been described that the association of the NPs with biological molecules such as bacterial endotoxins can strongly affect the immune response towards these materials.[10] Moreover, the diverse surface molecules, such as dextrans, citrate, synthetic polymers or phospholipids, used as an attempt to improve the biocompatibility of the NPs, result in highly differing physico-chemical properties and interactions of the particles with the cells.[11]

It should be also taken into account that alterations in the properties of the NPs can also occur when these compounds come in contact with the body or biological entities present in the environment, modifying the nanomaterial and causing dissolution, aggregation or coating of the NPs. The results of these potential alterations in the NPs range from free ions and chemicals released in the body to micrometre-sized aggregates.[5,12] In this regard, it is a well-known fact that an NP introduced into a biological system may rapidly adsorb proteins, forming a "protein corona", which surrounds the individual NPs and is responsible for the inter-particle aggregation.[13,14]

The direct toxicity of NPs in different human cells in vitro has been addressed in several papers,[15–22] but studying the interaction of the NPs with the immune system is particularly relevant in the case of NPs used for biomedical purposes, since these compounds are often injected into the blood stream and in direct contact with many immune cell types.[23]

The first cells to come in contact with injected NPs are leukocytes, such as lymphocytes and monocytes/macrophages, and in the case of ingested or inhaled NPs, the response of non-professional defence cells (such as human gut and lung epithelial cells) can be considered as representative of the real life situation.[24]

In this review, we summarise the information available about the effects of metallic and metal oxide NPs in the immune response of the host, ranging from the apoptosis (programmed cell death) of single cells to the recruitment of immune cells to the lungs and the associated cytokine production, cell proliferation, activation of intracellular signalling pathways and genotoxicity.

12.2 Effect of Different Metal and Metal Oxide Nanoparticles on the Immune Response

12.2.1 Titanium Dioxide Nanoparticles

Titanium dioxide (TiO_2) NPs are being increasingly used for energy and environmental applications, in pigments (makeup) and medical implants. TiO_2 is manufactured in large-scale production plants, thus resulting in risks for accidental high exposures to humans. Inhalation of high doses of TiO_2 potentially leads to acute and long-term adverse effects in the immune system. In this regard, Gustafsson and co-workers have recently found that a dynamic response to TiO_2 NPs takes place in the lungs of Dark Agouti rats (highly susceptible to develop long-lasting immune-mediated disorders). According to the authors, the response was long-lasting (90 days after NP instillation), beginning with activation of eosinophils, neutrophils, dendritic and natural killer (NK) cells (from the innate immune system), followed by a long-lasting lymphocyte activation and recruitment to the lung, predominantly of CD4+ helper T-cells.[20] Accordingly, previous works stated that lung exposure to TiO_2 NPs in mice caused inflammation, by activation of helper T-cells, participating both in immune responses and allergic sensitisation.[25,26] Moreover, an accumulation of titanium NPs in lymph nodes of rats upon inhalation has been reported, probably through uptake of the NPs by migratory antigen presenting cells.[27] Nevertheless, it is not clear if such immune activation can be generalised, since differences between species in the pulmonary effect of the NPs have been observed.[28] The immunotoxicity of TiO_2 NPs in rat pulmonary alveolar macrophages has been reported by several groups.[29,30] In addition, it seems that TiO_2 NPs can be phagocytosed by lung epithelial cells, not clearing but rather accumulating inside those cells, thus increasing their concentration and facilitating the interaction of the NPs with intracellular proteins like microtubules, with potential effects in vital cell functions such as cellular transport.[31,32]

12.2.2 Zinc Oxide Nanoparticles

Zinc oxide (ZnO) NPs are already being widely used in the cosmetic and sunscreen industry, as food additives, antimicrobial agents, for drug delivery, bioimaging probes, and cancer treatment.[33] Nevertheless, it seems clear from the literature that ZnO NPs have potentially harmful effects, due to their high surface area, unique physiochemical properties, and increased reactivity of the material's surface. In this regard, numerous reports have demonstrated the toxicity of ZnO NPs in various cellular systems,[12,17,33–38] while the bulk micron-sized materials remain nontoxic. However, the information about the impact of ZnO NPs in immune cells remains largely incomplete. In that context, Heng and coworkers have recently described a differential cytotoxic response between human immune cell subsets, lymphocytes being the most resistant and monocytes the most susceptible to ZnO nanoparticle-induced toxicity.[33] In addition, ZnO NPs seem to exhibit differential toxicity towards primary human cells depending upon their proliferation potential, with normal T lymphocytes that are stimulated to divide by signalling through the T-cell receptor (TCR) displaying significantly greater toxicity than quiescent nonproliferating cells of identical lineage. When these studies were extended to immortalised T leukemic and lymphoma cells, even greater sensitivity to NP-induced toxicity was observed.[39] According to the authors, susceptibility to NP-induced cytotoxicity appears to be related to the proliferative capacity of the cell and may also be affected by other physiologically relevant parameters, including cell–NP electrostatic interactions and inherent differences in cellular endocytic/phagocytic processes that facilitate NP uptake (ZnO NPs showed a preference for monocytic cells, which would explain, in part, the greater susceptibility of those cells to the nanomaterial). In this context, an inverse relationship between nanoparticle size and cytotoxicity, as well as nanoparticle size and reactive oxygen species (ROS) production was observed. The production of the proinflammatory cytokines, IFN-γ, TNF-α, and IL-12 was also detected upon ZnO NP exposure.[33] Accordingly, in a recent work, Roy and colleagues have observed that cytotoxicity and uptake of ZnO oxide NPs lead to enhanced inflammatory cytokine levels in murine macrophages, and it is suggested that the small size of the NPs may help them in evading the macrophage response.[40] An interesting point that should be taken into account is that ZnO NPs can also interfere with zinc ion homeostasis to cause cytotoxicity, and, although it is usually suggested that toxicity of ZnO-NPs may be related to their dissolution (excess of zinc ions in mammalian cells is toxic, although a trace amount of them serves as an effective antioxidant and a critical structural and functional component of zinc-binding proteins), the mechanism for that remains obscure, being probably related to programmed cell death (apoptosis).[12,41]

12.2.3 Iron Oxide Nanoparticles

Iron oxide NPs (Fe_3O_4 NPs) have magnetic properties, and are widely applied in clinical settings as contrasting agents, to enhance magnetic resonance imaging. These particles are also very promising for cell labelling, cancer therapy and drug delivery.[42] When administered systemically to mice, iron oxide NPs are rapidly engulfed by macrophages, the liver and spleen being the main distribution sites for the particles.[43] Macrophages constitute the central cellular compartment of the mononuclear phagocytic system. They are unique among immune cells in that they can enter any tissue and reside there as tissue macrophages. These cells scavenge for dead cells and foreign particles, engulfing them. Activated macrophages would then produce inflammatory molecules, which act as signals for other immune cells. NPs delivered in vivo by the systemic route or in a local compartment would undoubtedly be intercepted by macrophages as foreign bodies to be phagocytosed. Accordingly, it has been described that exposure of primary human macrophages to iron oxide NPs in culture results in a marked induction of oxidative stress and apoptosis.[44] Furthermore, in vivo studies with mice have revealed that intratracheal instillation with these NPs induced a marked infiltration of inflammatory cells in the lungs and elevated levels of proinflammatory cytokines, including interleukin (IL)-1, IL-6 and TNF-α in the bronchoalveolar lavage fluid (BAL). These results clearly demonstrated that the functionality of macrophages was modulated by iron oxide NPs in vitro and in vivo.[45]

In addition to the effect in macrophages, a recent study has shown that T-cells are also sensitive to iron oxide NPs, increasing the number of both CD4+ and CD8+ T lymphocytes, as well as the serum levels of IL-2 and interferon (IFN)-γ, two critical cytokines predominantly released by T-cells.[46]

Zhu and co-workers have recently described another potentially harmful effect of iron oxide NPs related to the immune system. The authors claim that an endothelial dysfunction and inflammation are induced by exposure to these metallic NPs, acting as risk factors for early atherosclerosis. The effects could be mediated by monocyte phagocytosis of the NPs and the subsequent activation of these cells. Accordingly, adhesion of monocytes to the endothelium was significantly enhanced as a consequence of the upregulation of intracellular cell adhesion molecule-1 (ICAM-1) and interleukin-8 (IL-8) expression, considered as early steps of atherosclerosis. Phagocytosis and dissolution of iron oxide NPs by monocytes simultaneously provoked oxidative stress and, once internalised and dissolved inside the monocytes, mediated severe endothelial toxicity, impacting the endothelial cells as free iron ions.[47]

Microglia are macrophage-like cells which play a key role in the innate immune responses in the central nervous system (CNS). Interestingly, it has been recently described that, upon iron oxide NP exposure, microglial activation, recruitment and phagocytosis of these nanomaterials take place in the CNS. NPs could then induce cell proliferation, phagocytosis and generation of free oxygen radicals (ROS), but no significant release of proinflammatory factors (such as interleukin 1β, interleukin 6 and tumour necrosis factor-α (TNF-α)) was observed. Microglial activation would rather act as an alarm and defence system against the entering and storage of NPs in the brain.[48]

A factor that must be taken into account when assessing NP toxicity is that the different NP coatings commonly used as an attempt to increase the biocompatibility of the nanomaterials can also damage cell physiology, especially for positively charged coating molecules.[11] Accordingly, Yang and colleagues, using the murine macrophage cell line RAW 264.7, have recently proposed that ferucarbutran—a clinically approved iron oxide NP coated with carboxydextran—was endocytosed by macrophages *via* the clathrin pathway, and this dose-dependent NP uptake affected the cellular behaviour of the cells, with an increase in cell proliferation.[49] These results, however, are in contradiction with a previous report by Hsiao et al., in which no changes in macrophage proliferation or viability were observed upon addition of ferucarbutran. In the same work, the authors state that the phagocytic capacity of the cells decreased when exposed to low doses of ferucarbutran, whereas other macrophage functions such as migration and production of the cytokine tumour necrosis factor-α (TNF-α) increased at higher concentrations of the NP.[50] In the same context, mice sensitised with the T-cell-dependent antigen ovalbumin (OVA) and exposed to iron oxide NPs, showed alterations not only in the production of antigen-specific antibodies, but also in T-cell functionality.[51] Thus, a decrease in Ag-specific antibodies was observed in these mice, along with an absence of IFNγ and IL-4, cytokines produced by T helper (Th) cells, which play a critical role in the activation and differentiation of B lymphocytes. In addition to T-cells, other targets and/or mediators may be affected by iron oxide NPs, and contribute to the impaired humoral immunity. In this regard, a recent study showed that the capability of dendritic cells to process the antigens and to stimulate T-cell cytokine expression was suppressed by exposure to iron oxide NPs in vitro.[52] In the same line of work, Naqvi and co-workers have described that iron oxide NPs coated with the surfactant Tween 80 induce significant toxicity only at high concentrations (300–500 μg/ml) and prolonged exposure time (6 h.) in the macrophagic murine cell line J774, being the cell damage dose- and time-dependent, due to induction of oxidative stress, and producing a marked reduction in cell viability (55–65%), mainly by apoptosis.[53]

The ample use of iron oxide NPs in biomedical research as magnetic resonance imaging (MRI) contrast agents, as mentioned above, means that the lack of cytotoxicity must be accompanied by a highly efficient internalisation, as MRI requires high levels of contrast agents to clearly depict signal alterations. Nevertheless, little information is available so far about the mechanisms and intracellular pathways involved in the potential detrimental effects of iron oxide NP internalisation. In a recent report, Soenen and colleagues stated that high intracellular levels of iron NPs affected the viability and physiology of human endothelial cells. The particles diminished cellular proliferation and affected the actin cytoskeleton and the microtubule architecture (a more compact microtubule network, closer to the cell nucleus, was formed), as well as focal adhesion formation and maturation, with potential decrease on cell migration efficiency. The extent of these effects correlated with the intracellular levels of the NPs, which accumulated in the perinuclear area.[54]

The concentration-dependent effects may indicate that cells have an intrinsic limit of the amount of non-degradable nanomaterial that they can incorporate without any significant alterations occurring. In the same context, although internalisation studies are mostly confined to cells that actively internalise foreign material, such as macrophages and dendritic cells,[55,56] iron oxide NP internalisation by non-phagocytic cells, like T lymphocytes, has been also addressed.[57]

Using the T-cell line Jurkat, the authors claim that the extent to which T-cells internalise iron oxide particles is not only dependent on particle size, but also on cell loading conditions, NP surface charge, particle concentration and incubation time.

12.2.4 Cerium Oxide Nanoparticles

Cerium oxide (CeO_2) NPs have great potential as antioxidant and radioprotective agents for applications in cancer therapy, but the interaction of these nanomaterials with cells, their uptake mechanism and subcellular localisation are still poorly understood, and the works addressing the effects of these nanomaterials on the immune system of humans or mice are still very scarce. The use of cerium compounds as diesel fuel catalyst results in the emission of cerium oxide NPs which can be inhaled, and thus be potentially harmful for human health. In this regard, Ma and co-workers have recently found that exposure to cerium oxide NPs in rats induce inflammation and functional changes in alveolar macrophages. Accordingly, an increased interleukin-12 (IL-12) production and alveolar macrophage apoptosis (through activation of the caspases 9 and 3) were observed.[58]

Vascular endothelial cell inflammation is critical in the development of cardiovascular pathology. In a recent work, Gojova and co-workers claim that direct exposure to CeO_2 NPs causes very little inflammatory response in human aortic endothelial cells (HAECS), as measured by different inflammatory markers such as intercellular adhesion molecule 1 (ICAM-1), interleukin (IL)-8, and monocyte chemotactic protein (MCP-1), even at the highest doses of the NPs. Therefore, cerium oxide NPs appear to be rather harmless, compared to other metallic NPs, such as ZnO NPs.[59]

12.2.5 Gold Nanoparticles

Gold NPs have attracted enormous scientific and technological interest because they are easy to synthetise, chemically stable, and have unique optical properties. They are widely used in chemical applications, biological imaging, drug delivery, and cancer treatment, so the in-depth knowledge about their potential toxicity and health impact is essential before these nanomaterials can be used in clinical settings.[60]

Gold NPs are "nontoxic" according to some reports. Using a human leukaemia cell line, gold nanospheres with diverse cappings were found to be harmless.[61] Accordingly, exposure to gold NPs did not induce cytotoxic effects and did not elicit secretion of the proinflammatory cytokines TNF-α and interleukin-1β (IL-1β) in the murine macrophage cell line RAW 264.7. These nanomaterials were internalised inside the macrophages as lysosomal bodies disposed in perinuclear zones.[62] Therefore, according to these reports, gold NPs are noncytotoxic and nonimmunogenic.

On the contrary, a work by Yen and colleagues presents opposing results. The authors claim that macrophage treatment with gold NPs of different sizes dramatically decreases the population and increases the size of these cells. The NPs enter the macrophages, up-regulating the levels of proinflammatory factors like interleukin-1 (IL-1), interleukin-6 (IL-6) and tumour necrosis factor-α (TNF-α). According to transmission electron microscopy assays, gold NPs would remain trapped in vesicles in the cytoplasma of the cells, organised into a circular pattern. The authors also hypothesise that maybe part of the negatively charged gold NPs might adsorb serum proteins and enter macrophages via endocytosis, not phagocytosis, which would result in cytotoxicity and harmful immunological responses.[63] Accordingly, another recent report by Gosens and co-workers described a mild inflammatory reaction after intratracheal instillation of gold NPs in the rat lung, as indicated by small increases in inflammatory cells and pro-inflammatory cytokine production. The NPs were taken up by macrophages.[64]

Dendritic cells (DC) are professional antigen presenting cells (APC), able to initiate the specific immune response, and, by their capacity to secrete various cytokines, also able to activate cells on the innate immune response, such as natural killer (NK) cells. In a recent report, Villiers and colleagues described the effect of gold NPs on DCs produced in culture from mouse bone marrow progenitors. They found that these NPs were not cytotoxic, even at high concentrations. Furthermore, the phenotype of the DC was unchanged after the addition of gold NPs, suggesting that there was no activation of the cells. Nevertheless, an in-depth analysis at the intracellular level revealed that important amounts of NPs accumulated in endocytic compartments, likely due to the phagocytic ability of these cells. Such accumulation of NPs inside the DCs, especially in compartments dedicated to antigen processing, may be a handicap for antigen presentation. Moreover, a significant change in cytokine secretion was also observed, indicating a potential perturbation of the immune response. Accordingly, the levels of the cytokine interleukin-12p70 (IL-12p70), an active complex formed by the association of IL12p35 and IL12p40, directly involved in T lymphocyte activation and thus in the regulation of the antigen-specific immune response, were significantly reduced in the presence of gold NPs, whereas the amount of IL-6, involved in the induction of inflammation, was not affected.[65]

12.2.6 Silver Nanoparticles

Silver NPs (Ag NPs) are among the most commercialised nanomaterials, due to their antimicrobial potential. Nanosilver is widely used for treatments of wounds, burns, water or air disinfection, or as coatings on diverse textiles. Nevertheless, important issues, such as the interactions of these NPs with cells, uptake mechanisms, distribution, excretion, toxicological endpoints and mechanism of action are not completely understood.[66,67]

The potential toxic effects of Ag NPs concerning the immune system have been addressed in several works. Foldbjerg and colleagues have described that these nanomaterials induced apoptosis and necrosis in the monocytic cell line THP-1, being these effects dose-and exposure-dependent. A drastic increase in ROS levels was also detected, suggesting that oxidative stress is an important mediator of cytotoxicity caused by Ag NPs.[68]

It should be also taken into account that Ag ions may be released from the NPs, in a way similar to ZnO NPs, and could contribute to the toxicity of the nanomaterials. In this regard, both the NPs and the free Ag ions induced cytotoxicity in Jurkat, a human T lymphocyte cell line. Both Ag NPs and Ag ions induced similar levels of cellular reactive oxygen species (ROS) during the initial exposure, and after 24 hours, they were increased on exposure to Ag NPs, compared to Ag ions. Moreover, the activation of some intracellular signalling molecules, such as p38 mitogen-activated protein kinase, upon Ag NP exposure was also reported, which induced DNA damage, cell cycle arrest and apoptosis in the lymphocytes.[69]

Nevertheless, a recent report by Stebounova and co-workers showed that minimal inflammatory response or toxicity was found following exposure to nanosilver in in vivo studies. Mice were exposed to silver NPs by inhalation, and NP toxicity was evaluated, among other assays, by measuring the levels of inflammatory cytokines in bronchoalveolar lavage fluid (BAL) of the animals. Measured cytokines included interleukin (IL)-6, IL-12(p40), tumour necrosis factor (TNF)-α, granulocyte macrophage colony stimulating factor (GM-CSF), keratinocyte-derived cytokine (KC), monocyte chemotactic protein (MCP)-1, and macrophage inflammatory protein (MIP)-1α. An increase in the number of macrophages and a slight inflammatory response was observed in NP-exposed animals. Accordingly, only two of the cytokines tested, IL-12(p40) and KC showed slight concentration changes upon exposure to the nanomaterial. IL-12(p40), which exhibited the most elevated concentration, is produced by dendritic cells and macrophages, and directly involved in T lymphocyte activation and thus in the regulation of the antigen-specific immune response, as previously mentioned in this review. Keratinocyte-derived cytokine (KC), involved in chemotaxis and cell activation of neutrophils, was also slightly elevated.[67]

12.2.7 Silica Nanoparticles

Silica (SiO_2) NPs are used as a polishing agent and remineralisation promoter for teeth in the oral care field. They are also employed in cosmetics and foods. Regarding the potential effects of these nanomaterials on the immune system, several reports have described macrophage toxicity and stimulation of inflammatory responses.[16,70–72] Accordingly, Nabeshi and colleagues have also recently reported cytotoxic effects on the murine macrophage cell line RAW 264.7 upon exposure to several types of silica NPs, with or without coating with amine or carboxyl groups. They found that the NPs without any coating were toxic to the cells, decreasing their proliferation rate, whereas the coated nanomaterial failed to display any detectable cytotoxicity up to concentrations of 1000 μg/ml, suggesting that the cytotoxic effect of silica NPs can be reduced by surface modification.[22] On the subject of the mechanisms underlying the toxicity of silica NPs, several explanations have been suggested, including membrane damage, caspase activation and cell death via apoptosis. Nevertheless, the precise trigger for these silica NP-induced cellular effects is uncertain. Some possible explanations would be lysosomal destabilisation,[73] or mitochondrial membrane damage.[74,75] It is, however, feasible that cytotoxicity by silica NPs depends on a combination of these effects. In this regard, Wang and colleagues have reported that silica NPs induced apoptosis on lung alveolar macrophages in mice through an increase in the expression of p53, a tumour suppressor, which acts as a key transcription factor, regulating many important apoptosis-related genes.[73] In the same line of work, Thibodeau and co-workers have also reported that silica NPs have apoptotic effects on the mouse alveolar macrophage cell line MH-S, through mitochondrial membrane depolarisation and activation of the caspases 3 and 9 (proteins interacting in a coordinated cascade, to propagate cell signalling pathways leading to eventual apoptosis).[76] In another work by the same group, a role for lysosomal enzymatic activity in the silica-induced apoptosis of mouse alveolar macrophages is described. Accordingly, MH-S cells pretreated with pepstatin A, an inhibitor of lysosomal cathepsin D, showed decreased caspase 3 and 9 activation, and were less apoptotic. Cell treatment with an inhibitor of lysosomal acidic sphingomyelinase also showed decreased caspase activation. Thus, according to the authors, silica NPs induce a lysosomal injury that precedes cell apoptosis, and the apoptotic signalling pathway involved includes cathepsin D and acidic sphingomyelinase.[75]

12.2.8 Copper Nanoparticles

Although widely synthesised and used as metal catalysts, heat transfer fluids in machine tools, semiconductors and even in antimicrobial preparations, information about the potential immunotoxic effects of copper NPs is very scarce. A report by Chen and co-workers suggested that copper NPs were very toxic in vivo, affecting the kidney, liver and spleen,[77] and in a recent article, Kim and colleagues found that Cu NPs induced strong inflammatory responses, with increased recruitment of total cells and neutrophils to the lungs.

Although copper NPs have anti-microbial activity in vitro, NPs may inhibit microbial clearance by inducing excessive neutrophil-mediated inflammation.[78,79] To address this question, the authors established a murine pulmonary infection model of *Klebsiella pneumoniae*, which causes infection and pneumonia in mammals, following NP exposure. They found that Cu-exposed mice showed increased inflammation, and that there was an upregulation of pro-inflammatory cytokines and recruitment of neutrophils to the lungs. Moreover, augmented levels of total protein and LDH (lactate dehydrogenase) activity in BAL fluid provided evidence of cytotoxicity. Inflammatory responses were dose-dependent.[80]

12.3 Metallic Nanoparticles and Their Interactions with Plasma Proteins

The immune complement (IC) is a cell-free protein cascade system, and the first element of the innate immune system to recognise foreign objects that enter the body. Activation of the complement cascade can be harmful if NPs inadvertently, or by design, enter the systemic circulation, increasing the risk of hypersensitivity reactions and anaphylaxis. It has also been suggested that activation of the complement system at tumour sites stimulates tumour-associated immune cells and promotes their conversion into a tumour-supportive phenotype, thereby stimulating cancer progression.[81,82] This type of response may impact the therapeutic efficacy of NP formulations intended for cancer diagnosis or therapy. Conversely, if particles are used for subcutaneous or intradermal administration, activation of the complement by the particles can benefit vaccine efficacy.[23] The information regarding the interactions between metallic NPs and the complement system is still scarce. A recent work by Hulander and colleagues showed that surface-bound hydrophilic gold NPs suppressed the activation of the complement system.[83] In a recent report by Lozano and colleagues, different metallic NPs (iron, zinc, cerium and titanium) were tested, but none of them seemed able to induce complement activation.[21]

On the other hand, it should be taken into account that, as previously mentioned in this review, NPs in a biological fluid (plasma, or otherwise) associate with proteins organised into the "protein corona", responsible for the inter-particle aggregation.[13,14] This adsorption of proteins onto NPs could alter the conformation state of the proteins, changing or losing their function and possibly presenting new antigenic peptides to the immune system, leading to auto reactivity against self-epitopes and thus to a persistent cell-mediated immune response.[84] It has been described that both size and surface properties of the NP play a very significant role in determining the protein corona, which may be loosely divided into a "soft" component, in which a rapid dynamical exchange of the biomolecules between medium and particles predominates, and a "hard" corona, with biological macromolecules with high affinity for the particle surface.[14]

Protein binding is also a key element that affects biodistribution of the NPs throughout the body.[85] In this regard, Dutta and colleagues have assessed the influence of proteins adsorbed onto the surface of carbon nanotubes in guiding nanomaterial uptake or toxicity in the macrophage mouse cell line RAW 264.7. Albumin was identified as the major foetal bovine or human serum/plasma protein adsorbed onto the nanotubes, and seems to target the nanotube-albumin complex to scavenger receptors.[71]

12.4 Intracellular Signalling Pathways Activated by Metallic Nanoparticles

According to the available literature, the main event implicated in NP-induced toxicity seems to be oxidative stress, which can activate a wide variety of cellular events such as cell apoptosis, inflammation and induction of antioxidant enzymes.[8] Indeed, reactive oxygen species (ROS) play important roles in cells, either by acting as second messengers leading to the activation of specific pathways and gene expression, or by causing cell death. The oxidative stress results from an excess of ROS, which overwhelms the antioxidant capacities of the cells. NPs have inherent abilities to produce ROS, especially due to their chemical composition and interactions with the cellular components. This model could explain the cell responses to diverse NPs.[86] Nevertheless, it is likely that oxidative stress is not enough to explain all the biological effects of the NPs.[87]

It is acknowledged that cell responses to NPs occur after the activation of different cellular pathways, mainly those involving MAP kinases (ERK (extracellular signal-regulated kinases), p38 mitogen-activated protein kinase and JNK (c-Jun N-terminal kinases), and redox-sensitive transcription factors such as NFκB and Nrf-2. The ability of NPs to interact with components of these signalling pathways could partially explain their cytotoxicity, and the induction of apoptosis is also closely related to the modulation of signalling pathways induced by NPs.[86]

It should also be taken into account that NPs do not interact directly with the cells, they do it through the "protein corona" that surrounds the NPs, as mentioned earlier in this review. The "corona" plays an essential role in the interaction of the nanomaterials with lipids or protein receptors of the cell membrane, is important for the NP uptake, and could also lead to the activation of specific signalling pathways.[14]

In a recent work by Kang and co-workers, the involvement of JNK/P38 pathway in the apoptosis induction by TiO_2 NPs has been demonstrated in phytohemagglutinin-stimulated human lymphocytes.[88] According to the authors, the NPs activate caspase-9 and caspase-3 and induce a loss of mitochondrial membrane potential, which suggests that titanium NPs induce apoptosis via a mitochondrial pathway. Moreover, two MAPKs, p38 and JNK were also activated by the NPs. Conversely, fullerene NPs selectively inhibit JNK-related apoptosis in cerebral microvasculature endothelial cells.[89]

Another interesting fact about magnetic NPs such as iron oxide NPs, is that they can be coated with specific ligands, enabling them to bind to receptors on the cell´s surface. When a magnetic field is applied, it pulls on the particles so that they deliver nanoscale forces at the ligand-receptor bond. This mechanical stimulation can activate cellular signalling pathways known as mechano transduction pathways. Integrin receptors, some ion channels, focal adhesions, and the cytoskeleton are key players in activating these pathways, but much of the information about how mechanosensors work is still lacking. It seems that applied forces at these structures can activate pathways involving calcium signalling, Src family protein kinase, MAPK and the small GTPase Rho.[90] Accordingly, it has been recently described by Soenen and colleagues that high intracellular iron oxide NP concentrations diminish cell proliferation in human endothelium and affect the actin cytoskeleton and microtubule network architectures, as well as focal adhesion formation and maturation. According to the authors, high levels of perinuclear localised iron oxide NPs diminish the efficiency of protein expression and sterically obstruct the mature actin fibres, with potential harmful effects on cell migration and differentiation.[54]

12.5 Genotoxic Studies on NMs

12.5.1 ZnO Nanoparticles

Cytotoxic effects of nanoparticulate ZnO have been well established in cell cultures.[91] DNA damage and induction of oxidative stress has been found in different cell lines as human epidermal cells A431, primary human keratinocytes[92,93] and human mucosa nasal cells.[94] All these data should be taken into account as nasal mucosa and skin are the first barriers of the body, which are exposed to the ZnO NPs used in industry and cosmetics.

In Chinese hamster ovarian (CHO) cells, genotoxicity of ZnO NPs can be slightly increased when cells are UV-irradiated during or before NPs exposition. The authors, however, propose that this result can be produced due to an increase of sensitiveness of the test system and do not necessarily represent a photo-genotoxic effect of ZnO NPs.[95]

Sub-acute long-term oral administration of ZnO NPs caused oxidative stress mediated DNA damage, evaluated by Fpg-modified Comet assay, and apoptosis in liver and kidney cells of exposed mice.

Genotoxic effects of ZnO NPs were also studied in plants. Kumari and co-workers have found an increase of chromosomal aberrations and micronuclei induction in root cells of *Allium cepa,*[96] but RAPD analysis in *Glycine max* soybean plants just shows minor modification in electrophoretical pattern when compared to control untreated plants.[97]

12.5.2 Aluminium Oxide

Aluminium oxide (Al_2O_3) NPs (30–40 nm) were found to be genotoxic in orally administrated dosages above 500 mg/kg, as the NPs caused micronuclei induction and positive comet assay in bone narrow and peripheral blood cells of exposed rats.[98,99]

Cytotoxic and genotoxic effects were found in in vitro studies, showing that those effects depend on the used cell line and on the dosage.[100] Wagner and co-workers have also found that the oxidised aluminium NPs are less toxic than metallic Al NPs in rat alveolar macrophages, but both of them can inhibit fagocitic function at a dosage that has no effect on cellular viability.[101]

12.5.3 TiO_2 Nanoparticles

In vivo studies with Titanium oxide NPs showed some controversial data: DNA damage, measured as increased tail in Comet assay, has not been found either in rodents, after inhalation or intratracheal instillation[102,103] nor in exposed aquatic animal models.[104]

On the other hand, Trouiller and co-workers have found increased 8-OHdG levels, gamma-H2AX foci, micronuclei induction and DNA deletions in a mice model treated with TiO_2 NPs supplemented water,[105] as well as previously described inflammatory response. Direct oral gavage of titanium oxide NPs also produced DNA damage—measured in comet assay—and micronuclei induction in bone marrow and liver cells of exposed mice.[106] These different results show that dosage and exposure method to NPs may be critical factors to control in toxicity studies.

Several studies have found genotoxic effects due to TiO_2 NPs on different types of cultured mammal cells using comet assay, micronuclei induction or HPRT assays.[100,107–112] Nevertheless, in a different study, nano-sized TiO_2 did not show any sign of genotoxicity in human peripheral blood lymphocytes with concentrations as high as 200 μg/ml, tested with comet assay.[113]

As most of the genotoxicity studies are short term experiments, some controversy about the genotoxic effects of long-term low dosage of TiO_2NPs has arisen. While Huang and co-workers have found long-term genotoxic effects in human fibroblasts that lead to genome instability,[114] other authors have found that long-term exposure produces a kind of adaptative response in treated cells with a reduction in long-term genotoxic effects measured with comet and HPRT assays in Chinese hamster ovary cells.[115]

Also, it has been proposed that titanium oxide genotoxicity is associated with ROS production, inflammatory response and caspase dependent apoptotic pathway activation.[107,116]

Regarding possible differential effects of the two common forms of titanium oxide, rutile and anatase, it has been found that the anatase form provokes greater DNA damage in human bronchial epithelium cells,[109] although both of them have been found to be genotoxic.

12.5.4 CeO_2 Nanoparticles

There are not many studies about the genotoxicity of nanoceria. Even though some studies propose that nanoceria could act as an ROS scavenger with anti-inflammatory properties,[117] there is some evidence that do not support that. Lin and co-workers have found that incubation with up to 23.5 μg/ml of nanoceria produce significant oxidative stress in human lung cancer cells reflected by reduced glutathione and alpha-tocopherol levels[118] and also slight genotoxic effects have been found in 72h 10–100 μg/ml exposed human lens epithelial cells.[119] In the same way, a study using environmental concentrations (1–100nM) of nanoceria led to decreased lifespan of treated nematode *Caenorhabditis elegans* and suggested that ROS generation is involved in the toxicity mechanism.[120]

Interestingly, only CeO_2 NPs—but neither SiO_2 nor TiO_2 NPs—led to DNA breaks measured with comet assay in two aquatic species widely used in biomonitoring, crustaceus *Daphnia magna* and larva of the aquatic midge *Chironomus riparius*,[104] when they were exposed to NPs in water (1 mg/ml).

In plants, the exposure to CeO_2 NPs treated water (2000–4000 mg/L) caused DNA damage measured as change in RAPD profile in soybean. Curiously, cerium oxide NPs showed higher genotoxic effect than zinc oxide NPs.[97]

12.5.5 Gold Nanoparticles

Gold NPs are widely used in bioimaging and diagnostic applications. In vivo studies showed that Au NPs only produce minor genotoxic effects in assayed systems. Geffroy and co-workers have found DNA damage as changes in RAPD patterns and also upregulation of genes related with DNA repair, mitochondrial metabolism, oxidative stress, apoptosis and detoxification process in low dose, 36–106 ng gold/day, treated zebrafish. At the cellular level, they also found dysfunctions in mitochondrial activity of brain and muscle.[121] Jacobsen and co-workers have described slight genotoxic effects in a study that compare gold NPs with carbon-based nanomaterials in mice.[122] On the other hand, Schulz and co-workers have not found any genotoxic effects in rats after tracheal instillation of NPs when these effects were measured using comet and micronuclei induction assays.[123]

Also, one specific in vitro assay shows induction of sixteen proteins related to oxidative stress, cell cycle, cytoskeleton and DNA repair, as well as positive comet and FISH assays in MRC-5 human lung fibroblasts treated with 1nM Au NPs for 72 h.[124] The use of different experimental conditions caused different results as Nelson and co-workers did not detect induction of up to four different types of DNA lesions HPLC/mass during spectroscopy measured in HepG2 exposed to 0.2 μg/ml of Au NPs.[125]

12.5.6 Silver Nanoparticles

In vivo studies suggest that Ag NPs can be genotoxic in some, but not in all of the studied systems. The impact of silver NPs in aquatic organisms is of special concern, as this NM can easily reach aquatic ecosystems. A long-term study using the aquatic worm *Nereis diversicolor* has found a small, but statistically significant, increase in tail of comet assay after ten days of 25–50 μg Ag NPs /g of dry sediment exposure.[126] In another aquatic environment related organism, *Chironomus riparius,* silver NPs exposure leads to chronic toxicity in development and reproduction, and also to changes in expression of different hormone and protein synthesis related genes.[127]

0.1–10mM Ag NPs caused a slight but significant increase in somatic mutation and recombination in *Drosophila melanogaster* fruitfly (SMART test), measured as loss of heterozigosity in multiple wing hairs and flare-3 recessive markers.[128] Ag NPs incubation induced cell death and upregulation of cell cycle checkpoint p53, DNA damage repair Rad51 proteins, and also phosporilated H2AX histone expression in two kinds of embryonic mouse cells.[129] On the other hand, Kim and co-workers have not found any statistically significant increase in the number of micronucleated polychromatic erythrocytes in 30–300 mg/kg silver NPs orally treated rats.[130]

The genotoxicity of Ag NPs has also been studied in several human cell lines. For example, in normal human lung fibroblast cells, the cytotoxicity of Ag NPs has been related to DNA damage, measured with single cell gel electrophoresis (SCGE) and with CBMN techniques, and also with the disruption of mitochondria function and ROS production.[66] Oxidative stress has also been proposed as the main cause of DNA damage provoked by silver NPs, as this can be reduced by antioxiodant pre-treatment in A549 human alveolar cell line[131] or by treatment with ROS scavengers as SOD2 enzyme in bronchial epithelial BEAS-2B cells.[132]

Regarding the relation between Ag NPs size and cytotoxicity, while most of the authors have found that small Ag NPs present more toxicity than the big ones,[133–135] others have found that 200 nm Ag NPs caused higher levels of DNA damage than 200 nm silver and 200 nm and 20 nm titanium NPs in exposed human testicular NT2 cell lines. None of these, however, caused significant DNA damage in mouse primary testicular cells, indicating that genotoxic effects are highly dependent on exposed cell line.[136]

12.5.7 Platinum Nanoparticles

Platinum has been used as an antitumoral agent for a long time, but the data about platinum NPs genotoxicity are limited. Pt NPs were found to cause DNA strand breaks in human colon carcinoma cells (HT29). Damage to DNA may be caused by Pt ions that are liberated from the particles during incubation.[137,138] DNA damage, apoptosis and P53 mediated growth arrest has also been reported in human cells treated with 5–8 nm platinum NPs.[139]

In vivo studies have also found that platinum NPs may induce inflammatory response in mice[140] as well as some phenotypic effects in zebrafish embryos, including pericardial effusion, abnormal cardiac morphology, circulatory defects and malformation of the eyes.[141] To our knowledge, however, there are no full studies reporting specifically genotoxic damage in vivo.

12.5.8 Fe_3O_4 Nanoparticles

Superparamagnetic iron oxide NPs (SPIONs) have great potential in biomedical applications as they can be directed by external magnetic fields. SPION physico-chemical properties as Fe redox status or size can highly modify their genotoxic effects.[142,143] Different surface modification or coating compounds can also reduce significantly the genotoxic effect of SPIONs, as it was shown when different coated NPs were tested in fibroblasts.[144,145] Magnetite Fe_3O_4 particles of a wide range of sizes have been found to provoke genotoxic effects when measured with cytokinesis block-micronucleus induction test and comet assay, as well as with ROS induction and c-Jun N-terminal kinases (JNK) induction in human alveolar A549 cells.[146] Again, there are no full studies regarding iron oxide NPs genotoxicity in vivo, although determining this is a critical step in future biomedical applications development for these NMs.

12.6 Conclusions

The increasing use of metal and metal oxide NPs in fields as diverse and present in our daily life as electronics, cosmetics and medicine offers undeniable advantages, mainly due to their unique physicochemical properties, different from the bulk form of the materials. Moreover, the fact that they are able of acting as drug carriers makes them potentially relevant as tools for cancer therapies.

Nevertheless, the possibility of unwanted interactions between NPs and the immune system, such as immunostimulation or immunosuppression, should be kept in mind as they may promote inflammatory or autoimmune disorders, or increase the host's susceptibility to infections and even cancer. Unintentional recognition of NPs as foreign by the immune cells may result in a multilevel immune response against them, eventually leading to toxicity in the host and/or lack of therapeutic efficacy. Therefore, understanding in detail the potential toxic effects of the different NPs and the signalling pathways activated by them in the cells is critical, so we can take full advantage of these promising materials without taking unnecessary risks.

There are, however, several inconsistencies in the literature, and some of these engineered nanomaterials can cause different genotoxic responses such as chromosomal fragmentation, DNA strand breaks, mutations and oxidative stress caused lesions in DNA. Some points should be taken into account in future research to understand potential risks for human health of genotoxic effects caused by nanomaterials.

While most of published studies are based on short term and high dosage experiments, assessing long-term and low-dosage exposition genotoxic effects of NPs is a must, as this exposition is more suitable for understanding the real risks to human health.

For in vitro and in vivo models based experiments are not enough to measure the genotoxic effects in different systems. It is even more important to try to understand the physical and chemical interactions of different nanomaterials with biological structures and their functions that lead to different results. Assessing occupational risk needs to be based not just on in vivo or ex vivo models, but also on long-term epidemiological studies.

References

1. Gheshlaghi, Z. N., Riazi, G. H., Ahmadian, S., Ghafari, M., Mahinpour, R. (2008) *Acta Biochim Biophys Sin (Shanghai)* 40, 777–782.
2. Schipper, M. L., et al. (2007) *J Nucl Med* 48, 1511–1518.

3. Guildford, A. L., et al. (2009) *J R Soc Interface* 6, 1213–1221.
4. Soenen, S.J, De Cuyper, M. *Contrast Media Mol Imaging* 6, 153–164 year 2011.
5. Oostingh, G. J., et al. (2011) *Part Fibre Toxicol* 8, 8.
6. Dobrovolskaia, M. A., Aggarwal, P., Hall, J. B., McNeil, S. E. (2008) *Mol Pharm* 5, 487–495.
7. Pfaller, T., et al. (2010) *Nanotoxicology* 4, 52–72.
8. Nel, A., Xia, T., Madler, L., Li, N. (2006) *Science* 311, 622–627.
9. Li, N., Xia, T., Nel, A.E. (2008) *Free Radic Biol Med* 44, 1689–1699.
10. Vallhov, H., et al. (2006) *Nano Lett* 6, 1682–1686.
11. Soenen, S. J., De Cuyper, M. (2009) *Contrast Media Mol Imaging* 4, 207–219.
12. Kao, Y. Y., Chen, Y. C., Cheng, T. J., Chiung, Y. M., Liu, P. S. (2011) *Toxicol Sci* , doi:kfr319 [pii] 10.1093/toxsci/kfr319.
13. Diaz, B., et al. (2008) *Small* 4, 2025–2034.
14. Lundqvist, M., et al. (2008) *Proc Natl Acad Sci U S A* 105, 14265–14270.
15. Hu, X., Cook, S., Wang, P., Hwang, H. M. I (2009) *Sci Total Environ* 407, 3070–3072.
16. Waters, K. M., et al. (2009) *Toxicol Sci* 107, 553–569.
17. De Berardis, B., et al (2010) Toxicol Appl Pharmacol, 246,3, 116-127.
18. Heng, B. C., et al. (2010) *Food Chem Toxicol* 48, 1762–1766.
19. Yang, X., et al. (2010) *Part Fibre Toxicol* 7, 1.
20. Gustafsson, A., Lindstedt, E., Elfsmark, L. S., Bucht, A. (2011) *J Immunotoxicol* 8, 111–121.
21. Lozano, et al (2011) Journal of Physics: Conference Series, 304, 1, article number 012046.
22. Nabeshi, H., et al. (2011) *Nanoscale Res Lett* 6, 93.
23. Zolnik, B. S., Gonzalez-Fernandez, A., Sadrieh, N., Dobrovolskaia, M. A. (2010) *Endocrinology* 151, 458–465.
24. Uboldi, C., et al. (2009) *Part Fibre Toxicol* 6, 18.
25. Larsen, S. T., Roursgaard, M., Jensen, K. A., Nielsen, G. D. (year 2010) *Basic Clin Pharmacol Toxicol* 106, 114–117.
26. Park, E. J., Yoon, J., Choi, K., Yi, J., Park, K. (2009) *Toxicology* 260, 37–46.
27. Ma-Hock, L., et al. (2009) *Inhal Toxicol* 21, 102–118.

28. Bermudez, E., et al. (2004) *Toxicol Sci* 77, 347–357.
29. Liu, R., et al. (2010) *J Nanosci Nanotechnol* 10, 8491–8499.
30. Liu, R., et al. (2010) *J Nanosci Nanotechnol* 10, 5161–5169.
31. Stearns, R. C., Paulauskis, J. D., Godleski, J. J. (2001) *Am J Respir Cell Mol Biol* 24, 108–115.
32. Rothen-Rutishauser, B., Muhlfeld, C., Blank, F., Musso, C., Gehr, P. (2007) *Part Fibre Toxicol* 4, 9.
33. Hanley, C., et al. (2009) *Nanoscale Res Lett* 4, 1409–1420.
34. Heng, B. C., et al. (2011) *Arch Toxicol* 85, 695–704.
35. Sharma, V., Anderson, D., Dhawan, A. (2011) *J Biomed Nanotechnol* 7, 98–99.
36. Sharma, V., Singh, P., Pandey, A. K., Dhawan, A. I (2011) *Mutat Res* doi:S1383-5718(11)00362-7 [pii] 10.1016/j.mrgentox.2011.12.009.
37. Taccola, L. (2011) *et al. Int J Nanomed* 6, 1129–1140.
38. Xie, Y., et al. (2011) *Toxicol Sci.* doi:kfr251 [pii] 10.1093/toxsci/kfr251.
39. Hanley, C., et al. (2008) *Nanotechnology* 19, 295103.
40. Roy, R., Tripathi, A., Das, M., Dwivedi, P. D. (2011) *J Biomed Nanotechnol* 7, 110–111.
41. Kim, Y. H., et al. (2010) *Am J Respir Crit Care Med* 182, 1398–1409.
42. Shen, C. C., Liang, H. J., Wang, C. C., Liao, M. H., Jan, T. R. (2011) *Int J Nanomed* 6, 2791–2798.
43. Wang, J., et al. (2010) *Int J Nanomed* 5, 861–866.
44. Lunov, O., et al. (2010) *Biomaterials* 32, 547–555.
45. Cho, W. S., et al. (2009) *Toxicol Appl Pharmacol* 239, 106–115.
46. Chen, B. A., et al. (2010) *Int J Nanomed* 5, 593–599.
47. Zhu, M. T., et al. (2011) *Toxicol Lett* 203, 162–171.
48. Wang, Y., et al. (2011) *Toxicol Lett* 205, 26–37.
49. Yang, C. Y., et al. (2011) *PLoS One* 6, e25524.
50. Hsiao, J. K., et al. (2008) lLabeling. *NMR Biomed* 21, 820–829.
51. Shen, C. C., Wang, C. C., Liao, M. H., Jan, T. R. (2011) *Int J Nanomed* 6, 1229–1235.
52. Blank, F., et al. (2011) *Nanotoxicology* 5, 606–621.
53. Naqvi, S., et al. (2010) *Int J Nanomed* 5, 983–989.
54. Soenen, S. J., Nuytten, N., De Meyer, S. F., De Smedt, S. C., De Cuyper, M. (2010) *Small* 6, 832–842.

55. de Vries, I. J., et al. (2005) *Nat Biotechnol* 23, 1407–1413.

56. Montet-Abou, K., Montet, X., Weissleder, R., Josephson, L. (2007) *Mol Imaging* 6, 1–9.

57. Thorek, D. L., Tsourkas, A. (2008) *Biomaterials* 29, 3583–3590.

58. Ma, J. Y., et al. (2011) *Nanotoxicology* 5, 312–325.

59. Gojova, A., et al. (2009) *Inhal Toxicol* 21 Suppl 1, 123–130.

60. Alkilany, A. M., Murphy, C. J. (2010) *J Nanopart Res* 12, 2313–2333.

61. Connor, E. E., Mwamuka, J., Gole, A., Murphy, C. J., Wyatt, M. D. (2005) *Small* 1, 325–327.

62. Shukla, R., et al. (2005) *Langmuir* 21, 10644–10654.

63. Yen, H. J., Hsu, S. H., Tsai, C. L. (2009) *Small* 5, 1553–1561.

64. Gosens, I., et al. (2010) *Part Fibre Toxicol* 7, 37.

65. Villiers, C., Freitas, H., Couderc, R., Villiers, M. B., Marche, P. (2010) *J Nanopart Res* 12, 55–60.

66. AshaRani, P. V., Low Kah Mun, G., Hande, M. P., Valiyaveettil, S. (2009) *ACS nano* 3, 279–290.

67. Stebounova, L. V., et al. (2011) *Part Fibre Toxicol* 8, 5.

68. Foldbjerg, R., et al. (2009) *Toxicol Lett* 190, 156–162.

69. Eom, H. J., Choi, J (2010) *Environ Sci Technol* 44, 8337–8342.

70. Cho, W. S., et al. (2007) *Toxicol Lett* 175, 24–33.

71. Dutta, D., et al. (2007) *Toxicol Sci* 100, 303–315.

72. Sayes, C. M., Reed, K. L., Warheit, D. B. (2007) *Toxicol Sci* 97, 163–180.

73. Wang, L., et al. (2005) *Am J Physiol Lung Cell Mol Physiol* 288, L488–496.

74. Fubini, B., Hubbard, A. (2003) *Free Radic Biol Med* 34, 1507–1516.

75. Thibodeau, M. S., Giardina, C., Knecht, D. A., Helble, J., Hubbard, A. K. (2004) *Toxicol Sci* 80, 34–48.

76. Thibodeau, M., Giardina, C., Hubbard, A. K. (2003) *Toxicol Sci* 76, 91–101.

77. Chen, Z., et al. (2006) *Toxicol Lett* 163, 109–120.

78. Shvedova, A. A., et al. (2008) *Am J Respir Cell Mol Biol* 38, 579–590.

79. Shvedova, A. A., Kagan, V. E., Fadeel, B. (2010) *Annu Rev Pharmacol Toxicol* 50, 63–88.

80. Kim, J. S., Adamcakova-Dodd, A., O'Shaughnessy, P. T., Grassian, V. H., Thorne, P. S. (2011) *Part Fibre Toxicol* 8, 29.

81. Markiewski, M. M., et al. (2009) *Nat Immunol* 9, 1225–1235.

82. Markiewski, M. M., Lambris, J. D. (2009) *Trends Immunol* 30, 286–292.
83. Hulander, M., et al. (2011) *Int J Nanomed* 6, 2653–2666.
84. Lundqvist, M., Nygren, P., Jonsson, B. H., Broo, K. (2006) *Angew Chem Int Ed Engl* 45, 8169–8173.
85. Aggarwal, P., Hall, J. B., McLeland, C. B., Dobrovolskaia, M. A., McNeil, S. E. (2009) *Adv Drug Deliv Rev* 61, 428–437.
86. Marano, F., Hussain, S., Rodrigues-Lima, F., Baeza-Squiban, A., Boland, S. (2010) *Arch Toxicol* 85, 733–741.
87. Donaldson, K., Borm, P. J., Castranova, V., Gulumian, M. (2009) *Part Fibre Toxicol* 6, 13.
88. Kang, S. J., Kim, B. M., Lee, Y. J., Hong, S. H., Chung, H. W. (2009) *Biochem Biophys Res Commun* 386, 682–687.
89. Lao, F., et al. (2009) *ACS Nano* 3, 3358–3368.
90. Sniadecki, N. J. (2010) *Endocrinology* 151, 451–457.
91. Kim, I.-S., Baek, M., Choi, S.-J. (2010) *J Nanosci Nanotechnol* 10, 3453–3458.
92. Sharma, V. *et al.* (2009) *Toxicol Lett* 185, 211–218.
93. Sharma, V., Singh, S. K., Anderson, D., Tobin, D. J., Dhawan, A. (2011) *J Nanosci Nanotechnol* 11, 3782–3788.
94. Hackenberg, S., et al. (2011) *Toxicol in vitro* 25, 657–663.
95 Dufour, E. K., Kumaravel, T., Nohynek, G. J., Kirkland, D., Toutain, H. (2006) *Mutat Res* 607, 215–224.
96. Kumari, M., Khan, S. S., Pakrashi, S., Mukherjee, A., Chandrasekaran, N. (2011) *J Hazard Mater* 190, 613–621.
97. López-Moreno, M. L., et al. *(2010) Environ Sci Technol* 44, 7315–7320.
98. Balasubramanyam, A. (2009) *et al. Mutat Res* 676, 41–47.
99 Balasubramanyam, A., et al. (2009) *Mutagenesis* 24, 245–251.
100. Di Virgilio, a L., Reigosa, M., Arnal, P. M., Fernández Lorenzo de Mele, M. (2010) *J Hazard Mater* 177, 711–718.
101. Wagner, A. J., et al. (2007) *J Phys Chem B* 111, 7353–7359.
102. Naya, M., et al. (2011) *Regul Toxicol Pharmacol.* doi:10.1016/j.yrtph.2011.12.002.
103. Lindberg, H. K., et al. (2011) *Mutat Res.* doi:10.1016/j.mrgentox.2011.10.011.
104. Lee, S.-W., Kim, S.-M., Choi, J. (2009) *Environ Toxicol Pharmacol* 28, 86–91.

105. Trouiller, B., Reliene, R., Westbrook, A., Solaimani, P., Schiestl, R. H. (2009) *Cancer Res* 69, 8784–8789.

106. Sycheva, L. P., et al. (2011) *Mutat Res* 726, 8–14.

107 Shukla, R. K., et al. (2011) *Nanotoxicology*.doi:10.3109/17435390.2011.629747.

108 Jugan, M.-L., et al. (2011) *Nanotoxicology*.doi:10.3109/17435390.2011.587903.

109. Falck, G. C. M., et al. (2009) *Human Exp Toxicol* 28, 339–352.

110. Lu, Z.-X., Liu, L.-T., Qi, X.-R. (2011) *Int J Nanomed* 6, 1661–1673.

111. Rahman, Q., et al. (2002) *Environ Health Perspect* 110, 797–800.

112. Wang, J. J., Sanderson, B. J. S., Wang, H. (2007) *Mutat Res* 628, 99–106.

113. Hackenberg, S., et al. (2011) *Environ Mol Mutagen* 52, 264–268.

114. Huang S, Chueh PJ, Lin YW, Shih TS, Chuang SM Toxicol Appl Pharmacol (2009), 241, 182-194.

115. Wang, S., Hunter, L. A., Arslan, Z., Wilkerson, M. G., Wickliffe, J. K. (2011) *Environ Mol Mutagen* 52, 614–622.

116. Shukla, R. K., et al. (2011) *Toxicol in vitro* 25, 231–241.

117. Hirst, S. M., et al. (2009) *Small (Weinheim an der Bergstrasse, Germany)* 5, 2848–2856.

118. Lin, W., Huang, Y.-W., Zhou, X.-D., Ma, Y. *Int J Toxicol* 25, 451–457 year 2006.

119. Pierscionek, B. K., Li, Y., Schachar, R. A., Chen, W. (2011) *Nanomed: Nanotechnol Biol Med* (2011).doi:10.1016/j.nano.2011.06.016.

120. Zhang, H., et al. (2011) *Environ Sci Technol* 45, 3725–3730.

121. Geffroy, B., *et al.* (2011) *Nanotoxicology*. doi:10.3109/17435390.2011.562328.

122. Jacobsen, N. R., et al. (2009) *Particle Fibre Toxicol* 6, 2.

123. Schulz, M., *et al.* (2011) *Mutat Res.* doi:10.1016/j.mrgentox.2011.11.016.

124. Li, J. J., et al. (2011) *Biomaterials* 32, 5515–5523.

125. Nelson, B. C., et al. (2011) *Nanotoxicology*. doi:10.3109/17435390.2011.626537.

126. Cong, Y., et al. (2011) *Aquat Toxicol (Amsterdam, Netherlands)* 105, 403–411.

127. Nair, P. M. G., Park, S. Y., Lee, S.-W., Choi, *J.* (2011) *Aquat Toxicol (Amsterdam, Netherlands)* 101, 31–37.

128. Demir, E., Vales, G., Kaya, B., Creus, A., Marcos, R. (2010) *Nanotoxicology* 5, 417–424.
129. Ahamed, M., et al. (2008) *Toxicol Appl Pharmacol* 233, 404–410.
130. Kim, Y. S., et al. (2008) *Inhal Toxicol* 20, 575–583.
131. Foldbjerg, R., Dang, D. A., Autrup, H. (2011) *Arch Toxicol* 85, 743–750.
132. Kim, H. R., Kim, M. J., Lee, S. Y., Oh, S. M., Chung, K. H. (2011) *Mutat Res* 726, 129–135.
133. Park, M. V. D. Z., et al. (2011) *Biomaterials* 32, 9817–9810.
134. Carlson, C., et al. (2008) *J Phys Chem B* 112, 13608–13619.
135. Liu, W., et al. (2010) *Nanotoxicology* 4, 319–330.
136. Asare, N., et al. (2012) *Toxicology* 291, 65–72.
137. Pelka, J., et al. (2009) *Chem Res Toxicol* 22, 649–659.
138. Gehrke, H., et al. (2011) *Arch Toxicol* 85, 799–812.
139. Asharani, P. V., Xinyi, N., Hande, M. P., Valiyaveettil, S. (2010) *Nanomedicine (London, England)* 5, 51–64.
140. Park, E.-J., Kim, H., Kim, Y., Park, K. (2010) *Arch Pharm Res* 33, 727–735.
141. Asharani, P. V., Lianwu, Y., Gong, Z., Valiyaveettil, S. (2011) *Nanotoxicology* 5, 43–54.
142. Singh, N., et al. (2012) *Biomaterials* 33, 163–170.
143. Zuzana, M., Alessandra, R., Lise, F., Maria, D. (2011) *J Biomed Nanotechnol* 7, 20–21.
144. Auffan, M., et al. (2006) *Environ Sc Technol* 40, 4367–4373.
145. Hong, S. C., et al. (2011) *Int J Nanomed* 6, 3219–3231.
146. Könczöl, M., et al. (2011) *Chem Res Toxicol* 24, 1460–1475.

Chapter 13

Analogies in the Adverse Immune Effects of Wear Particles, Environmental Particles, and Medicinal Nanoparticles

Eleonore Fröhlich
Center for Medical Research, Medical University of Graz, Stiftingtalstr. 24, 8010 Graz, Austria
eleonore.froehlich@medunigraz.at

13.1 Introduction

Nanoparticles (NPs) offer revolutionary new possibilities in various fields of nanomedicine. Biodegradable nanoparticles like Doxil®, DaunoXome® and Depocyt® were already approved by the Food and Drug Administration (FDA) in the last century, and together with Abraxane®, which was approved in 2005 for metastatic breast cancer, are standard medications in the treatment of cancer. Non-biodegradable NPs are less widely used, but iron oxides in magnetic resonance imaging (MRI) and silver NPs for topical antimicrobiosis also have an established place in medical diagnosis and treatment. As they are not biodegradable and may accumulate in the body, their effects upon chronic exposure are of major importance. As most

Horizons in Clinical Nanomedicine
Edited by Varvara Karagkiozaki and Stergios Logothetidis

ISBN 978-981-4411-56-1 (Hardcover), 978-981-4411-57-8 (eBook)
www.panstanford.com

particles accumulate in organs of the reticuloendothelial system, in lymph nodes, liver and spleen, immune effects are of particular importance. For the exposure to non-biodegradable medical NPs no systematic studies are available so far, but the data on biological effects of wear particles from orthopaedic implants and the action of environmental NPs on the respiratory system could serve as an indication for potential immune effects of medical NPs. This review compares the effects on the immune system of non-biodegradable NPs used in nanomedicine to those of wear particles and of environmental particles in order to identify common principles.

Studies on wear particles mostly originate from long-term exposures of humans with rather low levels of particles compared to animal exposures, which often use high doses of particles for short exposure times.

13.2 Orthopaedic Implants

Orthopaedic implants are medical devices that replace or provide fixation of bone or replace articulating surfaces of a joint. Major indications include degenerative joint disease, inflammatory arthropathy, avascular necrosis, and complicated fractures, whereas less common indications are bone tumours. Joint arthroplasty may replace hip, knee, shoulder and elbow joints and is the most frequently performed orthopaedic procedure after fracture fixation. Complications are infections, loosening, small particle disease/osteolysis, periprosthetic fracture, implant fracture or dislocation, and recurrent disease, especially in patients with tumours. While the rate of hip arthroplasties has increased between 2000/2001 and 2005/2006 slightly from 3.5 to 4.0 per 1000 beneficiaries, knee joint replacements increased from 6.0 to 8.8 procedures per 1000 beneficiaries. Compared to that shoulder joint replacements with 0.5 and 0.8 procedures per 1000 beneficiaries is not very common [1].

13.2.1 Common Joint Replacements

Joint implants have the goal to restore normal function of the joint and to relieve pain caused by inflammation or malfunction of the joint. Pioneering work was performed for hip replacement; other common joint replacements include knee, shoulder and elbow implants (Fig. 13.1). Replacements of ankle, finger and wrist are performed less frequently.

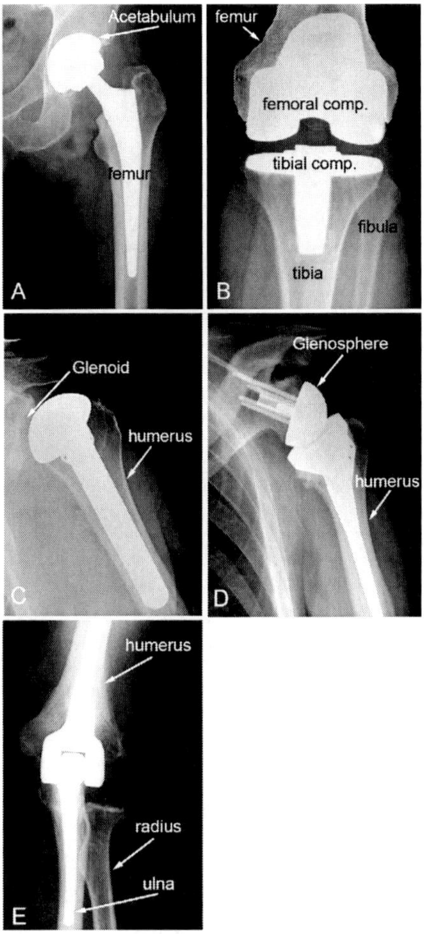

Figure 13.1 Examples for implants commonly used in total replacement of hip (A), knee (B), shoulder (C, D, normal and reverse) and elbow (E). A: The acetabular shell for the head of the femur is implanted into the acetabulum of the pelvis. The shell consists of metal and may be coated with plastic (UHMWPE) or ceramic. B: Total prosthesis of the knee with femoral component and tibial component separated by a spacer consisting of UHMWPE. C: total shoulder arthroplasty where the glenoid is replaced by a socket and D, where the glenoid is replaced by a glenosphere and the humerus carries the cavity for the articulation with the glenosphere. UHMWPE usually separates the articulating surfaces. E: Total elbow joint with humoral component and ulnar component and plastic between the articulating surfaces.

13.2.1.1 Hip replacement implants

The history of implants started in 1840, when wooden blocks inserted between the damaged ends of the hip joint were used for arthroplasty [2]. Later, a bell-shaped glass was put over the end of the femur to replace the damaged femur ball. The first solution did not produce acceptable results due to high wear, the second did not withstand the mechanical stress and failed quickly. With the use of cobalt-chromium alloys (Vitallium), mechanical properties greatly improved but surface quality was inadequate and pain relief less than expected. For total hip replacement (THR) the first surgeons used ivory, rubber and Vitallium as femur short stem prosthesis until, in 1950, Austin Moore developed the long-stem femur prosthesis. In the 1960s, this prosthesis was introduced into THR as McKee-Farrar prosthesis. A major improvement in THR was achieved by Charnley who established the use of poly(methyl methacrylate) as fixation for the implant. High Molecular Weight Polyethylene (HMWPE) replaced Teflon as socket material and since then a variety of implants, metal-on-metal, alumina and zirconia cement-on-metal, poly-tetra-fluoroethylene-on-metal, polyethylene-on-metal, ultra-high HMWPE (UHMWPE)-on-metal and ultra HMWPE-on-ceramic has been implanted. Most common indications for THR are arthritis and hip fractures.

13.2.1.2 Knee replacement implants

Knee replacements have been performed since 1860, but until 1973, when the total condylar prosthesis was invented, were quite unsuccessful because the mechanic of the knee joint is very complex. The introduction of methylmethacrylate as fixation material and of polyethylene as surface coating enabled the development of the total knee replacement. While Gunston and others worked on the anatomical approach—the exact reproduction of the knee joint—the Freeman-Swanson and other prostheses followed the functional approach. Unrestricted movements and high contact stress limited the survival of knee prostheses but with the use of composite matrix materials and the maintenance of the collateral ligaments, it may be possible that total knee replacements reach the survival of THRs. Arthritis is the most common indication for total knee replacements.

13.2.1.3 Shoulder

Replacements of the upper extremity, shoulder and elbow, were developed later and are performed less frequently. In 1951, Neer invented the hemiarthroplasty of the shoulder and, in combination with glenoid resurfacing, the total shoulder was developed in 1973. In the 1990s second-generation shoulder arthroplasty with modular construction was used. The total shoulder is indicated for patients with rheumatoid arthritis, with humoral head necrosis and with tumours. Reverse shoulder replacement, where the normal socket of the glenoid is replaced with an artificial ball and the normal ball is replaced with an implant that has a socket into which the artificial ball rests, is mainly used when the rotator cuff of muscles for the articulation of the joint is absent.

13.2.1.4 Elbow

Replacements of the equivalent to the knee were tried between 1940 and 1960 with poor outcomes. Compared to the hip and knee joints, the elbow is relatively small and its stability depends greatly on ligamentous integrity. As with the knee, major progress was made for the elbow with the new materials, methyl-methacrylate and polyethylene. This helped perform linked total elbow replacements for older patients and unlinked total elbow replacements for younger patients with competent soft tissue. The main indication for elbow replacements is inflammatory, mostly rheumatoid arthritis.

13.2.2 Failure of Implants

The survival of all joint implants is compromised by infection, instability fracture, nerve injury, loosening and anaesthesia problems. Failure of joint arthroplasty over the long-term consists of >75% aseptic osteolysis, 7% infection, 6% recurrent dislocation, 5% periprosthetic failure and 3% surgical error [3].

13.2.2.1 Infection

Periimplant osteolysis may be caused by infection, and in the absence of bacteria, infection of the implant site is favoured by the implant itself. The presence of a foreign body reduces the minimum inoculation of Staphylococcus aureus, one of the main bacteria involved in infected joints, by more than a factor of 100,000 [4].

13.2.2.2 Aseptic loosening

Just as loosening of the implant took place in the absence of infection, studies identified wear particles as inducers of inflammation. Extensive studies were performed in the physicochemical and biological characterisations of wear particles because they were regarded as a most important parameter in aseptic loosening of the prosthesis, periimplant osteolysis, also called particle disease [5]. Aseptic osteolysis is defined as bone loss around the implant that is observable radiographically.

13.2.2.3 Generation of particles

The mechanical theory for aseptic loosening of implants claims that the total artificial hip cannot adapt to cyclic stresses imposed on the interlock between the prosthesis and the patients' skeleton during everyday activities. The area where the bone attaches to the total hip is exposed to 70 times higher stresses in the implanted hip than in the normal one. Since the hip is denervated by the implantation, protection against the stress is minimal.

The biological theory relies on the generation of wear particles, which are created in enormous amounts with each step. These particles get into the surrounding tissue and generate inflammation. In the most common type of THR with polyethylene-on-metal, particles consisting of polyethylene are produced.

The production of wear particles is highly dependent on the type of implant. Ceramic on ceramic implants produce the lowest amount of particles, but their low fracture toughness limits their use for THR. Metal on metal implants have better mechanical properties and produce much less particles than HMWPE-on-metal and UHMWPE-on-metal (0.004 mm^3/y vs. 0.9 mm^3/y), but metal particles appear to show less favourable biological effects [2].

During the lifetime of the prosthesis, high amounts of particles are released by wear and tear. These particles are transported via the joint fluid to the effective joint space, where macrophages and bone cells get exposed to them. This transport is facilitated by high joint pressure.

13.2.2.4 Unspecific and specific immune response

For a better understanding of the immune effects of wear particles, the relevant cell types are cursorily described (Fig. 13.2).

Tissue macrophages carry a variety of membrane receptors and differentiate according to the stimulus into M1 (classical activation) and M2 a-c (alternative activation) subtypes. The differentiation into M1 macrophages is induced by bacteria and endotoxin and also by wear particles. M1 cells recruit Th17, cytotoxic T-cells and neutrophilic granulocytes and mainly cause inflammation [6]. M1 macrophages induce Th1 responses, tissue destruction, killing of intracellular parasites and promote tumour resistance. Macrophages differentiate into M2 cells by binding of IL-4 and IL-13 (M2a), of IgG complexes and ligands of toll-like receptors (M2b), or by IL-10 and glucocorticoids (M2c). M2a and M2c are immunoregulatory and act weakly microbicidal. M2 macrophages mediate allergic responses via stimulation of Th2 cells, eosinophilic granulocytes and naïve T-cells and, in addition, induce tissue remodelling [7,8]. M2 macrophages also participate in parasite encapsulation, angiogenesis and tumour promotion. The different subtypes can interconvert and appear to represent functional states of macrophages.

Dendritic cells (DCs) are key regulator of the specific immune response. Binding of pathogens to toll-like receptors (TLRs) and the endocytic pattern recognition receptor (PRRs, mannose receptor) of DCs activates naïve T-cells via the CD80/CD86-CD28 and the MHC class II-T-cell receptor pathway, respectively. Naïve T (T0) cells differentiate into Th1 cells to mediate immunity to intracellular pathogens and cause autoimmunity. Alternatively, they may differentiate into Th2 cells to provide protection against extracellular parasites by production of antibodies and recruitment of eosinophils. Th17 cells, which are involved in resistance to extracellular bacteria and fungi, generate a pro-inflammatory response by recruiting neutrophilic granulocytes. Follicular Th cells, located in the lymph nodes, stimulate B-cells to produce antibodies. Regulatory T-cells inhibit the activity of most of the effector T-cells and act as immunoregulator. In addition to these main classes, distinct T-cell subsets, Th9 cells and Th22 cells, have been identified. Th9 cells, a subclass of Th2 cells, may be involved in immune-mediated diseases ranging from autoimmunity to asthma. Their existence is not yet unanimously accepted. Th22 belonging to Th17 cells are able to regulate epidermal responses in inflammatory skin and lung diseases.

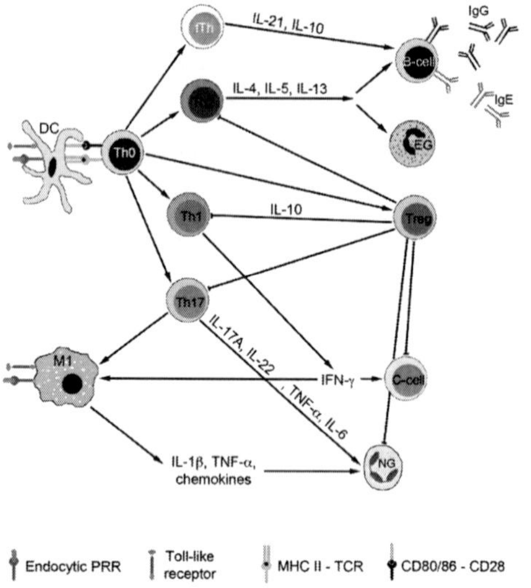

Figure 13.2 Simplified sketch on the interactions of unspecific and specific immune system. M1 macrophages (M1) secrete cytokines for the recruitment of neutrophilic granulocytes (NG). They interconnect with Th1 and Th17 cells of the specific immune system. Dendritic cells (DC) initiate maturation of the naïve T-cells (Th0) to different types of mature T-cells. Th1 cells secrete mainly IFNγ and TNFα and promote an immune response against intracellular pathogens for instance by activation of cytotoxic T-cells. Th2 cells secrete IL-4, IL-5 and IL-13 and stimulate eosinophilic granulocytes (EG). Th9 cells, a subgroup of Th2 cells with similar functions, are not shown in the figure. Follicular (fTh) cells are found in the lymph nodes in close association to B cells and are characterized by the secretion of IL-21. The fTh cells stimulate antigen specific B cells to secrete high antibody levels. Th17 secrete IL-17 and recruit neutrophils and activate of macrophages. Th22 cells are a subgroup of TH17 cells with similar function to Th2 and are not represented in the figure. Regulatory T-cells (Treg) inhibit immune response and inflammation by blocking the activity of effector, helper and/or antigen presenting cells.

13.2.2.4.1 *Immune effects of wear particles*

Most implants contain UHMWPE coating. Wear particles from this material are 0.5–10 μm large and are more active than larger (>10 μm) and smaller (<0.5 μm) ones. Particles >10 μm cause only giant cell formation but no osteolysis, whereas smaller particles induce cytokine secretion, which activate the resorption of bone by osteoclasts. Irregularly shaped particles are more reactive in this respect than regularly shaped ones.

The periimplant tissue contains mainly (60–80%) macrophages with 5–10% of T-lymphocytes; Th1 and Th2 in equal proportion [9]. Activation of macrophages with differentiation into M1 macrophages is the main mechanism for osteolysis in most patients (Fig. 13.3). M1 macrophages promote differentiation and protease activity of osteoclasts and fibroblasts. On the other hand, they inhibit osteoblast activity. PMMA wear particles and UHMWPE particles activate macrophages directly via PRR/TLR2 signalling [10,11]. Specific macrophage functions such as cytokine release and nitric oxide generation are stimulated [12]. Cytokines involved in aseptic osteolysis include PGE2, TNF-α, IL-1, IL-8, IL-10 and IL-6. Also, endotoxin bound to Ti particles may trigger the activation of macrophages and increase the recruitment of white blood cells [13].

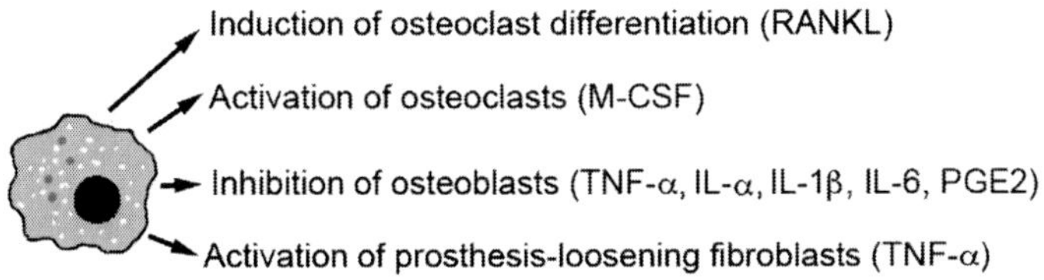

Figure 13.3 Wear particle-related alterations of the unspecific immune system. Monocytes attracted by particles from the blood stimulate the secretion of RANKL by stromal precursor cells of osteoclasts. Maturation and activation of the osteoclast precursors is promoted by secretion of the macrophage colony stimulating factor, whereas inhibition of osteoblasts occurs through secretion of a variety of inflammatory cytokines, such as TNF-α, IL-1, IL-6 and PGE2. The secretion of matrix degrading proteases by activated fibroblasts is induced.

Particles released from conventional metal-on-metal implants are CoCrMo or CrO_2 particles with varying amounts of Co and Cr, and Ti and Ti alloy particles with oval or needle-shaped morphology. The greatest biological effects were observed by 1.5–4 μm Ti particles [14]. Phagocytosis of the wear particles was not a precondition for inflammatory response of macrophages [15]. Binding of complement to CD11b/CD18 receptors induced the release of cytokines by macrophages [16], where 0.1–1 μm UHMWPE particles were most effective. Particle number is also critical for 0.49–4.2 μm particles: a high ratio (100 particles/cell) cause more inflammation than a lower ratio (10 particles/cell). Hydroxyl apatite particles induced the secretion of pro-inflammatory cytokines in human bone marrow cells dose-dependently; cells with more ingested particles produced more cytokines [17].

Few effects are known from Co in the nano-sized form [18]. It appeared that Co NPs decrease the antibacterial response by changed expression of TLRs and decreased CD14 expression, in addition to changes in cytokine secretion.

Wear particles from metal-on-metal bearings may also act on the adaptive immune system by causing lymphocyte-dominated granuloma like lesions in the bone around the implant. This reaction is supposed to be due to metal hypersensitivity caused by complexes of metals with native proteins because the solid particles did not markedly influence the lymphocyte function. Ti NPs, for instance, did not alter the proliferation of lymphocytes to pokewood mitogen but inhibited IgG production in endotoxin-stimulated B-cells [19]. Although lymphocytes play only a small role in osteolysis, commercial Ti particles may increase antigen-specific IgG and IgE levels in OVA-immunised mice [20].

As data on wear particles are mostly generated in humans, they reflect the specific situation of the human immune system. Wear particles have a constant and relatively well-defined composition because the material is produced in a standardised way. The particles do not have to cross (external) epithelial barriers of the body to reach the blood circulation, a situation more similar to NPs in medicine, which are mostly applied intravenously. A limitation of these studies is that the size range of wear particles is broad and many particles are not in the nano-size range.

13.3 Environmental Particles

Ambient or environmental particles are also termed airborne particulate matter and are classified according to their aerodynamic diameter. The diameter of PM10 is less than 10 μm, PM2.5 are particles smaller than 2.5 μm and PM0.1 smaller than 0.1 μm. Sources for ambient air particles are power generation plants, road traffic and ships with residual oil fly ash particles and diesel exhaust particles as main components. The two types of particles have a different composition: residual oil fly ash contains high proportions of water-soluble sulphate and metals such as vanadium, nickel and iron [21] while the main contaminations of diesel exhaust particles (DEP) are elemental and organic carbon species with the aromatic polycyclic hydrocarbon phenanthrene as the main compound. The composition of DEP varies according to the engine, engine load and type of diesel fuel.

The negative influence of air pollution on human health was already identified in the 1970s and has been confirmed in many studies. In evaluations based on hospital admissions and interviews, however, neither the exact effective concentration nor the composition of the inhaled air is known. One problem in the evaluation of the immune effects of PM, therefore, is the heterogeneity of the samples. Studies that compared the effects of washed, or stripped, DEP and carbon black to unpurified diesel exhaust reached contradictory conclusions, regarding the contribution of contaminations and carbonaceous nuclei for the observed immune reactions [22,23].

Compared to wear particles, only few data on PM are based on human studies; mainly exposures of rodents to relatively high concentrations of particles were analysed. The immune system and also the respiratory system, however, show major differences between mice and men. Anatomical differences include a larger nasal surface area, fewer, less symmetric airway branches, the absence of respiratory bronchioles, few mucous or serous cells, and no submucosal glands below the trachea in mice. In addition, mice are obligate nose breathers and possess a weak cough reflex. In terms of the reaction to pathogens mice differ from humans in cellular expression of TLRs 3,5,9 and in ligand specificity of TLRs 2,4,9 and human nucleotide binding and oligomerisation domain protein 1. The pattern of lysozyme secretion is different and mice also lack the antimicrobial hypothiocyanite, present in the human respiratory tract [24]. When used as models to study allergy, additional problems arise because immunisation with ovalbumin (OVA) as antigen, which is usually employed to induce asthma in mice, is not representative for humans as it is seldom implicated in human asthma. The BALB/c mouse asthma model, which is the preferred one due to its robust Th2 response, presents several anatomical and physiological differences to human asthma. The relevance for the human situation is further reduced by the strain-specificity of the immune response in mice because pronounced differences in the degree of airway inflammation and antibody production between the mouse strains have been reported [25].

13.3.1 Immune Effects of Environmental Particles

The few studies performed with healthy volunteers show an increase in the number of neutrophils in bronchial and alveolar fractions after two hours exposure to concentrated ambient particles [26] and mild inflammation in the lower respiratory tract but no effect on immune phenotype or macrophage function [27]. In these studies, soluble components of PM were linked to the observed effects [28].

As the composition of PM is subjected to large regional and seasonal variations, and contaminants are supposed to contribute more to the immune effects than the carbonaceous nucleus, studies with more homogeneous materials like DEP and carbon black will be mentioned in the following. Also significant is the fact that DEP have a heterogeneous composition and consist of 20–40% of adsorbed organic compounds, sulphates, nitrates and metals.[1]

13.3.1.1 Diesel exhaust particles

13.3.1.1.1 *Human exposure*

Data obtained by human inhalation exposures with relatively high doses of DEP for short duration report an increase in cytokines and white blood cells in bronchoalveolar lavage fluid and in bronchial mucosal biopsies [29,30]. The recruitment of neutrophilic granulocytes and macrophages as indication for a stimulation of the unspecific immune system was increased, but phagocytosis of macrophages suppressed [31]. This effect was only slightly reduced when particles were removed from the exhaust, suggesting that contaminations of DEP mainly caused this effect. Various studies investigated local effects of DEP at the nasal mucosa in the absence and presence of additional challenge. Increases in cytokine levels,

[1]US Environmental Protection Agency (EPA). Health effects assessment document for diesel engine exhaust. EPA/600/8-90/057F. National Center for Environmental Assessment, Office of Research and Development. Washington, DC: US EPA, 2002.

switch to the IgE isotype and sensitisation to neoallergens were observed [32–34]. In pre-disposed (allergic rhinitis) individuals, challenge with DEP and influenza virus reduced virus clearance and increased allergic inflammation [35].

13.3.1.1.2 *Animal exposure*

Inhalative exposure of rodents to DEP resulted in stimulation of the unspecific immune system. Increases in pro-inflammatory (TNF-α, MIP-2, IL-6, IL-8) and immune (INF-γ, IL-13) cytokines, in addition to increased IgE levels, in the absence of specific antigen and recruitment of leukocytes were observed [36,37]. Also, several groups reported maturation and activation of dendritic cells upon exposure to DEP in the absence of a specific antigen [38–41].

On the other hand, an impairment of specific functions of macrophages, such as phagocytosis, was seen after exposure to DEP. This depression in phagocytosis appears to be caused by stiffening of the cytoskeleton with consequent reduction in intracellular motion of phagosomes [42]. Prominent changes in cytokine secretion were also observed. Exposure to DEP decreased the cytokine production of splenocytes to endotoxin [43] and decreased the endotoxin-elicited inflammation by suppression of LPS-induced MCP-1 expression [44]. Pro-inflammatory cytokines TNF-α, IL-1β and IL-12 were reduced and anti-inflammatory cytokines IL-10 augmented in rat alveolar macrophages upon pre-exposure to DEP before challenge with *Listeria monocytogenes* [45].

A higher propensity to allergic reaction was seen upon exposure to particles alone and for combinations of DEP and pathogens. IgE levels were increased in DEP-exposed animals [37] and exposure to DEP amplifies the allergen-induced allergic inflammation evidenced by high levels of cytokines like IL-4 and IL-5, and by high antigen-specific IgE production [46–48]. Mice primed with DEP were more likely to show virus-induced exacerbation of allergic inflammation after exposure to influenza virus [49].

It is proposed that DEP stimulate M1 and M2 effects of macrophages. The M1 effects induce inflammation by secretion of TNF-α, IL-1β, IL-6 and IL-8, whereas the M2 effects induce tissue remodelling by secretion of TGF-β. The predisposition to allergic response is induced in combination with the respiratory epithelium and induces the recruitment of eosinophilic granulocytes (Fig. 13.4).

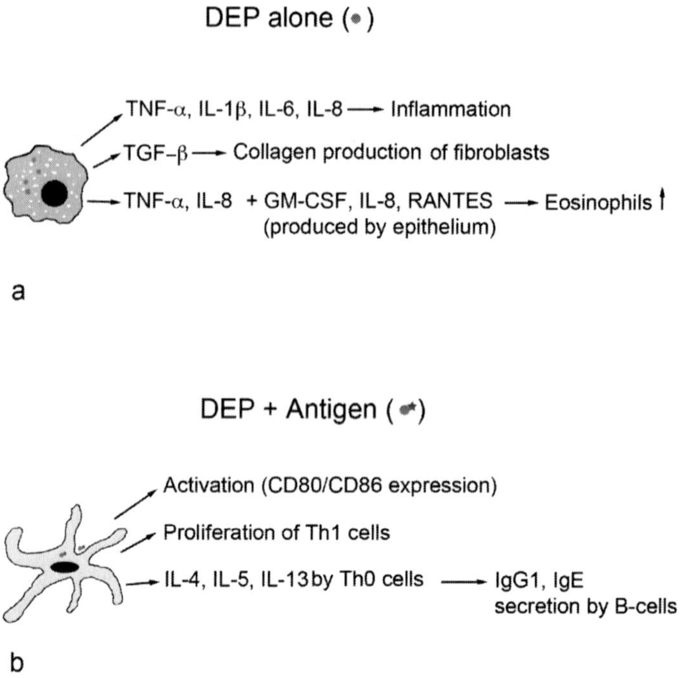

Figure 13.4 Role of macrophages (a) and dendritic cells (b) in the immune effects caused by diesel exhaust particles (DEP) alone and in combination with an antigen (DEP + antigen). a: In the lung, macrophages may induce unspecific inflammation or induce tissue remodelling by stimulation of collagen production by fibroblasts. In synergism with the alveolar epithelium, they cause sensitisation with increase in eosinophilic granulocytes. b: DCs activated by the contact with DEP + antigen increase the expression of CD80 and CD86. They induce proliferation of Th1 cells and induce the production of theTh2 cytokines IL-4, IL-5 and IL-13 by naïve T-cells, thereby stimulating the production of IgE by plasma cells.

13.3.1.2 Carbon black

Carbon black (CB) represents only the carbonaceous nucleus of DEP and, therefore, is a good control particle for the discrimination between particle induced and contaminant-induced effects of DEP.

Upon exposure to CB, maturation and activation of bone-marrow derived monocytic cells has been observed [50] and significant upregulation of CD86 and of the dendritic cell receptor DEC205 has been reported upon intratracheal instillation of CB [51]. Mice pre-exposed to CB reacted to infection with respiratory syncytial virus by allergic rather than antimicrobial response [52] In combination with OVA, CB acted as adjuvant and promoted the production of OVA-specific Th2 cytokine release and the local and systemic allergic reaction to OVA [53–55].

The pro-inflammatory effects upon instillation to BALB/c mice were size-dependent because MIP-1 protein and mRNA levels, IL-1β, IL-6 and TNF-α levels and numbers of macrophages, lymphocytes and neutrophilic granulocytes in BAL were significantly higher induced by 14 nm CB than by 95 nm particles [56]. Similarly, thymus and activation regulated chemokine was induced by 14 nm CB to a higher extent than by 56 nm CB [44].

13.3.1.3 Role of contaminants in the immune reaction to diesel exhaust particles

Most studies aimed at evaluating the role of contamination in the DEP in the immune effects conclude that the majority of the observed effect is caused by the contaminations, and not by the carbonaceous core. The following studies observed a much lower effect of CB or purified DEP, compared to DEP. Eosinophilic infiltration in the lung, activation of dendritic cells and the promotion of Th2 cytokine response were absent upon stimulation with carbon black [57]. In the absence of an additional challenge such as endotoxin, bacteria, or allergen, CB did not increase the secretion of IL-8, GM-CSF and IL-1β by airway cells [39]. DEP, but not the organic-stripped DEP, depressed pulmonary clearance and reduced IL-1β, TNF-α and ROS response to endotoxin challenge in-vitro [58]. Organic contaminants of DEP alone produced similar effects to DEP regarding the challenge of mice with endotoxin, whereas carbon black had no effects [59]. Simultaneous application of DEP and the allergen OVA augmented lymph node weight, cell number and proliferation of lymphocytes compared to either OVA or DEP alone, while CB displayed only little effects [60]. A stronger effect on eotaxin and cytokine expression and on OVA-specific IgG1 levels was reported upon challenge with a combination of OVA and not washed DEP, compared to the combination of OVA and washed DEP preparations [61].

In other studies, however, the removal of the contaminants had only a small effect. Yanagisawa et al. observed similar increases of neutrophilic granulocytes, pulmonary oedema and IL-1β, MIP-1a and MCP1 expression upon exposure to a combination of endotoxin with washed and not washed DEP, suggesting that only the carbonaceous nucleus caused these reactions [62]. No differences between the immune reaction of DEP and carbon black on OVA-induced sensitisation were seen either in the study by Al-Humadi et al. [63]. It is also possible that because both types of particles act by different mechanisms by the choice of one read-out parameter instead of other(s), a difference is detected. While DEP in combination with OVA induce mainly Th2 responses, CB stimulates both Th1 and Th2 responses [64,65].

13.4 Medical Nanoparticles

13.4.1 Use of Non-Biodegradable Nanoparticles in Medicine

NPs intended for use in nanomedicine include gold, silver, iron oxide, silica and carbon nanotubes. Gold NPs could be used in diagnostic imaging, biosensing and cancer therapeutic techniques; silver NPs act as antimicrobial agents, and applications for iron oxide NPs range from imaging, drug delivery to hyperthermia treatment and tissue repair. Silica particles and carbon nanotubes could be carriers for drug and gene delivery [66].

13.4.2 Immunotoxicity and Effects of Medical Nanoparticles on the Immune System

The evaluation of compounds intended for medical use requires cytotoxicity screening as an initial step. Cytotoxicity of NPs on immune cells can be different from that on epithelial cells because immune cells grow in suspension whereas epithelial cells are adherent. The growth in suspension can lead to higher cytotoxicity (e.g., for polystyrene particles [67]) because a larger plasma membrane area is exposed to NPs, but may also result in lower cytotoxicity because sedimentation of NP aggregates prevents the contact with the cells in suspension. The immune effects described in the following were detected by using medical NPs in non-cytotoxic concentrations.

13.4.2.1 Iron oxide nanoparticles

Microglial BV2 cells increased phagocytosis, reactive oxygen and nitric oxide generation and proliferation upon stimulation with iron oxide [68] whereas phagocytosis in RAW 264.7 cells was suppressed by iron oxide NPs. Although uptake of poly(vinylalcohol) coated Fe2O3 NPs did not alter the surface marker expression of dendritic cells nor their antigen uptake, it reduced their capacity for antigen processing, for T-cell activation and for cytokine induction [69].

In animal experiments, pro-inflammatory effects as evidenced by increased cytokine levels and raised CD4 and CD8 cells in the peripheral blood are seen after intravenous and intratracheal application [70–72]. The production of pro-inflammatory cytokines, of Th1 cytokines, of Th2 cytokines and of IgE were induced suggesting an activation of several types of immune responses. Antigen-specific antibody levels were decreased by intravenous injection of iron oxide particles and appeared to be due to the abnormal presentation of the antigen by antigen-presenting cell like DC [73].

13.4.2.2 Gold nanoparticles

Gold particles activate DCs and macrophages in-vitro according to cytokine release and nitric oxide production, but inhibit LPS-induced nitric oxide production [74–76]. In contrast to these data, Villiers et al. [77] excluded any activation of mononuclear cells by gold particles.

In-vivo, gold particles accumulated in organs of the reticuloendothelial system. Although accumulation in the spleen was only observed for 30 nm particles, relative spleen weight and immune cell numbers were increased for 5, 10 and 60 nm gold particles [78]. For 5 and 10 nm gold particles accumulation in the liver was seen and 10 and 60 nm gold particles caused slight liver damage. These data show that organ damage may occur in the absence of obvious accumulation. Another group reported inflammatory liver damage by 13 nm gold NPs [79].

13.4.2.3 Silver nanoparticles

Immune effects on silver particles include increases in stress gene expression and IL-8 secretion of macrophages by 5 nm, not by 100 nm particles, and a decline in nitric oxide production in murine peritoneal macrophages [80,81].

Silver NPs, in vivo, showed pro-inflammatory effects and repeated, oral, application increased Th1 and Th2 responses [82]. In inflamed or diseased tissue, however, also anti-inflammatory effects may occur [83].

13.4.2.4 Silica nanoparticles

Silica NPs have been shown to act size-dependent on the immune system. 270 nm silica particles stimulated DCs less than 2.5 μm ones at the same concentration [84]. NPs functionalised with sheep IgG interact intensely with macrophages but not with lymphocytes [85]. No interference, however, with the bactericidal activity of macrophages was detected.

Colloidal silica acts as immunogenic sensitiser and induced contact hypersensitivity [86]. Silica NPs can penetrate into Langerhans cells of the skin and aggravate atopic dermatitis [87]. Intraperitoneal injection of silica particles caused increased blood levels of IL-1β and TNF-α [88]. Inflammatory liver damage was size-dependent: 70 nm particles caused an effect whereas 300 nm and 1000 nm particles did not [89].

13.4.2.5 Carbon nanotubes

Single-walled carbon nanotubes activated natural killer cells and monocytes and impaired specific functions of the immune system like proliferation of T-cells, chemotaxis and phagocytosis in human monocyte-derived macrophages [90–92]. Inflammatory cytokine mRNA expression was increased in mouse macrophages [93] and induction of these cytokines was also observed in A549 cells exposed to single- and multi-walled carbon nanotubes [94,95].

Similar effects were also observed in-vivo. The pro-inflammatory effect was seen after inhalation and subcutaneous application to mice and resulted in an increase of neutrophilic granulocytes, monocytes and cytokine secretion [93,96–98]. Other effects included the impairment of immune function, such as, for instance, the decreased function of splenocytes and monotonic systemic immune suppression [99,100]. A predisposition to allergy upon intratracheal instillation of multi-walled carbon nanotubes was evidenced by higher increases in Th2 cytokines than in Th1 cytokines [101]. In tumour-bearing animals, the promotion of cytokine secretion, complement activation and increased phagocytosis reduced tumour growth [102]. In combination with a defined antigen, such as OVA, allergy occurred [103,104].

13.4.2.6 Size-dependency of the immune effects

Studies on wear particles, environmental particles and medical particles have the disadvantage that size-dependent effects cannot be studied in a systematical way. Given the high surface reactivity of NPs and the preference of phagocytes for particles between 500 and 2000 nm, size-dependent effects appear likely. Such effects are commonly studied using polystyrene particles because the bulk material is inert, lacks metal contamination and the particles can be obtained in various sizes from commercial providers.

Particles of 40 nm appeared to possess a greater stimulating effect than smaller (20 nm) and larger (100, 500, 1000 and 2000 nm) ones upon intradermal application [105]. The 40, 49 and 67 nm particles caused higher IFN-γ levels whereas 93, 101 and 123 nm ones acted by increases of IL-4 [106]. 25 nm and 50 nm latex particles upon intratracheal instillation had a stronger effect than 100 nm particles [107]. The effect as adjuvant in combination with OVA was higher for 58 nm and 202 nm polystyrene particles than for larger particles up to 11.14 μm [108]. On simultaneous intratracheal instillation of *Listeria monocytogenes* and polystyrene beads, viable bacteria numbers were reduced to a higher extent by 64 nm and 202 nm than by larger particles [109]. When the challenge with bacteria followed particle exposure, 64 nm, 202 nm and 4646 nm sized particles reduced bacterial colonisation of the lung and only 1053 nm particles showed no reduction. Studies on polystyrene also showed that the response to particles is strain-specific because, after exposure to 100 nm polystyrene particles, cytokine responses were obtained in some but not all mouse strains [108]. Taken together, the effects of smaller polystyrene particles were stronger than that of larger ones. One important size limit appears to be 100 nm for particles alone. In combination with bacteria or antigens, no size-dependency is obvious.

13.4.2.7 Comparison of immune effects of different particles

Hussain et al. [110] mentioned immune effects caused by a variety of NPs, such as CeO_2, TiO_2, Al_2O_3, polystyrene particles, carbon nanotubes, Si O_2, DEP, CB as examples for non-biodegradable NPs.

These effects are presented in Fig. 13.5 and are compared to the summary of the immune effects of wear particles, DEP, CB and medical NPs reviewed in this chapter (Table 13.1).

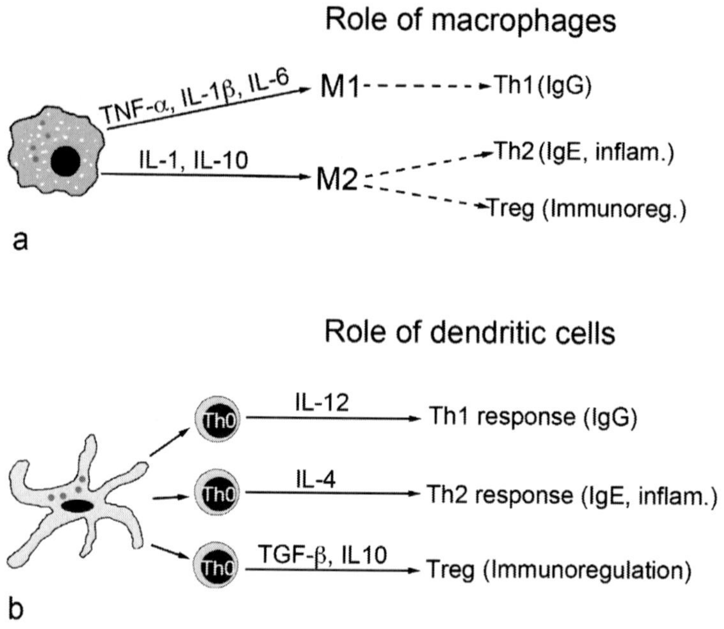

Figure 13.5 Role of macrophages (a) and dendritic cells (b) in the potential immune effects of non-biodegradable NPs. Inflammatory and allergic pathways can be activated by contact with non-biodegradable NPs and are mediated via polarization of macrophages into M1 and M2 types, where M1 stimulate the Th1 response and M2 increase Th2 and Treg cells. Increases in Th1, Th2 and Tregs are also caused by dendritic cell induced cytokine secretion of naïve T (T0) cells.

As Fig. 13.5 suggests, all non-biodegradable NPs may stimulate Th1 and Th2 pathways through activation and polarisation of macrophages and activation of DCs. Due to the heterogeneity of the collected data, it is not clear to which extent these effects are material-, size- and application-specific. Using similar settings, material-related differences between metal and metal oxide NPs were reported by Lucarelli et al. [18]. They showed that Co NPs decreased CD14 expression, while TiO_2 NPs increased it. CD14 is an important regulator of the responsiveness of the immune system to endotoxin. Whereas particles had different effects on the expression of TLRs and IL-1β secretion, they all caused anti-inflammatory effects in macrophages exposed to endotoxin. According to these data, it is not likely that all non-biodegradable NPs cause similar immune effects. The release of ions from metal particles and co-precipitates of metal oxides particles with other ions are some of the reasons that may explain the different actions.

Table 13.1 Action of wear particles, environmental particles and medical NPs on immune cells

Action	WP	DEP	CB	Fe_2O_3	Au	Ag	SiO_2	CNTs
Particles alone								
Proinflamm. Cytokines	↑	↑	↑	↑	↑	↑	↑	↑
Macrophage function*	↑	↓		↑, ↓	↑	↓	↑	↓
Leukocyte recruitment	↑		↑		↑			↑
Particles + pathogen/immune challenge								
Macrophage function	↓	↓			↓			↓
Th2 response	↑	↑	↑	↑	↑		↑	↑
Th1 response	↑	↑		↑	↑			↑
Size effect (nm)	100–1000		14, not 95		5–60	5, not 100	70, not 300	—

Abbreviations: WP, wear particles, ↑, increase, ↓, decrease; *, function other than production of cytokines.

Table 13.1 indicates common modes of action and particle-specific effects are seen. General effects (effects detected in ≥ 3 different particle types) include the stimulation of pro-inflammatory cytokines, the recruitment of leukocytes, a negative effect on the antibacterial defence of macrophages and stimulation of both Th1 and Th2 pathways. In contrast, the effect on macrophage function of particles alone appears to be particle-specific.

13.5 Conclusions

The immune system is one of the major targets of particles in the micro- and in the nano-size. NPs accumulate in immune cells and cause general and material/particle specific effects. Pro-inflammatory effect, mitigation of the reaction to bacteria and predisposition for allergic reactions appear to be general effects caused by particles. Historically, NP effects were mainly deduced from environmental data, but studies on wear particles can also provide valuable information on immune effects of medical NPs.

References

1. Fisher, E., Bell, J., Tomek, I., Esty, A. and Goodman, D. (2010) *Trends and Regional Variation in Hip, Knee, and Shoulder Replacements*Ed. (Robert Wood Johnson Foundation, Princeton, NJ).
2. Pramanik, S., Agarwal, A. and Rai, K. (2005). Chronology of Total Hip Joint Replacement and Materials Development, *TIBAO*, **19** (1), 1526.
3. Caicedo, M., Jacobs, J. and Hallab, N. (2010). Aseptic loosening of implants limits long-term success. Inflammatory bone loss in joint replacements: The mechanisms, *J. Musculoskeletal Med.*, **27** (6), 6.
4. Campoccia, D., Montanaro, L. and Arciola, C. R. (2006). The significance of infection related to orthopedic devices and issues of antibiotic resistance, *Biomaterials*, **27** (11), 2331–2339.
5. Drees, P., Eckardt, A., Gay, R. E., Gay, S. and Huber, L. C. (2007). Mechanisms of disease: Molecular insights into aseptic loosening of orthopedic implants, *Nat Clin Pract Rheumatol*, **3** (3), 165–171.
6. Murray, P. J. and Wynn, T. A. (2011). Protective and pathogenic functions of macrophage subsets, *Nat Rev Immunol*, **11** (11), 723–737.
7. Sica, A., Larghi, P., Mancino, A., Rubino, L., Porta, C., Totaro, M. G., Rimoldi, M., Biswas, S. K., Allavena, P. and Mantovani, A. (2008). Macrophage polarization in tumour progression, *Semin Cancer Biol*, **18** (5), 349–355.
8. Benoit, M., Desnues, B. and Mege, J. L. (2008). Macrophage polarization in bacterial infections, *J. Immunol.*, **181** (6), 3733–3739.
9. Arora, A., Song, Y., Chun, L., Huie, P., Trindade, M., Smith, R. L. and Goodman, S. (2003). The role of the TH1 and TH2 immune responses in loosening and osteolysis of cemented total hip replacements, *J. Biomed. Mater. Res. A*, **64** (4), 693–697.

10. Maitra, R., Clement, C. C., Scharf, B., Crisi, G. M., Chitta, S., Paget, D., Purdue, P. E., Cobelli, N. and Santambrogio, L. (2009). Endosomal damage and TLR2 mediated inflammasome activation by alkane particles in the generation of aseptic osteolysis, *Mol. Immunol.*, **47** (2–3), 175–184.

11. Pearl, J. I., Ma, T., Irani, A. R., Huang, Z., Robinson, W. H., Smith, R. L. and Goodman, S. B. (2011). Role of the Toll-like receptor pathway in the recognition of orthopedic implant wear-debris particles, *Biomaterials*, **32** (24), 5535–5542.

12. Shanbhag, A. S., Macaulay, W., Stefanovic-Racic, M. and Rubash, H. E. (1998). Nitric oxide release by macrophages in response to particulate wear debris, *J. Biomed. Mater. Res.*, **41** (3), 497–503.

13. Sundfeldt, M., Carlsson, L. V., Johansson, C. B., Thomsen, P. and Gretzer, C. (2006). Aseptic loosening, not only a question of wear: a review of different theories, *Acta Orthop*, **77** (2), 177–197.

14. Miyanishi, K., Trindade, M. C., Ma, T., Goodman, S. B., Schurman, D. J. and Smith, R. L. (2003). Periprosthetic osteolysis: induction of vascular endothelial growth factor from human monocyte/macrophages by orthopaedic biomaterial particles, *J. Bone Miner. Res.*, **18** (9), 1573–1583.

15. Nakashima, Y., Sun, D. H., Trindade, M. C., Chun, L. E., Song, Y., Goodman, S. B., Schurman, D. J., Maloney, W. J. and Smith, R. L. (1999). Induction of macrophage C-C chemokine expression by titanium alloy and bone cement particles, *The Journal of bone and joint surgery. British volume*, **81** (1), 155–162.

16. Dean, D. D., Schwartz, Z., Liu, Y., Blanchard, C. R., Agrawal, C. M., Mabrey, J. D., Sylvia, V. L., Lohmann, C. H. and Boyan, B. D. (1999). The effect of ultra-high molecular weight polyethylene wear debris on MG63 osteosarcoma cells in vitro, *J. Bone Joint Surg. Am.*, **81** (4), 452–461.

17. Wilke, A., Endres, S., Griss, P. and Herz, U. (2002). [Cytokine profile of a human bone marrow cell culture on exposure to titanium-aluminium-vanadium particles], *Z Orthop Ihre Grenzgeb*, **140** (1), 83–89.

18. Lucarelli, M., Gatti, A. M., Savarino, G., Quattroni, P., Martinelli, L., Monari, E. and Boraschi, D. (2004). Innate defence functions of macrophages can be biased by nano-sized ceramic and metallic particles, *Eur Cytokine Netw*, **15** (4), 339–346.

19. Goodman, S. B. (2007). Wear particles, periprosthetic osteolysis and the immune system, *Biomaterials*, **28** (34), 5044–5048.

20. Mishra, P. K., Wu, W., Rozo, C., Hallab, N. J., Benevenia, J. and Gause, W. C. (2011). Micrometer-sized titanium particles can induce potent Th2-

type responses through TLR4-independent pathways, *J. Immunol.*, **187** (12), 6491–6498.

21. Dreher, K. L., Jaskot, R. H., Lehmann, J. R., Richards, J. H., McGee, J. K., Ghio, A. J. and Costa, D. L. (1997). Soluble transition metals mediate residual oil fly ash induced acute lung injury, *J. Toxicol. Environ. Health*, **50** (3), 285–305.

22. Granum, B. and Lovik, M. (2002). The effect of particles on allergic immune responses, *Toxicol. Sci.*, **65** (1), 7–17.

23. Braun, A., Bewersdorff, M., Lintelmann, J., Matuschek, G., Jakob, T., Gottlicher, M., Schober, W., Buters, J. T., Behrendt, H. and Mempel, M. (2010). Differential impact of diesel particle composition on pro-allergic dendritic cell function, *Toxicol. Sci.*, **113** (1), 85–94.

24. Mizgerd, J. P. and Skerrett, S. J. (2008). Animal models of human pneumonia, *Am. J. Physiol. Lung Cell Mol. Physiol.*, **294** (3), L387–398.

25. Ichinose, T., Takano, H., Miyabara, Y., Yanagisawa, R. and Sagai, M. (1997). Murine strain differences in allergic airway inflammation and immunoglobulin production by a combination of antigen and diesel exhaust particles, *Toxicology*, **122** (3), 183–192.

26. Ghio, A. J., Kim, C. and Devlin, R. B. (2000). Concentrated ambient air particles induce mild pulmonary inflammation in healthy human volunteers, *Am. J. Respir. Crit. Care Med.*, **162** (3 Pt 1), 981–988.

27. Harder, S. D., Soukup, J. M., Ghio, A. J., Devlin, R. B. and Becker, S. (2001). Inhalation of PM2.5 does not modulate host defense or immune parameters in blood or lung of normal human subjects, *Environ. Health Perspect.*, **109** (Suppl 4), 599–604.

28. Huang, Y. C., Ghio, A. J., Stonehuerner, J., McGee, J., Carter, J. D., Grambow, S. C. and Devlin, R. B. (2003). The role of soluble components in ambient fine particles-induced changes in human lungs and blood, *Inhal. Toxicol.*, **15** (4), 327–342.

29. Salvi, S. S., Nordenhall, C., Blomberg, A., Rudell, B., Pourazar, J., Kelly, F. J., Wilson, S., Sandstrom, T., Holgate, S. T. and Frew, A. J. (2000). Acute exposure to diesel exhaust increases IL-8 and GRO-alpha production in healthy human airways, *Am. J. Respir. Crit. Care Med.*, **161** (2 Pt 1), 550–557.

30. Stenfors, N., Nordenhall, C., Salvi, S. S., Mudway, I., Soderberg, M., Blomberg, A., Helleday, R., Levin, J. O., Holgate, S. T., Kelly, F. J., et al (2004). Different airway inflammatory responses in asthmatic and healthy humans exposed to diesel, *Eur. Respir. J.*, **23** (1), 82–86.

31. Rudell, B., Blomberg, A., Helleday, R., Ledin, M. C., Lundback, B., Stjernberg, N., Horstedt, P. and Sandstrom, T. (1999). Bronchoalveolar

inflammation after exposure to diesel exhaust: comparison between unfiltered and particle trap filtered exhaust, *Occup Environ Med*, **56** (8), 527–534.

32. Diaz-Sanchez, D., Garcia, M. P., Wang, M., Jyrala, M. and Saxon, A. (1999). Nasal challenge with diesel exhaust particles can induce sensitization to a neoallergen in the human mucosa, *J. Allergy Clin. Immunol.*, **104** (6), 1183–1188.
33. Diaz-Sanchez, D., Tsien, A., Casillas, A., Dotson, A. R. and Saxon, A. (1996). Enhanced nasal cytokine production in human beings after in vivo challenge with diesel exhaust particles, *J. Allergy Clin. Immunol.*, **98** (1), 114–123.
34. Fujieda, S., Diaz-Sanchez, D. and Saxon, A. (1998). Combined nasal challenge with diesel exhaust particles and allergen induces In vivo IgE isotype switching, *Am. J. Respir. Cell Mol. Biol.*, **19** (3), 507–512.
35. Noah, T. L., Zhou, H., Zhang, H., Horvath, K., Robinette, C., Kesic, M., Meyer, M., Diaz-Sanchez, D. and Jaspers, I. (2012). Diesel exhaust exposure and nasal response to attenuated influenza in normal and allergic volunteers, *Am. J. Respir. Crit. Care Med.*, **185** (2), 179–185.
36. Gowdy, K., Krantz, Q. T., Daniels, M., Linak, W. P., Jaspers, I. and Gilmour, M. I. (2008). Modulation of pulmonary inflammatory responses and antimicrobial defenses in mice exposed to diesel exhaust, *Toxicol. Appl. Pharmacol.*, **229** (3), 310–319.
37. Lee, C. C., Liao, J. W. and Kang, J. J. (2004). Motorcycle exhaust particles induce airway inflammation and airway hyperresponsiveness in BALB/C mice, *Toxicol. Sci.*, **79** (2), 326–334.
38. Ohtoshi, T., Takizawa, H., Okazaki, H., Kawasaki, S., Takeuchi, N., Ohta, K. and Ito, K. (1998). Diesel exhaust particles stimulate human airway epithelial cells to produce cytokines relevant to airway inflammation in vitro, *J. Allergy Clin. Immunol.*, **101** (6 Pt 1), 778–785.
39. Boland, S., Baeza-Squiban, A., Fournier, T., Houcine, O., Gendron, M. C., Chevrier, M., Jouvenot, G., Coste, A., Aubier, M. and Marano, F. (1999). Diesel exhaust particles are taken up by human airway epithelial cells in vitro and alter cytokine production, *Am. J. Physiol.*, **276** (4 Pt 1), L604–613.
40. Reibman, J., Hsu, Y., Chen, L. C., Bleck, B. and Gordon, T. (2003). Airway epithelial cells release MIP-3alpha/CCL20 in response to cytokines and ambient particulate matter, *Am. J. Respir. Cell Mol. Biol.*, **28** (6), 648–654.
41. Provoost, S., Maes, T., Willart, M. A., Joos, G. F., Lambrecht, B. N. and Tournoy, K. G. (2010). Diesel exhaust particles stimulate adaptive

immunity by acting on pulmonary dendritic cells, *J. Immunol.*, **184** (1), 426–432.

42. Karavitis, J. and Kovacs, E. J. (2011). Macrophage phagocytosis: effects of environmental pollutants, alcohol, cigarette smoke, and other external factors, *J. Leukoc. Biol.*, **90** (6), 1065–1078.
43. Inoue, K., Takano, H., Yanagisawa, R., Sakurai, M., Ueki, N. and Yoshikawa, T. (2007). Effects of diesel exhaust particles on cytokine production by splenocytes stimulated with lipopolysaccharide, *J. Appl. Toxicol*, **27** (1), 95–100.
44. Inoue, K., Takano, H., Yanagisawa, R., Hirano, S., Sakurai, M., Shimada, A. and Yoshikawa, T. (2006). Effects of airway exposure to nanoparticles on lung inflammation induced by bacterial endotoxin in mice, *Environ. Health Perspect.*, **114** (9), 1325–1330.
45. Yin, X. J., Dong, C. C., Ma, J. Y., Roberts, J. R., Antonini, J. M. and Ma, J. K. (2007). Suppression of phagocytic and bactericidal functions of rat alveolar macrophages by the organic component of diesel exhaust particles, *J. Toxicol. Environ. Health*, **70** (10), 820–828.
46. Takano, H., Yoshikawa, T., Ichinose, T., Miyabara, Y., Imaoka, K. and Sagai, M. (1997). Diesel exhaust particles enhance antigen-induced airway inflammation and local cytokine expression in mice, *Am. J. Respir. Crit. Care Med.*, **156** (1), 36–42.
47. Fujimaki, H., Saneyoshi, K., Shiraishi, F., Imai, T. and Endo, T. (1997). Inhalation of diesel exhaust enhances antigen-specific IgE antibody production in mice, *Toxicology*, **116** (1–3), 227–233.
48. Inoue, K., Koike, E., Yanagisawa, R. and Takano, H. (2008). Impact of diesel exhaust particles on th2 response in the lung in asthmatic mice, *J. Clin. Biochem. Nutr.*, **43** (3), 199–200.
49. Jaspers, I., Sheridan, P. A., Zhang, W., Brighton, L. E., Chason, K. D., Hua, X. and Tilley, S. L. (2009). Exacerbation of allergic inflammation in mice exposed to diesel exhaust particles prior to viral infection, *Part. Fibre Toxicol.*, **6,** 22.
50. Koike, E., Takano, H., Inoue, K., Yanagisawa, R. and Kobayashi, T. (2008). Carbon black nanoparticles promote the maturation and function of mouse bone marrow-derived dendritic cells, *Chemosphere*, **73** (3), 371–376.
51. Koike, E., Takano, H., Inoue, K. I., Yanagisawa, R., Sakurai, M., Aoyagi, H., Shinohara, R. and Kobayashi, T. (2008). Pulmonary exposure to carbon black nanoparticles increases the number of antigen-presenting cells in murine lung, *Int. J. Immunopathol. Pharmacol.*, **21** (1), 35–42.

52. Lambert, A. L., Trasti, F. S., Mangum, J. B. and Everitt, J. I. (2003). Effect of preexposure to ultrafine carbon black on respiratory syncytial virus infection in mice, *Toxicol. Sci.*, **72** (2), 331–338.

53. de Haar, C., Hassing, I., Bol, M., Bleumink, R. and Pieters, R. (2006). Ultrafine but not fine particulate matter causes airway inflammation and allergic airway sensitization to co-administered antigen in mice, *Clin. Exp. Allergy*, **36** (11), 1469–1479.

54. Inoue, K., Takano, H., Yanagisawa, R., Sakurai, M., Ichinose, T., Sadakane, K. and Yoshikawa, T. (2005). Effects of nano particles on antigen-related airway inflammation in mice, *Resp. Res.*, **6**, 106.

55. Inoue, K., Takano, H., Yanagisawa, R., Ichinose, T., Sakurai, M. and Yoshikawa, T. (2006). Effects of nano particles on cytokine expression in murine lung in the absence or presence of allergen, *Arch. Toxicol.*, **80** (9), 614–619.

56. Shwe, T. T., Yamamoto, S., Kakeyama, M., Kobayashi, T. and Fujimaki, H. (2005). Effect of intratracheal instillation of ultrafine carbon black on proinflammatory cytokine and chemokine release and mRNA expression in lung and lymph nodes of mice, *Toxicol. Appl. Pharmacol.*, **209** (1), 51–61.

57. Bezemer, G. F., Bauer, S. M., Oberdorster, G., Breysse, P. N., Pieters, R. H., Georas, S. N. and Williams, M. A. (2011). Activation of pulmonary dendritic cells and Th2-type inflammatory responses on instillation of engineered, environmental diesel emission source or ambient air pollutant particles in vivo, *J. Innate Immun.*, **3** (2), 150–166.

58. Siegel, P. D., Saxena, R. K., Saxena, Q. B., Ma, J. K., Ma, J. Y., Yin, X. J., Castranova, V., Al-Humadi, N. and Lewis, D. M. (2004). Effect of diesel exhaust particulate (DEP) on immune responses: contributions of particulate versus organic soluble components, *J. Toxicol. Environ. Health*, **67** (3), 221–231.

59. Pacheco, K. A., Tarkowski, M., Sterritt, C., Negri, J., Rosenwasser, L. J. and Borish, L. (2001). The influence of diesel exhaust particles on mononuclear phagocytic cell-derived cytokines: IL-10, TGF-beta and IL-1 beta, *Clin. Exp. Immunol.*, **126** (3), 374–383.

60. Lovik, M., Hogseth, A. K., Gaarder, P. I., Hagemann, R. and Eide, I. (1997). Diesel exhaust particles and carbon black have adjuvant activity on the local lymph node response and systemic IgE production to ovalbumin, *Toxicology*, **121** (2), 165–178.

61. Yanagisawa, R., Takano, H., Inoue, K. I., Ichinose, T., Sadakane, K., Yoshino, S., Yamaki, K., Yoshikawa, T. and Hayakawa, K. (2006).

Components of diesel exhaust particles differentially affect Th1/Th2 response in a murine model of allergic airway inflammation, *Clin. Exp. Allergy*, **36** (3), 386–395.

62. Yanagisawa, R., Takano, H., Inoue, K., Ichinose, T., Sadakane, K., Yoshino, S., Yamaki, K., Kumagai, Y., Uchiyama, K., Yoshikawa, T. and Morita, M. (2003). Enhancement of acute lung injury related to bacterial endotoxin by components of diesel exhaust particles, *Thorax*, **58** (7), 605–612.
63. Al-Humadi, N. H., Siegel, P. D., Lewis, D. M., Barger, M. W., Ma, J. Y., Weissman, D. N. and Ma, J. K. (2002). The effect of diesel exhaust particles (DEP) and carbon black (CB) on thiol changes in pulmonary ovalbumin allergic sensitized Brown Norway rats, *Exp. Lung Res.*, **28** (5), 333–349.
64. van Zijverden, M. and Granum, B. (2000). Adjuvant activity of particulate pollutants in different mouse models, *Toxicology*, **152** (1–3), 69–77.
65. van Zijverden, M., van der Pijl, A., Bol, M., van Pinxteren, F. A., de Haar, C., Penninks, A. H., van Loveren, H. and Pieters, R. (2000). Diesel exhaust, carbon black, and silica particles display distinct Th1/Th2 modulating activity, *Toxicol. Appl. Pharmacol.*, **168** (2), 131–139.
66. Sekhon, B. S. and Kamboj, S. R. (2010). Inorganic nanomedicine--part 1, *Nanomedicine: Nanotech. Biol. Med.*, **6** (4), 516–522.
67. Fröhlich, E., Meindl, C., Roblegg, E., Griesbacher, A. and Pieber, T. R. (2011). Cytotoxicity of nanoparticles is influenced by size, proliferation and embryonic origin of the cells used for testing, *Nanotoxicology*, **6** (4), 424–439.
68. Wang, Y., Wang, B., Zhu, M. T., Li, M., Wang, H. J., Wang, M., Ouyang, H., Chai, Z. F., Feng, W. Y. and Zhao, Y. L. (2011). Microglial activation, recruitment and phagocytosis as linked phenomena in ferric oxide nanoparticle exposure, *Toxicol. Lett.*, **205** (1), 26–37.
69. Blank, F., Gerber, P., Rothen-Rutishauser, B., Sakulkhu, U., Salaklang, J., De Peyer, K., Gehr, P., Nicod, L. P., Hofmann, H., Geiser, T., et al (2011). Biomedical nanoparticles modulate specific CD4+ T cell stimulation by inhibition of antigen processing in dendritic cells, *Nanotoxicology*, **5** (4), 606–621.
70. Cho, W. S., Cho, M., Kim, S. R., Choi, M., Lee, J. Y., Han, B. S., Park, S. N., Yu, M. K., Jon, S. and Jeong, J. (2009). Pulmonary toxicity and kinetic study of Cy5.5-conjugated superparamagnetic iron oxide nanoparticles by optical imaging, *Toxicol. Appl. Pharmacol.*, **239** (1), 106–115.
71. Chen, B. A., Jin, N., Wang, J., Ding, J., Gao, C., Cheng, J., Xia, G., Gao, F., Zhou, Y., Chen, Y., et al (2010). The effect of magnetic nanoparticles of Fe(3)

O(4) on immune function in normal ICR mice, *Int. J. Nanomedicine*, **5**, 593–599.

72. Park, E. J., Kim, H., Kim, Y., Yi, J., Choi, K. and Park, K. (2010). Inflammatory responses may be induced by a single intratracheal instillation of iron nanoparticles in mice, *Toxicology*, **275** (1–3), 65–71.
73. Shen, C. C., Wang, C. C., Liao, M. H. and Jan, T. R. (2011). A single exposure to iron oxide nanoparticles attenuates antigen-specific antibody production and T-cell reactivity in ovalbumin-sensitized BALB/c mice, *Int. J. Nanomedicine*, **6**, 1229–1235.
74. Moyano, D. F., Goldsmith, M., Solfiell, D. J., Landesman-Milo, D., Miranda, O. R., Peer, D. and Rotello, V. M. (2012). Nanoparticle hydrophobicity dictates immune response, *J. Am. Chem. Soc.*, **134** (9), 3965–3967.
75. Ma, J. S., Kim, W. J., Kim, J. J., Kim, T. J., Ye, S. K., Song, M. D., Kang, H., Kim, D. W., Moon, W. K. and Lee, K. H. (2010). Gold nanoparticles attenuate LPS-induced NO production through the inhibition of NF-kappaB and IFN-beta/STAT1 pathways in RAW264.7 cells, *Nitric Oxide*, **23** (3), 214–219.
76. Hutter, E., Boridy, S., Labrecque, S., Lalancette-Hebert, M., Kriz, J., Winnik, F. M. and Maysinger, D. (2010). Microglial response to gold nanoparticles, *ACS Nano*, **4** (5), 2595–2606.
77. Villiers, C., Freitas, H., Couderc, R., Villiers, M. B. and Marche, P. (2010). Analysis of the toxicity of gold nano particles on the immune system: effect on dendritic cell functions, *J. Nanopart. Res.*, **12** (1), 55–60.
78. Zhang, X. D., Wu, D., Shen, X., Liu, P. X., Yang, N., Zhao, B., Zhang, H., Sun, Y. M., Zhang, L. A. and Fan, F. Y. (2011). Size-dependent in vivo toxicity of PEG-coated gold nanoparticles, *Int. J. Nanomedicine*, **6,** 2071–2081.
79. Cho, W. S., Kim, S., Han, B. S., Son, W. C. and Jeong, J. (2009). Comparison of gene expression profiles in mice liver following intravenous injection of 4 and 100 nm-sized PEG-coated gold nanoparticles, *Toxicol. Lett.*, **191** (1), 96–102.
80. Lim, D. H., Jang, J., Kim, S., Kang, T., Lee, K. and Choi, I. H. (2012). The effects of sub-lethal concentrations of silver nanoparticles on inflammatory and stress genes in human macrophages using cDNA microarray analysis, *Biomaterials*, **33** (18), 4690–4699.
81. Shavandi, Z., Ghazanfari, T. and Moghaddam, K. N. (2011). In vitro toxicity of silver nanoparticles on murine peritoneal macrophages, *Immunopharmacol. Immunotoxicol.*, **33** (1), 135–140.
82. Park, E. J., Bae, E., Yi, J., Kim, Y. S., Choi, K., Lee, S. H., Yoon, J., Lee, B. C. and Park, K. (2010). Repeated-dose toxicity and inflammatory responses

in mice by oral administration of silver nanoparticles, *Environ. Toxicol. Pharmacol.*, **30**, 162–168.

83. Bhol, K. C. and Schechter, P. J. (2005). Topical nanocrystalline silver cream suppresses inflammatory cytokines and induces apoptosis of inflammatory cells in a murine model of allergic contact dermatitis, *Br J Dermatol*, **152** (6), 1235–1242.

84. Vallhov, H., Gabrielsson, S., Stromme, M., Scheynius, A. and Garcia-Bennett, A. E. (2007). Mesoporous silica particles induce size dependent effects on human dendritic cells, *Nano Lett.*, **7** (12), 3576–3582.

85. Kulikova, G. A., Parfenyuk, E. V., Ryabinina, I. V., Antsiferova, Y. S., Sotnikova, N. Y., Posiseeva, L. V. and Eliseeva, M. A. (2010). In vitro studies of interaction of modified silica nanoparticles with different types of immunocompetent cells, *J. Biomed. Mater. Res. A*, **95** (2), 434–439.

86. Lee, S., Yun, H. S. and Kim, S. H. (2011). The comparative effects of mesoporous silica nanoparticles and colloidal silica on inflammation and apoptosis, *Biomaterials*, **32** (35), 9434–9443.

87. Hirai, T., Yoshikawa, T., Nabeshi, H., Yoshida, T., Tochigi, S., Ichihashi, K., Uji, M., Akase, T., Nagano, K., Abe, Y., et al (2012). Amorphous silica nanoparticles size-dependently aggravate atopic dermatitis-like skin lesions following an intradermal injection, *Part. Fibre Toxicol.*, **9**, 3.

88. Park, E. J. and Park, K. (2009). Oxidative stress and pro-inflammatory responses induced by silica nanoparticles in vivo and in vitro, *Toxicol. Lett.*, **184** (1), 18–25.

89. Nishimori, H., Kondoh, M., Isoda, K., Tsunoda, S., Tsutsumi, Y. and Yagi, K. (2009). Silica nanoparticles as hepatotoxicants, *Eur. J. Pharm. Biopharm.*, **72** (3), 496–501.

90. Delogu, L. G., Venturelli, E., Manetti, R., Pinna, G. A., Carru, C., Madeddu, R., Murgia, L., Sgarrella, F., Dumortier, H. and Bianco, A. (2012). Ex vivo impact of functionalized carbon nanotubes on human immune cells, *Nanomedicine: Nanotech. Biol. Med.*, **7** (2), 231–243.

91. Tkach, A. V., Shurin, G. V., Shurin, M. R., Kisin, E. R., Murray, A. R., Young, S. H., Star, A., Fadeel, B., Kagan, V. E. and Shvedova, A. A. (2011). Direct effects of carbon nanotubes on dendritic cells induce immune suppression upon pulmonary exposure, *ACS Nano*, **5** (7), 5755–5762.

92. Witasp, E., Shvedova, A. A., Kagan, V. E. and Fadeel, B. (2009). Single-walled carbon nanotubes impair human macrophage engulfment of apoptotic cell corpses, *Inhal Toxicol*, **21** (Suppl 1), 131–136.

93. Chou, C. C., Hsiao, H. Y., Hong, Q. S., Chen, C. H., Peng, Y. W., Chen, H. W. and Yang, P. C. (2008). Single-walled carbon nanotubes can induce pulmonary injury in mouse model, *Nano Lett.*, **8** (2), 437–445.

94. Herzog, E., Byrne, H. J., Casey, A., Davoren, M., Lenz, A. G., Maier, K. L., Duschl, A. and Oostingh, G. J. (2009). SWCNT suppress inflammatory mediator responses in human lung epithelium in vitro, *Toxicol. Appl. Pharmacol.*, **234** (3), 378–390.

95. Ye, S. F., Wu, Y. H., Hou, Z. Q. and Zhang, Q. Q. (2009). ROS and NF-kappaB are involved in upregulation of IL-8 in A549 cells exposed to multi-walled carbon nanotubes, *Biochem. Biophys. Res. Commun.*, **379** (2), 643–648.

96. Park, E. J., Roh, J., Kim, S. N., Kang, M. S., Han, Y. A., Kim, Y., Hong, J. T. and Choi, K. (2011). A single intratracheal instillation of single-walled carbon nanotubes induced early lung fibrosis and subchronic tissue damage in mice, *Arch. Toxicol.*, **85** (9), 1121–1131.

97. Meng, J., Yang, M., Jia, F., Xu, Z., Kong, H. and Xu, H. (2011). Immune responses of BALB/c mice to subcutaneously injected multi-walled carbon nanotubes, *Nanotoxicology*, **5** (4), 583–591.

98. Shvedova, A. A., Fabisiak, J. P., Kisin, E. R., Murray, A. R., Roberts, J. R., Tyurina, Y. Y., Antonini, J. M., Feng, W. H., Kommineni, C., Reynolds, J., et al (2008). Sequential exposure to carbon nanotubes and bacteria enhances pulmonary inflammation and infectivity, *Am. J. Respir. Cell Mol. Biol.*, **38** (5), 579–590.

99. Mitchell, L. A., Gao, J., Wal, R. V., Gigliotti, A., Burchiel, S. W. and McDonald, J. D. (2007). Pulmonary and systemic immune response to inhaled multiwalled carbon nanotubes, *Toxicol. Sci.*, **100** (1), 203–214.

100. Mitchell, L. A., Lauer, F. T., Burchiel, S. W. and McDonald, J. D. (2009). Mechanisms for how inhaled multiwalled carbon nanotubes suppress systemic immune function in mice, *Nat. Nanotechnol.*, **4** (7), 451–456.

101. Park, E. J., Cho, W. S., Jeong, J., Yi, J., Choi, K. and Park, K. (2009). Pro-inflammatory and potential allergic responses resulting from B cell activation in mice treated with multi-walled carbon nanotubes by intratracheal instillation, *Toxicology*, **259** (3), 113–121.

102. Meng, J., Yang, M., Jia, F., Kong, H., Zhang, W., Wang, C., Xing, J., Xie, S. and Xu, H. (2010). Subcutaneous injection of water-soluble multi-walled carbon nanotubes in tumor-bearing mice boosts the host immune activity, *Nanotechnology*, **21** (14), 145104.

103. Nygaard, U. C., Hansen, J. S., Samuelsen, M., Alberg, T., Marioara, C. D. and Lovik, M. (2009). Single-walled and multi-walled carbon

nanotubes promote allergic immune responses in mice, *Toxicol. Sci.*, **109** (1), 113–123.

104. Yamaguchi, A., Fujitani, T., Ohyama, K., Nakae, D., Hirose, A., Nishimura, T. and Ogata, A. (2012). Effects of sustained stimulation with multi-wall carbon nanotubes on immune and inflammatory responses in mice, *J. Toxicol. Sci.*, **37** (1), 177–189.

105. Fifis, T., Gamvrellis, A., Crimeen-Irwin, B., Pietersz, G. A., Li, J., Mottram, P. L., McKenzie, I. F. and Plebanski, M. (2004). Size-dependent immunogenicity: therapeutic and protective properties of nano-vaccines against tumors, *J. Immunol.*, **173** (5), 3148–3154.

106. Mottram, P. L., Leong, D., Crimeen-Irwin, B., Gloster, S., Xiang, S. D., Meanger, J., Ghildyal, R., Vardaxis, N. and Plebanski, M. (2007). Type 1 and 2 immunity following vaccination is influenced by nanoparticle size: formulation of a model vaccine for respiratory syncytial virus, *Mol. Pharmaceutics*, **4** (1), 73–84.

107. Inoue, K., Takano, H., Yanagisawa, R., Koike, E. and Shimada, A. (2009). Size effects of latex nanomaterials on lung inflammation in mice, *Toxicol. Appl. Pharmacol.*, **234** (1), 68–76.

108. Nygaard, U. C., Samuelsen, M., Aase, A. and Lovik, M. (2004). The capacity of particles to increase allergic sensitization is predicted by particle number and surface area, not by particle mass, *Toxicol. Sci.*, **82** (2), 515–524.

109. Samuelsen, M., Nygaard, U. C. and Lovik, M. (2009). Particle size determines activation of the innate immune system in the lung, *Scand. J. Immunol.*, **69** (5), 421–428.

110. Hussain, S., Vanoirbeek, J. A. and Hoet, P. H. (2012). Interactions of nanomaterials with the immune system, *Wiley Interdiscip. Rev. Nanomed. Nanobiotechnol.*, **4** (2), 169–183.

Index